FOOD BIOTECHNOLOGY—1

FOOD BIOTECHNOLOGY—1

Edited by

R. D. KING

Department of Food Science and Technology, University of Reading, UK

and

P. S. J. CHEETHAM

PPF International Ltd, Ashford, Kent, UK

ELSEVIER APPLIED SCIENCE
LONDON and NEW YORK

ELSEVIER APPLIED SCIENCE PUBLISHERS LTD
Crown House, Linton Road, Barking, Essex IG11 8JU, England

Sole Distributor in the USA and Canada
ELSEVIER SCIENCE PUBLISHING CO., INC.
52 Vanderbilt Avenue, New York, NY 10017, USA

WITH 35 TABLES AND 73 ILLUSTRATIONS

British Library Cataloguing in Publication Data

Food biotechnology.
1. Food industry and trade
I. King, R. D. II. Cheetham, P. S. J.
664 TP370

Library of Congress Cataloging in Publication Data

Food biotechnology.

Includes bibliographies and index.
1. Food industry and trade. 2. Biochemistry.
I. King, R. D. (Richard D.) II. Cheetham, P. S. J.
TP370.5.F655 1986 664 86-19675

ISBN 1-85166-029-1

The selection and presentation of material and the opinions expressed are the sole responsibility of the author(s) concerned.

Printed in Great Britain by Galliard (Printers) Ltd, Great Yarmouth

Contents

List of Contributors

M. P. BROCKLEBANK
John Brown Engineers and Constructors Ltd, 1 Buckingham Street, Portsmouth PO1 1HN, UK

M. M. BURRELL
Biochemistry Department, Rothamsted Experimental Station, Harpenden, Herts AL5 2JQ, UK

RAFAEL JIMENEZ-FLORES
Department of Food Science and Technology, College of Agricultural and Environmental Sciences, University of California, Davis, California 95616, USA

M. KREIS
Biochemistry Department, Rothamsted Experimental Station, Harpenden, Herts AL5 2JQ, UK

IRVIN E. LIENER
Department of Biochemistry, University of Minnesota, 140 Gortner Laboratory, 1479 Gortner Avenue, St Paul, Minnesota 55108, USA

B. J. MIFLIN
Biochemistry Department, Rothamsted Experimental Station, Harpenden, Herts AL5 2JQ, UK

P. E. MILSOM
John & E. Sturge Ltd, Denison Road, Selby, North Yorkshire YO8 8EF, UK

V. J. MORRIS
AFRC Food Research Institute, Colney Lane, Norwich NR4 7UA, UK

THOMAS RICHARDSON
Department of Food Science and Technology, College of Agricultural and Environmental Sciences, University of California, Davis, California 95616, USA

P. D. RUSSELL
GRC Consultants, 105 Roundway, Waterlooville, Hampshire PO7 7QB, UK

P. R. SHEWRY
Biochemistry Department, Rothamsted Experimental Station, Harpenden, Herts AL5 2JQ, UK

Introduction

Biotechnology in its many guises has developed very considerably over the last few years. We now feel that it is appropriate for the publication of a series of books that discuss the technical aspects of biotechnology specifically as applied to foods, and in particular concentrating on new and emerging techniques, processes and products. Food is without doubt one of the oldest bioindustries; however, some of the new areas of biotechnology, such as diagnostic and health-care applications, are likely to mature much faster than applications in the food industry. Eventually, however, biotechnology must have a very great impact on a wide scale in the food industry, simply because of the size and diversity of the industry, and because most food products are substantially natural in origin and are therefore very suitable for processing by biocatalysts. Some of the ways in which the food industry is likely to be affected by developments in biotechnology include the following:

The modification of food components to give products with new and/or improved properties, for instance high fructose corn syrups, and by modifying the functional properties of proteins.

New methods of assaying food constituents, such as immobilized enzyme sensors.

New processes for the production of foods and food components, for instance the use of plant cell cultures for the production of flavours.

Many of these topics will be described in detail in this series of books. In this first volume some of the methods and techniques of biotechnology are considered in chapters on protein and carbohydrate functionality, genetic engineering, bio-reactor design and protein separation. Also some applications are covered including organic acid production and detoxifying enzymes. In subsequent volumes subjects such as enzyme technology, economic and commercial factors, enzymic synthesis of flavours, lipid modification and bioaffinity methods of analysis will be dealt with.

The crux of the recent advances being made in biotechnology is the application of new biological techniques to commercial applications. However, it should also be borne in mind that new advances in other competing areas such as chemical catalysis are still taking place; for instance, the inorganic catalysts developed recently for synthesis gas catalysis. Chemical catalysis is well established in industry and has a good track record in terms of successful translation to commercial-scale operation. Therefore at present biocatalysis tends to be restricted to those applications that chemical catalysts cannot perform, or carry out only poorly, for instance stereoselective reactions. However, even if chemical catalysis does prove to be the eventual method of manufacture, biocatalysis may have roles in producing preliminary trial samples, for removing impurities from the final product, or by providing a 'lead' catalyst which can then give a starting point for the development of a new chemical catalyst. However, the consumer-led movement towards more 'natural' food ingredients and additives may provide a unique opportunity for biotechnology processes to establish themselves in the food industry.

The scientific or technical advance from which a new commercial opportunity arises is obviously very important, but is only the first of many factors which determine whether commercial success is achieved or not. These other factors include ability to scale-up processes in a cost-effective and logistically convenient manner, obtaining patent protection, regulatory approval in the case of pharmaceuticals and foods, and advertising, tax incentives, etc. The final arbiter of success is invariably not scientific novelty but functionality and cost. Is the product or service sufficiently useful and cheap enough to persuade customers to buy it, and for it to compete with existing competing products or services?

In discussing new biotechnological advances as related to food applications, it is important to keep in mind the economic status of the food industry. As an example, in 1985 UK consumers spent £182,427,000,000,[1] of which about 20% was spent on foods. The output of the UK food and drinks industry was approximately £30,000,000,000, that is about 17·5% of the total gross manufacturing output, and employed

2·3% of the UK workforce. These values should be viewed against a consistent long-term decline in the proportion of total consumer spending accounted for by food purchases. About 70% of the food purchased had undergone some form of processing prior to sale.[2]

The food and drinks processing industry covers a wide variety of companies with disparate interests. This divergence of interests may be one of a number of impediments to innovation. Another is that in most developed countries food consumption is relatively constant with only qualitative changes due to 'fashion', changing social habits, changing age structure of the population, etc. This tendency is combined with an innate and strong conservatism on the part of many consumers. On a more technical level, most foods have great chemical and physical complexity, and there is a poor working knowledge about the relationship between a food's composition and its organoleptic properties, so there is a requirement to put traditional empirically based processes and products on a better and more easily understood basis. Technical advances are also hindered by the lengthy and expensive petitioning procedures that are required to obtain regulatory approval for new food products.

Government research is generally concentrated on agriculture and fisheries, while industry concentrates on processing operations such as grinding, heating, distilling, etc. Too little effort is directed to understanding the fundamentals of processes and the development of new processes. Also, much less money is spent on food research in universities than on many other sciences and technologies. Thus there is a need to carry out rather more innovative research than at present; this could be conducted more cost-effectively if common interests were identified and 'clubs' of companies formed to share costs, particularly when small companies that find it very difficult to carry out 'in-house' innovative R&D are involved.

The R&D activities of food companies tend to be relatively small, typically 0·2–0·4% of turnover.[3] Regrettably, this parameter has tended to fall with time. Research expenditure appears to be strongly correlated with company turnover. Thus many companies, particularly the small ones, do no genuine research at all, so their rate of uptake of new technology is likely to be relatively slow.

However, there is a considerable problem in identifying products and services that justify the research and development effort needed to bring them successfully to the market place. Generally, food manufacturers require cheaper ingredients and processing for the products. Currently the biggest potential area of impact of research is on improved food processing, e.g. to improve energy consumption, product quality and decrease costs etc. There are also opportunities to produce improved food processing equipment, to improve testing methodology and in consultancy services. Uptake of new biotech ingredients and biocatalytically based processes will be slower, and probably more difficult due to conservation, the need to obtain regulatory approval and the relatively low profit margins of the food industry, which makes it difficult to invest in new processes and products. Also, a tremendous amount of work needs to be done to improve our limited knowledge of the relationship between food structure, processing properties, function and perception, and to put research and product development on a more rational basis.

There is also need to make food manufacturers more innovation conscious and increase their level of R&D activities. Nevertheless, many companies have an excellent record in research, e.g. Unilever, Ranks Hovis McDougall, and Tate & Lyle. There is also an interesting tendency for chemical and pharmaceutical companies to produce food ingredients, e.g. ICI produce microbial protein and Pfizer and Searle have strong interests in high intensity sweeteners such as Aspartame.

Eventually, food may prove to be the area of greatest commercial importance to 'biotechnology'! We hope this series of books on Food Biotechnology will make some contribution.

R. D. King
P. S. J. Cheetham

REFERENCES

1. HMSO, *Annual Abstracts and Statistics*, Central Statistical Office, London, 1985.
2. Dunnill, P. and Rudd, M., *Biotechnology and British Industry*, a report to UK SERC Biotechnology Directorate, 1984.
3. *The UK Food Processing Industry: Opportunities for Change*, Technical Change Center, 1985.

1

Fermenter and Bio-reactor Design

P. D. RUSSELL

GRC Consultants, 105 Roundway, Waterlooville, Hants., UK

NOTATION

A, A'	Growth rate constants
D	Agitator diameter
E, E'	Activation energies for growth rate
G	Specific growth rate
G_{max}	Maximum specific growth rate
$K_L a$	Volumetric oxygen transfer rate
K	Viscosity constant (non-Newtonian)
K_s	Monod equation constant
M	Cell mass
M_w	Carbon source molecular weight
n	Viscosity constant (non-Newtonian)
N	Agitator speed
N_p	Agitator power number (ungassed)
Nu	Nusselt number
P	Product concentration
P_g	Gassed agitator power
P_0	Ungassed agitator power
Pr	Prandtl number
Q_a	Volumetric gas flow rate
Q_f	Heat evolution
$Q(O_2)$	Oxygen transfer rate
Re	Reynolds number
S	Substrate concentration
t	Time

T	Absolute temperature
V	Fermenter liquid volume
V_s	Superficial gas velocity
X	Cell concentration
y	Rate of product formation per unit cell mass
Y	Cellular yield per substrate mass consumed
γ	Shear rate
μ	Viscosity
μ_a	Apparent viscosity
μ_w	Viscosity of water
τ	Shear stress

1. FERMENTATION PROCESSES

Fermentation processes are generally understood to be those caused by the actions of micro-organisms, but there is some argument as to whether the term should refer only to the actions of live cells, or be extended to include the actions of extracted enzymes (the biocatalysts which govern the chemical processes within cells). However, since the requirements of live micro-organisms are markedly different from those of enzymes, it is convenient to separate the design of fermenters from the design of enzyme reactors which are discussed in detail in a later chapter.

The metabolism of micro-organisms is a complex system in which a series of chemical processes are carried out, usually simultaneously, within the cell confines. Each reaction is catalysed by a specific enzyme, and the rate of each step is governed by the cell's production of the various enzymes. By the use of triggers to promote and control the rate of enzyme production, the organism can control the different stages in equilibrium and can also respond to changes in its environment. Industrial fermentations take advantage of a particular part of the cell's metabolic process, and to this end a successful and economic fermentation process requires not only an efficient micro-organism and a suitable growth medium, but also a fermenter in which the optimum conditions for metabolism exist.

It is essential to recognise that, whereas many of the principles of chemical reactor design apply to fermenters, there are certain fundamental differences. Although some of the major advantages quoted for fermentation over 'conventional' chemical reactions are the moderate conditions of temperature, pressure and pH, an organism is normally only viable within a narrow range of conditions. More importantly, if different conditions exist within the fermenter, whether due to poor mixing, recycle loops or rigid flow conditions, then the metabolism may be altered due to the cell's response to the changing environment. A typical example would be a poorly aerated zone in a loop fermenter; if the organism is forced to 'hold its breath' for too long in the unaerated zone then the cells will become overstressed and stop production.

A further difference is the need to maintain sterile conditions within the fermenter. Whereas the main concerns in a chemical reactor are to prevent materials from leaking, a major concern of industrial fermentations is to prevent competing organisms from entering the fermenter since the contaminating organisms will consume part of the substrate to reduce the yield and may also contaminate the product. Containment of the fermenter system is only necessary in a small proportion of fermentations which utilise hazardous organisms.

The concentration of micro-organisms in a fermenter is usually low; typically cell weight may comprise only 10–15% of the fermenter contents, and the product weight is a small fraction of this. A typical industrial fermentation process comprises the different activities shown in Fig. 1. It is important that the fermenter design is not considered in isolation but as part of the whole system. The capital cost of the fermenter and its associated equipment will usually be only 40–50% of the total plant cost; the downstream separation recycle and feed preparation processes will make up the remainder. There is considerable scope, especially in large-scale processes, for optimisation between the fermenter conditions and the cost of the downstream and recycle processes.

The oldest uses of fermentation have been for the production of foodstuffs, and it is only in the recent past that fermentation has been utilised to make antibiotics and chemicals. An essential characteristic of these 'traditional' industries is that the flavour and appearance of the final product must be consistent from day to day, despite seasonal changes to the raw materials. For this reason, industries still rely largely on human judgement to control the fermentation, since analysis and control of the number of minor components that constitute 'flavour' are virtually impossible by other means. Fermenter design in the traditional industries has changed to allow more hygienic operation and more efficient practices, but batch operation is usually retained in order to permit control of the final product.

In large-scale processes, the use of continuous fermentation results in significant economies in capital, labour and operating costs compared to batch operation. The trend will be to the use of continuous processes for new bulk foodstuffs especially where the product must compete with

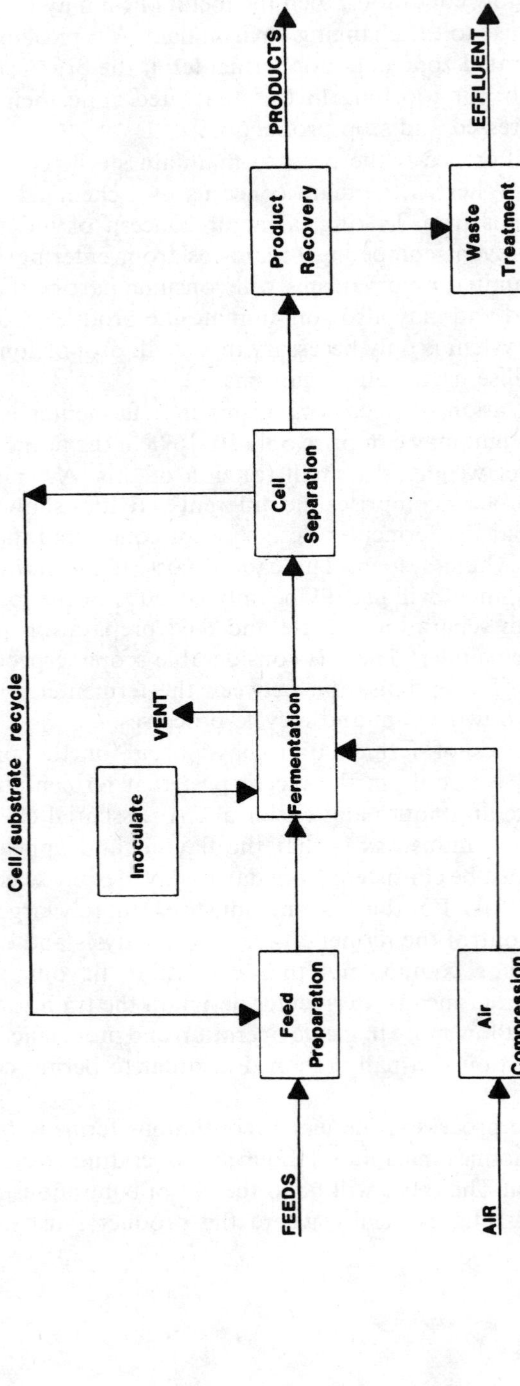

Fig. 1. Typical fermentation process.

existing products manufactured by other methods and flavour is not critical (e.g. SCP). Better understanding of the mechanisms involved will also allow some existing batch processes (such as beer production) to be carried out continuously.

2. BIOLOGICAL CONSTRAINTS

The primary aim of an industrial fermenter system is to provide the optimum conditions for the micro-organism, and a thorough understanding of the requirements and limitations of the organism is essential to the design of the fermenter. Much of the development effort that goes into the scale-up of new microbial processes is aimed at gaining an understanding of the ideal conditions for the system; this is further discussed in Section 5.

Although most fermentations are still performed in stirred, air-sparged tank fermenters, the wide range of designs that have been utilised or proposed for different fermentations arise from the need to satisfy the needs of the organisms as economically as possible. It is therefore useful to discuss these requirements before considering both the range of fermenter designs available and the methods used to design the systems.

2.1. Nutrient

Unlike a chemical reaction, where the reaction rate is a direct function of the concentrations of each of the reactants and products,[1] in microbial growth the relationships are more complex. The basic relationship between growth rate and concentration of each substrate is given in the Monod equation:[2]

$$G = G_{\text{max}}\left(\frac{S}{K_s + S}\right)$$

where G is the specific growth rate; G_{max} is the maximum specific growth rate; S is the substrate concentration; K_s is a constant, equal to the concentration when specific growth rate is half the maximum rate.

This expression is shown graphically in Fig. 2(a) and it is apparent that, when $S > 10K_s$, the specific growth rate becomes virtually constant. The value of K_s for growth of *S. cerevisiae* in glucose is about 25 mg/litre and therefore the specific growth rate is close to the maximum when the concentration is above about 0·25 g/litre.

At higher concentrations of nutrient the organism usually suffers inhibition and the specific growth rate is reduced, resulting in a growth

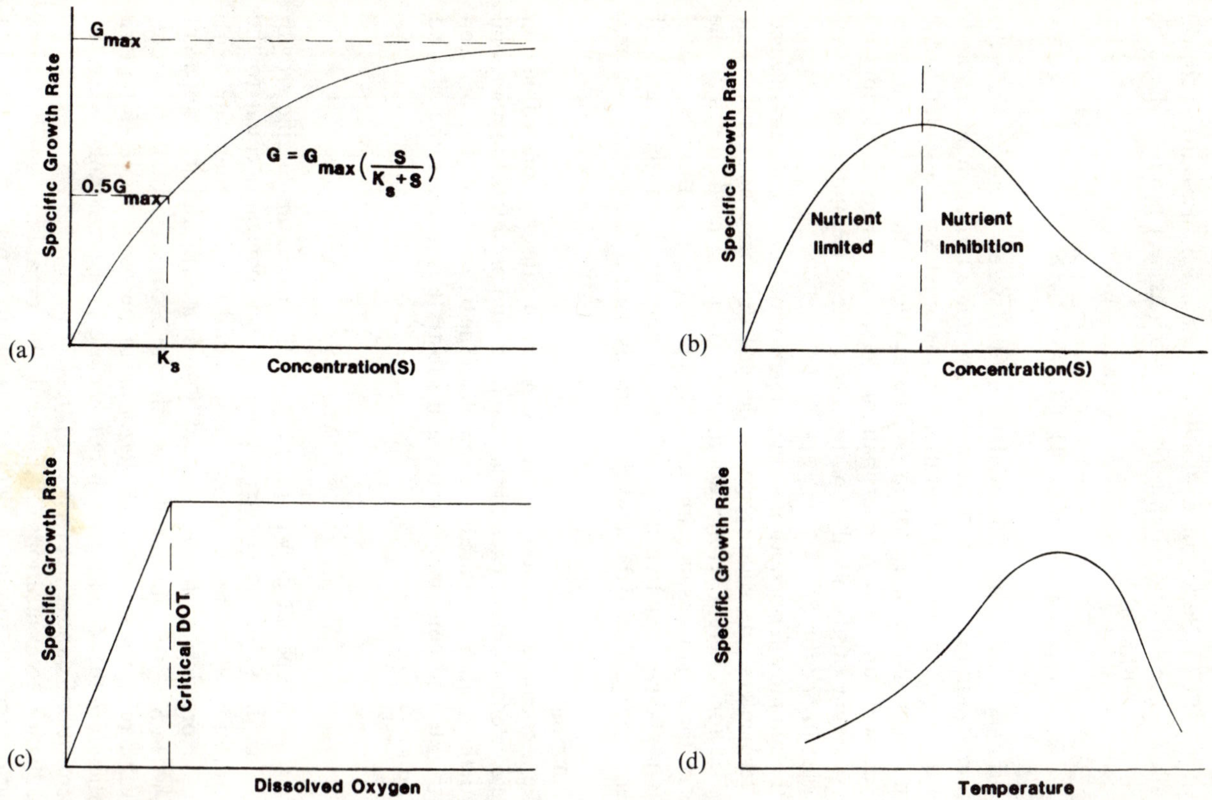

Fig. 2. Variation of specific growth rate. (a) Growth rate predicted by Monod equation. (b) The effect of substrate inhibition on growth rate curve. (c) The effect of dissolved oxygen tension on growth rate. (d) The effect of temperature on growth rate.

rate curve similar to Fig. 2(b). In some cases, inhibition is caused purely by the increasing osmotic pressure within the cell, but high concentrations of substrate may also have toxic effects on the organism or may trigger the organism to use a different metabolic route. The Monod model does not always predict the behaviour of an organism at low substrate concentrations, since low nutrient concentrations may also trigger a different metabolic path.

A further complication when using commercial substrates such as molasses and saccharified starches is that most organisms will consume one growth nutrient in preference to another, and will not start to utilise the second nutrient until the first has been depleted. While this would merely lengthen the cycle time for a batch fermentation, in a continuous fermenter the high loss of substrate may have serious economic implications. For this reason, most *Zymomonas* strains are unsuitable for economic ethanol production on mixed substrates, since they can only utilise a portion of the substrates.

The restraints on nutrient levels outlined above mean that optimisation of the medium is essential for economic fermentation. A high proportion of batch fermentations use the fed-batch technique, where nutrient is continually added during the course of the fermentation. This maintains the nutrient concentration at the optimum level and results in high specific growth rates and improved yields. In continuous fermenters, careful attention may be needed in the distribution of the substrate feed so as to avoid local high concentrations if the organism is sensitive to these.

2.2. Oxygen

The vast majority of current fermentation processes are aerobic in nature, and utilise oxygen to generate energy for growth, but micro-organisms in submerged culture are unable to use gaseous oxygen and can only absorb dissolved oxygen. Since oxygen is considerably less soluble in water than the other nutrients required by the organism (8 mg/litre at 30°C compared to 10 000–100 000 mg/litre for most nutrient salts), mass transfer of oxygen into the broth is normally the rate-limiting step in fermentations.

An expression for the oxygen required for cell growth was derived by Mateles[3] as:

$$\frac{\text{Mass oxygen}}{\text{Mass cells}} = \frac{32\text{C} + 8\text{H} + 16\text{O}}{YM_{\text{w}}} + 1{\cdot}0(\text{O}') - 2{\cdot}67(\text{C}') + 1{\cdot}7(\text{N}') - 8{\cdot}0(\text{H}')$$

where Y is the cellular yield/substrate used; M_{w} is the carbon source molecular weight; O′, C′, H′, N′ are fractions present in the cell; O, C, H, N are numbers of atoms/substrate molecule. Because the oxygen demand is inversely proportional to the cellular yield, it is apparent that the oxygen requirement of the fermentation is a good measure of the efficiency of substrate conversion.

The important measurement of oxygen availability to the organism is the dissolved oxygen tension (DOT), expressed either as mg O_2/litre or as a percentage of oxygen saturation. The relationship between DOT and cell growth rate, shown in Fig. 2(c), is similar to the Monod relationship; above a critical DOT the growth rate and formation of products are independent of the DOT. This critical DOT is normally low and will typically be less than 10% saturation. Below the critical DOT the growth rate falls rapidly.[4]

This behaviour reflects the mechanism for oxygen transfer, which has been found by numerous workers to be limited by transfer through the liquid film. If insufficient oxygen is available, the cell will absorb it from the substrate as rapidly as it is transferred from the gas phase, since little driving force is required for the transfer from liquid to the cell. The DOT in the substrate will therefore remain low. Immediately the rate of oxygen transferred into the liquor exceeds the cell's demand, then the DOT will rise significantly. As with nutrient concentration, the DOT may affect the metabolic pathway chosen by the organism, and the influence of both oxygen and glucose concentrations on yeast fermentation is well documented. At low glucose concentrations and high DOT, aerobic fermentation takes place resulting in cell growth, while at higher sugar concentrations the 'Crabtree effect' takes place even at high DOT and ethanol is preferentially produced.

2.3. Temperature

The optimal growth temperature arises from a balance between the growth rate of the organism, which is enhanced by increasing temperature, and the death rate which is also increased at higher temperatures. Both rates can be represented by the Arrhenius equation and so the overall growth rate becomes:

$$\frac{\mathrm{d}M}{\mathrm{d}t} = M(A\,\mathrm{e}^{-E/RT} - A'\,\mathrm{e}^{-E'/RT})$$

where M is the cell mass; A, A' are constants for growth and death; E, E' are activation energies for growth and death; T is the absolute temperature. A typical plot of the overall growth rate is shown in Fig. 2(d).

A complication in determining the optimum fermenter conditions is that the temperature at which maximum growth occurs may not be the same as that at which the best yield of product is obtained. In addition, increasing temperature will reduce the lag time in batch fermentation but may also increase the mutation rate of an organism. Data on the effect of temperature on a range of organisms has been given.[5]

Heat is evolved from all stages of fermentation processes, but the amount evolved depends on the balance between cell growth and cell maintenance requirements. Under conditions of cell growth, 50–60% of the energy available is converted to heat, while during cell maintenance virtually all the available combustion energy is released as heat to the environment.[6] This heat must be efficiently removed from the fermenter in order to maintain suitable conditions for fermentation; the mechanisms for heat removal are discussed in Section 5.4.

It has been shown that heat evolution is closely related to the efficiency of conversion of nutrients to cell mass. Because the rate of oxygen uptake increases with the degree of oxidation (and so the heat evolution), the heat evolution can be calculated from the oxygen uptake rate by the correlation developed by Cooney *et al.*[7]

$$Q_f = 0{\cdot}12Q(O_2)$$

where Q_f is the heat evolution (kcal/litre-h); $Q(O_2)$ is the oxygen uptake rate (mmol/litre-h). This correlation has been shown to be valid for a number of organisms and substrates, and is useful for initial design estimates. During the scale-up of a microbial process, the heat evolution should be confirmed from practical measurements.

A serious engineering problem is caused by the moderate temperatures required for most fermentations because the fermenter temperature is usually close to, or even below, that of the available cooling water supply. This results in the need for excessively large heat transfer areas in commercial fermenters, and refrigeration is frequently required. The potential economic benefit in developing thermophilic organisms for many commercial fermentations is great, and could be essential in developing new fermentation routes to existing products.

2.4. pH

As with temperature, most organisms will only retain viability within a limited range of pH. Ideal conditions for growth are at around pH 7 for most organisms, but cells will remain viable over a range of several pH units. The control of optimum pH is essential, either by buffering (on laboratory scale) or by a pH control system. In general, yeasts are more resistant to acid conditions than bacteria and other organisms, and this can be used to protect appropriate yeast fermentations from bacterial infection.

Specialised organisms may operate either at high acidity (e.g. lactic acid bacteria) or at high alkalinity. Several organisms are triggered to produce different metabolites at different pH conditions, and in these cases extremely close pH control is required to optimise the product yield.

2.5. Rheology

The effect of the fermentation broth rheology is often critical in fermenter design, and the maximum cell concentration achievable in a fermenter may be limited by the resulting viscosity.

The effect of cell concentration on viscosity varies depending on the form of the organism (ref. 5, p. 693); the viscosity of mycelial broths is significantly higher than those containing more spherical yeast or bacterial cells. At low concentrations, viscosity is roughly proportional to concentration, but at higher cell concentrations the viscosity rapidly increases.

Fermentation broths are highly non-Newtonian (i.e. the apparent viscosity is dependent on the shear rate), and most broths approximate to pseudo-plastic behaviour where:

$$\tau = K\gamma^n \qquad \text{or} \qquad \mu_a = K\gamma^{n-1}$$

where τ is the shear stress; γ is the shear rate; μ_a is the apparent viscosity; K, n are non-Newtonian viscosity constants.

As a fermentation progresses, the values of K and n would typically vary as shown in Fig. 3(a). As the viscosity increases, the transfer rate of both oxygen and nutrient to the cell will be reduced. The heat transfer rate will also fall with increased viscosity, resulting in the need for larger areas of cooling surfaces. In a highly non-Newtonian viscous broth, the degree of agitation decreases rapidly away from the agitator; see Fig. 3(b). Thus, aeration may only occur in the zone local to the impeller, and the rate of heat transfer through the vessel wall will be drastically reduced. All of these effects have a detrimental effect on fermentation, and fermenters to handle viscous broths require different designs for aeration and agitation from those handling more fluid broths.

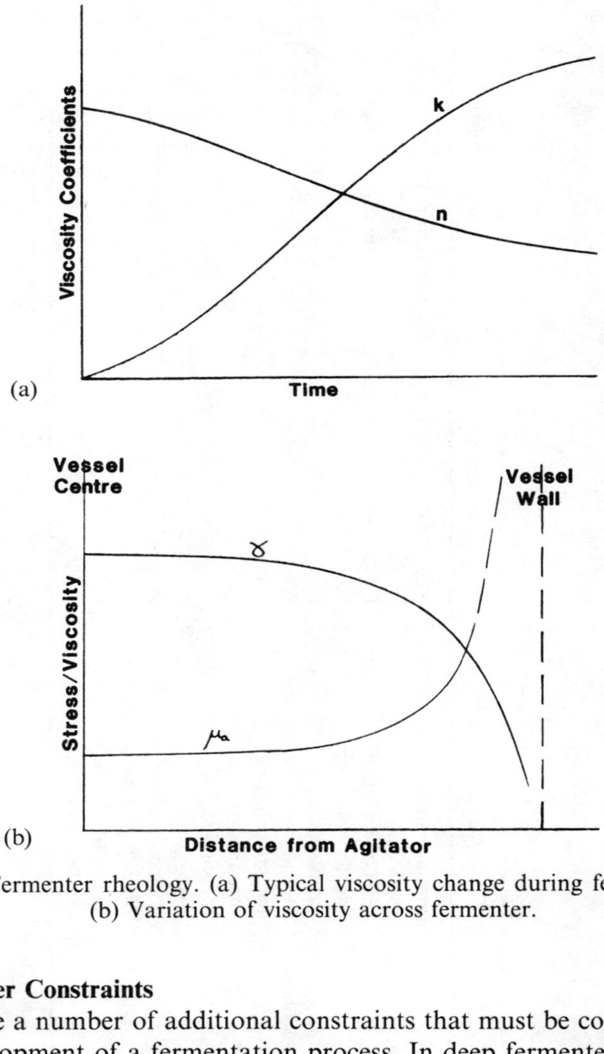

Fig. 3. Fermenter rheology. (a) Typical viscosity change during fermentation. (b) Variation of viscosity across fermenter.

2.6. Other Constraints

There are a number of additional constraints that must be considered in the development of a fermentation process. In deep fermenters or those which are pressurised to enhance oxygen transfer, the effects of pressure may become limiting. The cause may be either the increased hydrostatic pressure on the cell causing damage, or alternatively due to carbon dioxide inhibition. Carbon dioxide will inhibit most organisms, and at increased pressures the solubility of CO_2 is increased until it eventually reaches an inhibitory level.

Table 1
Fermentation Products

Product type	*Metabolic role*	*Examples*	*Availability*
Cellular matter	Material formed by cellular growth	SCP, baker's yeast	Intracellular
Enzymes	Produced by the cell to carry out the metabolic processes	Gluco-amylase, pectinase, invertase	Usually intracellular
Primary metabolic products	Materials used by the cell for growth, i.e. intermediates in the metabolism of micro-organism	Amino acids, vitamins, ethanol	Intracellular or extracellular
Secondary metabolic products	Not produced for growth; usually produced by the micro-organism to control its environment	Antibiotics, toxins	Usually extracellular

The effect of shear on an organism is related to the cell shape; bacteria and yeast tend to be more resistant to shear than mycelia. As the shear rate is increased in a mycelial fermentation, the productivity may initially increase as the filaments become separated and mass transfer is improved. Eventually, increasing shear will damage the cells and their viability will rapidly fall. The effect of shear is particularly severe on animal tissue cultures, where the cells are easily damaged by high-speed agitation, and special low-shear fermenter designs are required. Shear damage can be caused both by high-speed agitators and by external recycle pumps.

The mutation rate of an organism, especially a genetically engineered one, may limit the length of time for which a fermentation may be operated before the cell population reverts to a less favoured variant. If an organism is not stable for greater than 3 months, then continuous fermentation is unlikely to be economic.

In many processes, the economics may be improved by recycling unconsumed substrate from the downstream recovery systems. However, trace nutrients and by-products may have an inhibitory effect on the organism, and this will limit the amount of substrate that can be recovered by recycle from the downstream separation stages.

3. PRODUCT TYPE

The nature of the required product and its relationship to the growth phase of the micro-organism also have a fundamental bearing on the design of the fermenter system. As shown in Table 1, the required product may be contained within the cell or secreted from the cell, and may be formed either as part of the growth process or as a secondary metabolite.

3.1. Intracellular Products

The majority of fermentation products are retained within the cell by the organism, and sufficient cells must be grown to contain the yield of product. Unless the product is the whole cell (e.g. SCP) then a significant proportion of the substrate must be used in producing 'by-product' cellular material; indeed the nutrients needed for cell growth may be greatly different from those needed for production of the required product.

Production of intracellular product is usually an advantage in product recovery, because as the cells will not be recycled to the fermenter they may be separated from the broth in non-sterile equipment following the fermentation, and the cost of this initial product concentration equipment

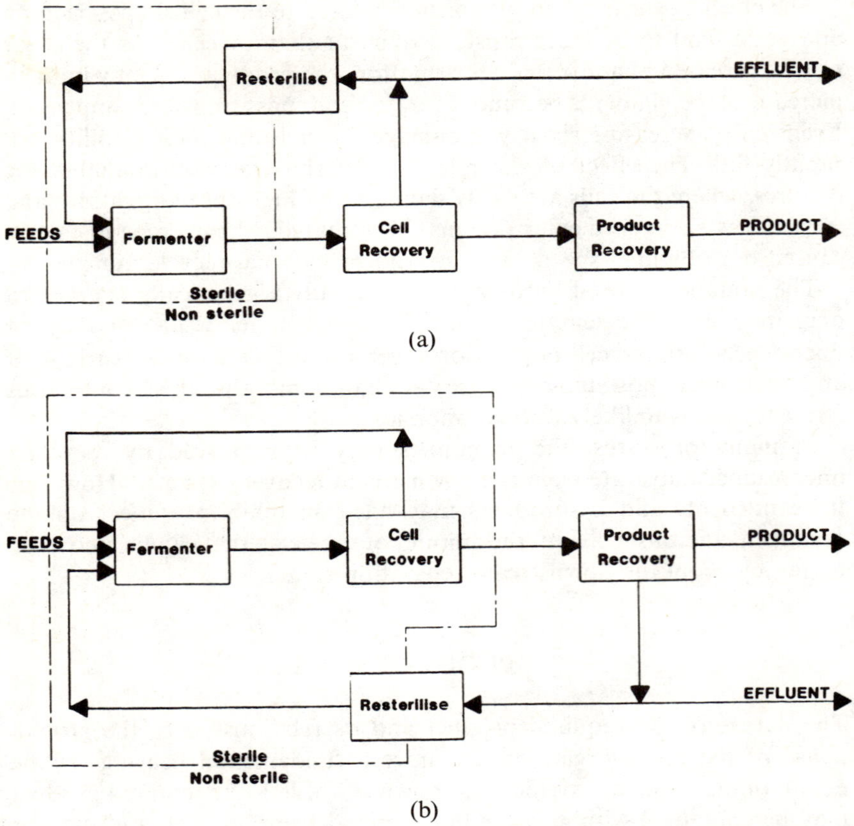

Fig. 4. (a) Fermenter system: intracellular product. (b) Fermenter system: extracellular product.

is relatively low; see Fig. 4(a). If substrate is to be recycled to improve the process economics, then this would need to be re-sterilised.

3.2. Extracellular Product

A minority of fermentation products are secreted by the organism, either because they are a waste product (ethanol) or because their role is to influence the environment of the organism (antibiotics). In this case, the product must be recovered from the broth after removal of the cells, and the opportunity exists to recycle cells back to the fermenter; see Fig. 4(b). This not only reduces the amount of nutrient required for cell growth, but also reduces the size of the fermenter since the cellular concentration (and so the volumetric yield) may be maintained at a high level.

Since the live cells may not be sterilised before return to the fermenter, the separation equipment must be operated aseptically and the cost of the separation equipment will increase significantly. Genetic selection of an organism which will secrete the product rather than retain it within the cell may in the future bring significant improvements in production costs of bulk products as better sterile separation equipment (e.g. ultrafiltration systems) is developed.

3.3. Secondary Metabolites

Secondary metabolites are materials that are not needed for growth of the micro-organism and are usually produced after the growth phase of the organism. Most secondary products are antibiotics and mycotoxins, and formation of the secondary metabolite requires deficiency of a key nutrient as well as the presence of a precursor to trigger production. The fermentation must therefore be carried out in two distinct phases, and the optimum conditions determined to maximise production of the required compound in the second phase.

Similar considerations apply to a number of other fermentation processes (such as ethanol and citric acid production) where, although a primary metabolite is formed, the product is generated after growth of the organism has ceased. In both cases, cellular growth is only encouraged to produce enough cells for efficient formation of the secondary product.

4. FERMENTER CATEGORIES

4.1. Overview

The range of fermenter designs that have been either successfully used or proposed reflects the large number of factors that can affect the design. Selection depends on factors such as the biological constraints of the organism, the scale of production, the level of technology available, economics and often the range of products that are to be considered.

Because of the many different forms of fermenters and the way in which they are operated, it is necessary to consider the design from a number of different aspects such as batch/continuous operation, the method of agitation, the use of free microbial flocs or immobilised systems, and whether the fermentation is aerobic or anaerobic.

Table 2
Examples of Fermentations for Food Products

Product	*Mode*	*Aerobic*	*Organism*	*Batch time (h)*	*Maximum fermenter size (m^3)*	*Type of fermenter*	*Major substrates*
Baker's yeast	Fed batch	Aerobic	Yeast	8–20	200	Stirred tank Draft tube Air agitated Air lift	Molasses Sugars
Beer	Batch	Initial saturation only	Yeast	170	320	Unagitated	Carbohydrates
SCP	Batch Continuous	Aerobic	Yeast Bacteria	$D = 0{\cdot}1$ to $0{\cdot}2\,h^{-1}$	1 500	Stirred tank Air lift Draft tube Loop	Carbohydrates Methanol Hydrocarbons
Citric acid	Batch Semi-continuous	Aerobic	Mycelia	72–350	240	Tray Stirred tank	Carbohydrates
Vinegar	Batch Semi-continuous	Aerobic	Bacteria	36		Stirred tank Trickle tower	Ethanol
Enzymes	Batch	Aerobic	Yeast Bacteria Mycelia	days	200	Stirred tank Trays Deep bed	Various depending on product/organism
Vitamin B_{12}	Batch	Aerobic Mixed	Bacteria	35–160		Stirred tank	Molasses Corn steep liquor
Xanthan gum	Batch	Aerobic	Bacteria	80		Stirred	Starch/glucose
Cheese	Batch	Minimal	Bacteria	5–24	20	Vats	Milk
Yoghurt	Batch	Minimal	Bacteria	3–36	10	Vats	Milk

For example, many baker's yeast processes are batch, aerobic fermentations involving free organisms in a stirred tank fermenter, while some waste treatment plants using nitrifying bacteria have anaerobic continuous systems with immobilised organisms, agitated in a fluidised bed. A summary of some of the different fermentation methods and fermenters used in the food industry is given in Table 2 to show the range of possibilities that have been used to date.

It is apparent from the table that the vast majority of commercial fermenters are stirred vessels containing free aerobic organisms and are batch operated. The reasons for this are discussed in the following sections, but as fermentation technology improves a wider range of designs will become used.

The design of fermenter is also affected by whether the commercial plant is to be dedicated for a single product or to be adaptable for multi-product operation. A batch, stirred tank fermenter will almost always be used for a multi-product plant since the design is adaptable for different conditions.

Bulk products will normally be produced in single-product plants where the fermenter will be designed specifically to provide the most efficient oxygen transfer, volumetric efficiency and heat removal for that process in order to minimise the production costs. In these cases, other designs of fermenter will often be more suitable. Good examples of this trend are the development of the air-lift fermenter for SCP production, and the various fermenter designs in use for fuel alcohol production.

4.2. Batch/Continuous Fermenters

4.2.1. Batch Operation

In batch fermentation, conditions within the fermenter change during the fermentation cycle, with the product and cell concentrations increasing as the substrate is depleted; see Fig. 5(a).

Traditionally, fermentations have been carried out in batch operation and this design has many advantages which are still valid and lead to its continued use in many applications.

(1) Each batch retains its discrete identity, and this is essential for pharmaceuticals and other products which require FDA-type batch logging. Each batch can also be individually controlled to take account of variations in feeds or other conditions.
(2) The fermenter and substrate can be sterilised between batches, ensuring that unwanted organisms are minimised. Once the

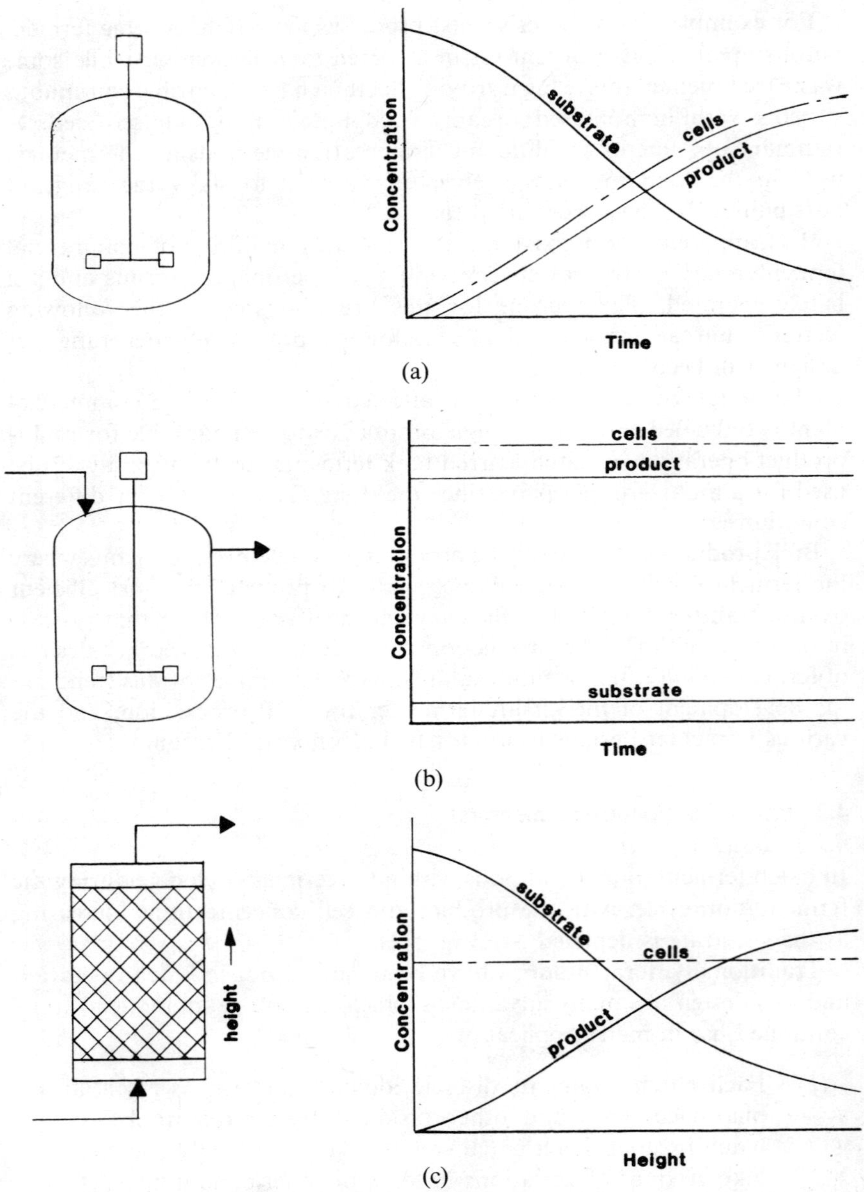

Fig. 5. Fermenter profiles. (a) Batch fermenter. (b) Continuous stirred tank fermenter. (c) Immobilised packed bed.

organism has been established in the fermenter, the effect of infections is usually small since a competing organism has little time to become dominant. In addition, the loss of an occasional batch through contamination may be less expensive than the effects of contamination of a continuous fermenter, where the system would need to be drained, sterilised, re-inoculated and restarted.

(3) Conditions within the fermenter can be changed during the time course of the fermentation for production of e.g. secondary metabolites and other growth-unrelated products such as enzymes.

(4) Scale-up from laboratory and pilot-plant scale is relatively straightforward.

(5) Some suitable batch fermentations may be carried out using unsophisticated technology, and without continuous supervision. For this reason, batch plants are used for much fuel alcohol production in countries where controlled operation under hygienic conditions is not feasible.

(6) Production may be easily adjusted to suit demand, availability of seasonal feedstocks etc. The same fermenters may often be used for different products.

(7) Fed batch operation may be used to maintain the concentration of critical nutrients within acceptable limits by addition of substrate throughout the fermentation rather than solely at the start.

A major disadvantage is that only a proportion of a batch fermentation cycle is available for fermentation; a significant part of the cycle is taken up by sterilisation, filling and emptying, and inoculation. In addition to the main fermenter(s), a seed fermenter is also required to grow the inoculum, and this normally operates at a higher level of sterility than the production vessel.

4.2.2. Continuous Operation

In a continuous fermentation, the conditions in the fermenter remain constant during operation (apart from the start-up period) so that fermentation is controlled with the same cell and substrate concentrations, temperature etc. throughout; see Fig. 5(b). Continuous operation is normally chosen for production of low value, high volume products because of the cost benefits. However, there are a number of other potential advantages to be gained by its use. The main advantages are listed below.

(1) The productivity based on fermenter volume is better since the fermenter can operate at a constant high cell concentration and peak fermentation rate. In addition, there is no downtime between batches as in a batch fermentation.
(2) The fermenter conditions can be optimised to maintain the best conditions for yield throughout operation.
(3) Higher productivity results in reduced capital costs since fewer fermenters are required, and the ancillary seed fermenter can be small since it is only required for occasional start-up duty.
(4) Labour costs are reduced since there is no labour-intensive turnaround of fermenters between batches.
(5) Control of the fermenter is simplified since it operates at steady-state conditions.

There are, however, disadvantages inherent in continuous operation which may make it inappropriate for some fermentations:

(1) 'Wash out' of the fermenter will occur if the rate of cell removal is greater than that of cell growth. In a continuous stirred tank fermenter (CSTF) the minimum fermenter volume is limited by this effect, while in a plug flow type fermenter cells must always be recycled to the inlet in order to maintain a high cell concentration.
(2) The cell population has a spread of ages, all of which operate in the same environment. If a product is required that is not produced in the growth phase, then a second fermenter stage will be required, operating at the necessary conditions for product formation.
(3) Continuous operation is only economic if the period of continuous operation can be extended for at least several months. This requires that the organism be stable over the period (i.e. unlikely to mutate or degenerate) and that sterile conditions are maintained for the same lengthy period.

The increased use of continuous fermentation has led to improvements in methods for aeration and heat transfer so that high fermentation rates can be used, and to the maintenance of aseptic conditions to enable monoseptic operation over extended periods.

4.3. Agitated Fermenters

Most fermenters employ free organisms which are suspended in the fermentation broth either as individual cells or as flocs. The fermenter broth is not usually homogeneous and the organisms are subject to any variations in the system due to cooling, aeration or mass transfer.

4.3.1. Stirred Fermenter

The different forms of fermenters used for free-floating organisms (either individually or as flocs) differ mainly in the means used for aeration and agitation (Fig. 6). The most commonly used fermenter form is the stirred vessel with air introduced via a sparger under the agitator. To achieve the best transfer rate of oxygen into the broth, a radial flow turbine agitator is normally used, and there may be two or three impellers on a single shaft in tall fermenters. The fermenter is fitted with a jacket and/or internal coils for removing the heat evolved from fermentation. Both top and bottom entry agitators are commonly used. The design features of this fermenter are discussed in more detail in Section 5.

The major advantages of stirred fermenters are that the scale-up methods are fairly well understood, and scale-up is relatively simple from laboratory equipment especially if a specialist agitator manufacturer is involved during development. The fermenter can also be readily adapted for multi-product use. The stirred fermenter design is less suitable for high viscosity broths (Fig. 3(b)), and even with broths of normal viscosity the circulation pattern within the fermenter leads to significant changes in conditions throughout the fermenter.

Many manufacturers are now offering mixed axial/radial turbines or better hydrodynamically shaped agitators to improve the mixing efficiency within the fermenter. These designs can lead to significant improvements in fermentation efficiency by ensuring that fermentation conditions are more constant throughout the fermenter. Because heat must be removed through a vessel jacket or internal coils, fermenters above 120–150 m^3 require an additional external cooling circuit.

Stirred fermenters are used at a number of different sizes, ranging normally from 1 litre laboratory units to 150–200 m^3 commercial fermenters, although vessels over 1000 m^3 have been installed for SCP production.[8]

4.3.2. Air-agitated Fermenter

Air-agitated fermenters use an excess of air fed into the vessel through air spargers covering the whole of the base area. The method was used historically for production of baker's yeast, and is still utilised for some mycelial fermentations in which a conventional impeller would be inefficient due to the high broth viscosity. Normally, air agitation is

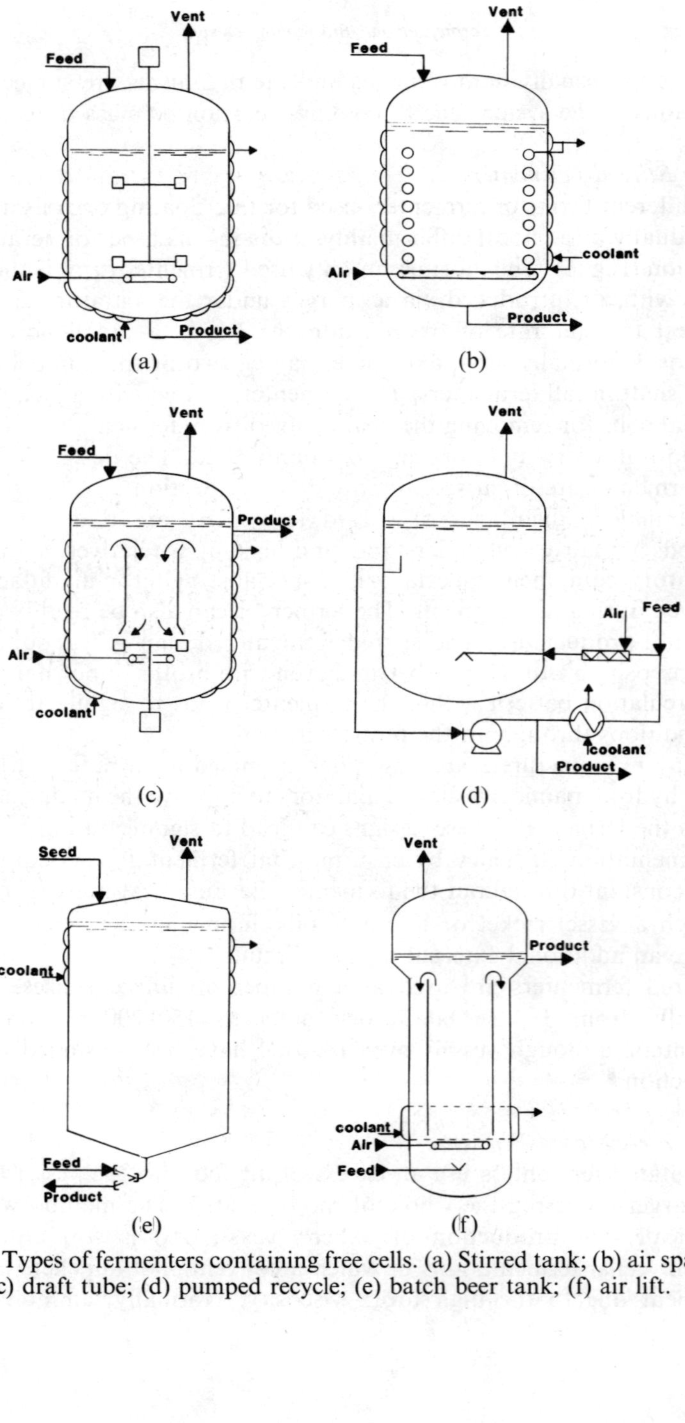

Fig. 6. Types of fermenters containing free cells. (a) Stirred tank; (b) air sparged; (c) draft tube; (d) pumped recycle; (e) batch beer tank; (f) air lift.

considerably less efficient than a stirred tank and the cost of the large volumes of sterile air required is prohibitive. Foaming is also a problem.

In some simpler fermenters, the agitation caused by the carbon dioxide evolved in the fermentation provides an adequate degree of agitation; this is used in beer fermenters. The intensity of agitation increases according to the underlying depth of fermenting broth, so that agitation is more intense at the top than the bottom, and this encourages the use of tall fermenters.

4.3.3. Internal Circulation

Various configurations have been proposed to improve the efficiency of the stirred fermenter by including draft tubes or similar devices to give a pumped internal circulation. The best known is the Waldhorf fermenter developed for baker's yeast production in which the flow rate past the air sparger is increased by placing the draft tube around the agitator. The pumped circulation ensures that once the broth has been aerated it will not immediately be drawn back into the aeration zone but will circulate round the vessel until the oxygen has been depleted. In this way the organism is subjected to a cyclic variation of dissolved oxygen tension (and, in a continuous fermenter, nutrient level) and the design must ensure that the critical level is always exceeded.

At high fermentation rates, the area required for cooling may be greater than that available from the combination of fermenter jacket and internal coils, and the broth must be circulated via an external cooler. In this case the pumped recycle may be utilised for agitating the vessel and for aeration. However, the use of an external circulation loop introduces new problems in maintaining sterile conditions in the pump, heat exchanger and piping of the outside loop.

4.3.4. Air-lift Fermenter

The air-lift fermenter creates an internal recirculation system by efficient use of sparged air. The fermenter is a tall vessel divided internally into a riser and downcomer section, usually by a concentric tube. Air is sparged into the base of the riser section, and the decrease in density caused by the gas–liquid mixture results in circulation of the liquid. Cooling is by internal coils, and internal baffles are fitted to improve mass transfer. The high rate of internal circulation allows the substrate concentration to be maintained at the optimum level. However, the downcomer is unaerated and care must be taken to avoid oxygen starvation in this zone.

By using the power available from air compression, the air-lift fermenter can achieve the high energy input required for aeration and mass transfer in large fermenters without the problems associated with massive agitators. Air-lift fermenters are most economical at larger sizes, and have been installed up to 1500 m^3 (for SCP production).

4.4. Immobilised Systems

Because micro-organisms are small and have a density close to that of water, it is difficult to retain them preferentially within the fermenter. The equipment for separation from the outgoing broth must be kept sterile and is large in relation to the weight of cells to be returned. Immobilised systems are utilised to overcome this problem, and allow the product yield to be increased compared to the cellular growth. Immobilised systems are also used to retain enzymes (which have been extracted from the micro-organism) within a fermenter system.

Immobilised cell systems are not economic for batch fermentation with live cells but the high value of enzymes encourages the use of immobilisation to recycle enzymes in batch systems. The ideal applications are in processes such as waste treatment, citric acid production and ethanol production where the product is formed under conditions different from those needed for growth, and research has been carried out into the use of immobilised systems for production of beer, wine and various solvents.[9-12]

Immobilisation is carried out separately from the main fermenter and involves either absorbing the live cells onto the surface of an inert carrier or combining them within a permeable support such as silica or polysaccharide gels. Alternatively, in some systems which are only mildly agitated (e.g. waste treatment), the presence of an inert support within the vessel enables a film of micro-organisms to form which attains an equilibrium between growth of new cells and removal of dead cells.

Various fermenter configurations can be used to take advantage of immobilised cells, and the main forms are shown in Fig. 7. Most are variations of the packed-bed fermenter in which the immobilised cells are supported on a fixed bed which may be of random packing material, fixed grids, fibres or sheets. Substrate is passed over the bed in plug flow and the concentration of substrate and product change is shown in Fig. 5(c), from a high substrate concentration at the inlet to high product level at the outlet. Substrate may be recycled if necessary to improve efficiency and reduce the bed height.

Other examples of packed beds are packed tower fermenters, trickle bed fermenters (in which the flow is downwards over the packing) and

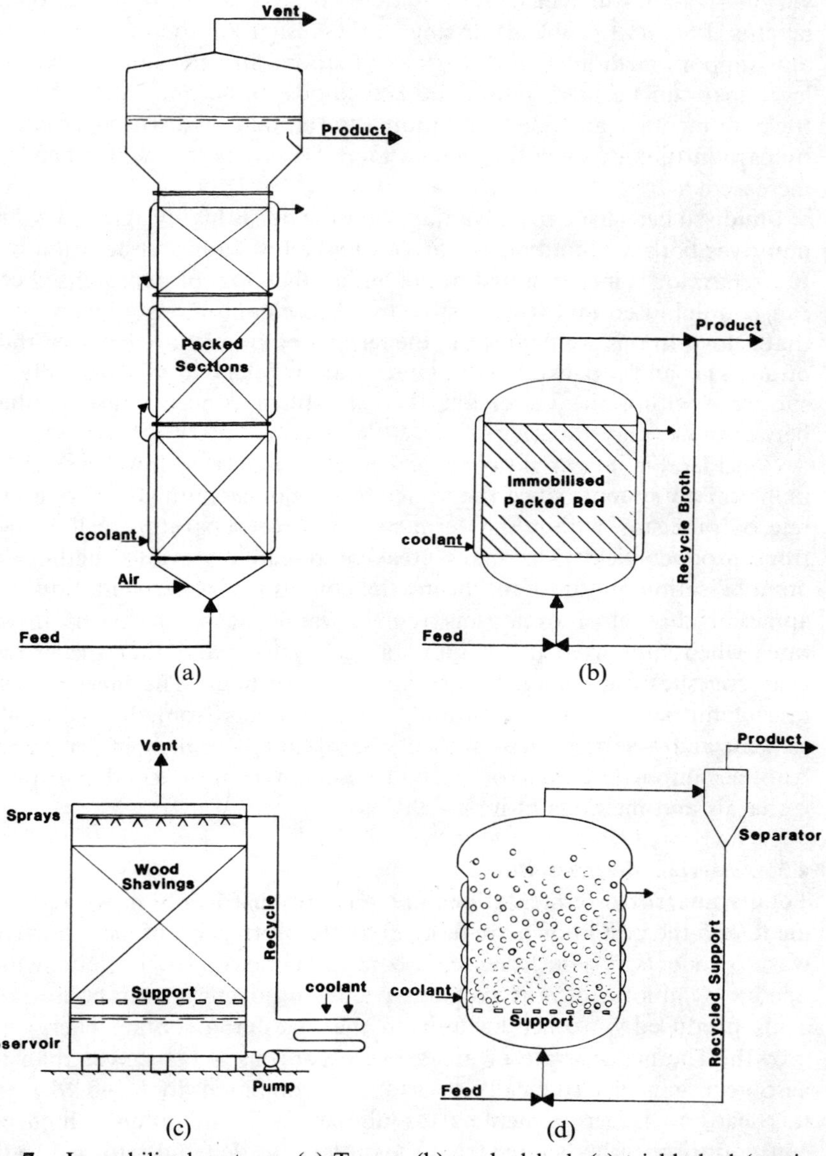

Fig. 7. Immobilised systems. (a) Tower; (b) packed bed; (c) trickle bed (acetic acid); (d) fluidised bed.

various reactors in which the immobilised cells are held within piping lengths. Potential problems arising in the design are the need to replace any support medium that is lost by attrition, and the limited aeration level that can be used before the bed floods or becomes disturbed. In tower fermenters, an added limitation is the amount of disturbance caused by carbon dioxide evolution which increases as the fermenter height is increased.

Fluidised beds have the advantage that the bed is highly agitated, which improves both heat and mass transfer, and allows a higher aeration rate. If a separator is incorporated in the outlet, then loss of immobilised cells can be minimised and it is feasible to replace continuously any support that is lost. In this configuration, the fermenter kinetics are between those of a CSTF and a packed bed; agitation is sufficient to mix partially the substrate within the vessel but there is still a concentration gradient between inlet and outlet.

Considerable research work is being carried out into the use of immobilised systems since they offer the major advantages of retaining micro-organisms within the fermenter and of separating cell growth from product yield. The main areas of research are into methods of immobilisation and into optimising the conditions for fermentation. It is apparent that most organisms remain viable at much lower growth rates when supported in an immobilised system, and that the growth characteristics are changed from those in free flocs. The mass transfer rate of nutrients into the cell and waste products from the cell is also critical, and all these factors affect the optimum fermenter conditions. Another important area of current research is into suitable support materials and methods of immobilisation.

4.5. Anaerobic Fermentations

Totally anaerobic fermentations are generally used only in waste treatment, and the best known are fermenters for production of methane from waste products. These fermenters contain a mixture of bacteria which operate symbiotically; the methane-producing bacteria metabolise fatty acids produced by other bacteria in the digester to obtain energy for growth. The growth rate of anaerobic organisms is far slower than for aerobic organisms (typically 0·1–0·2/day compared to 0·1–0·2/h) and this leads to larger fermenters (or digesters). To maintain a high cell concentration, either sludge from the outlet is settled and returned to the digester, or immobilised systems such as packed or fluidised beds are used.

Few plants have been built since they require a high capital investment and depend for economic operation on the value of the methane by-product. The main applications are for high volume wastes with relatively low concentrations of fermentable material, such as from distilleries, breweries and paper plants.[13]

4.6. Novel Fermenters

Although the majority of commercial fermenters are conventional stirred vessels, a wide range of configurations has been proposed and tested, usually at laboratory or pilot scale, and some of these designs will be used commercially. The aims of these novel designs are variously to improve oxygen transfer, to increase the cell concentration, to give better or more suitable agitation or heat transfer, and to improve sterility. Reviews of different systems have been published.[14,15]

Much work is being carried out into immobilised systems to increase cell concentration without an external separation system. In addition, different designs have been proposed for mammalian cell culture where the requirements are rather different from those of micro-organisms and much lower shear rates are required. Another approach to increasing cell concentration is the use of membrane fermenters which use filtration membranes in the form of sheets or hollow fibres to separate secreted product from cellular material. Fermenter systems with gravity settlers are used with flocculating yeasts in ethanol production to enable a high degree of cell recycle under sterile conditions.[16]

Various forms of air-lift fermenter have been suggested which aim at increased oxygen transfer efficiency at a high level of sterility. Another fermenter design giving the same benefits is the high turbulence plunging jet[17] in which aeration is achieved by the action of a coherent jet of recycled broth.

5. DESIGN PROCEDURES

5.1. Scale-up

The information needed to design a fermenter system for a new process is normally obtained by scale-up from small-scale trials. Although a considerable number of correlations exist it is necessary to carry out experimental work to generate accurate scale-up methods and to optimise the fermentation at successive scales of equipment. In addition, confidence

Table 3
Summary of Scale-up Procedure

	Laboratory scale	*Pilot plant scale*	*Semi-scale*	*Production*
Size	1–3 litres	25–100 litres	0·5–10 m^3	100–1000 m^3
Aims	Operating conditions Kinetics Materials balance Rheology	Additional kinetics Heat transfer Aeration and agitation Effects of recycle, pressure, sterilisation	Confirm scale-up Test production process Asepsis Economic data	Produce product for sale economically
Raw materials	Reagent grade	Reagent grade	Commercial grade	Commercial grade
Product	Small quantities Recover by lab. methods High purity	Initial test material Varying purity	Quality as final process	For sale
Fermenter sterile level	Highly aseptic	Highly aseptic Test lower levels	As final process	Appropriate to process
Containment/safety	High level if genetically engineered organisms are used	As laboratory scale	Depends on regulatory bodies and organism	Appropriate to organism
Other features		Try different combinations of equipment Prove feasibility of larger scale operation	Prove production process	

in the scale-up techniques is acquired at each successive stage, and any pitfalls are hopefully discovered before erection of a full-scale plant.

The number of stages in the development of a process from the laboratory to full-scale production will depend on the amount of data necessary to ensure confidence in the design and costing of the production unit. In general, the importance of obtaining scale-up information is increased by the use of continuous processes, novel organisms, above average size fermenters, and by the use of fermenters other than 'conventional' agitated vessels. Depending on the size of the final production fermenters, four stages of scale-up may be required following initial laboratory shake-flask experiments, and these are outlined in Table 3. The size of each stage is about an order of magnitude greater than the preceding stage.

At the initial laboratory phase, the main aims are to collect kinetic data, to arrive at the optimum fermenter conditions for nutrient, temperature, pH etc., and to determine the materials balance. The final development of the organism is normally carried out at the same time. It is necessary to develop the organism, the fermenter and the overall process design together to ensure not only an economic engineering design but also that research is carried out into the most important aspects. It is frequently the case in the development of fermentation processes that research work is centred solely on the fermenter performance without considering the effects on upstream and downstream processing. The features of a hypothetical ideal organism given in Table 4 are not achieved by any current commercial micro-organism, but the list shows the number of factors that contribute to a successful commercial fermentation process.

Several laboratory fermenters are required to investigate the range of fermenter conditions and obtain the necessary data, and the work may

Table 4
The Ideal Organism

High product concentration
High conversion efficiency
Stable
Low sterility needs
Tolerant of substrate and oxygen variations
High-purity product
Ferments above 45°C
Insensitive to salt level
Non-toxic

still continue at least throughout the pilot plant operation, since the cost of the equipment and the raw materials is relatively low. The next stage of development is pilot plant operation, where the aims are to obtain engineering data for aeration, agitation and heat transfer, to investigate alternative fermenter configurations, and to develop the necessary methods for feed preparation and product recovery.

Operation of the semi-scale plant is intended to prove that the overall process works, including any recycle streams and the maintenance of product quality and asepsis. Further optimisation will be carried out at this stage and the plant will be expected to produce commercial grade product for extensive testing. In many cases, the initial production will also be carried out on this plant.

Apart from the microbiological aspects (e.g. organism, substrate optimisation), the main engineering data obtained for scale-up of a typical aerobic fermentation relate to fermentation kinetics, oxygen transfer rate and agitation, and heat transfer.

5.2. Kinetics

A knowledge of the kinetics of a fermentation is necessary to size the fermenter and its associated equipment, and this information is normally obtained from laboratory experimentation with 1–3 litre fermenters. In batch fermentations, the kinetic model provides information to predict the rate of cell mass or product generation, while in continuous fermentation it will predict the rate of product formation under given conditions. The optimum fermenter size and fermentation conditions can then be calculated for the highest or most economic yield of product by modelling the fermenter system. In a continuous fermenter system, such an economic model might include the capital cost of fermenters with the associated equipment for sterilisation, aeration and initial product recovery, the cost of utilities for sterilisation, air compression, cooling etc., the cost of substrate and nutrients, and manpower.

Since all fermentations involve an interrelated series of enzymic reactions, the kinetic model will usually be greatly simplified compared to the actual series of reactions that occur. The form of kinetics for product formation will vary dependent on whether the product is cellular material, primary metabolite or secondary metabolite.

Detailed treatment of fermentation kinetics is outside the scope of this article, because of the variety of metabolic processes and the complexity of the different fermenter types, but a full treatment may be found in various texts.[5,6,18,19]

5.3. Oxygen Transfer

5.3.1. Oxygen Demand

The oxygen demand for established fermentations may be well documented or alternatively an initial estimate may be predicted by the Mateles equation given in Section 2.2. However, in the development of a new process it is necessary to determine the oxygen requirement experimentally by materials balance techniques before designing the fermenter aeration and agitation.

5.3.2. Basis of Scale-up

Much work has been carried out into predicting the rate of oxygen transfer into stirred fermenters, but most of these correlations are based on experiments in which the mass transfer is determined by chemical methods such as sulphite oxidation (which do not necessarily predict the rate in a biological system). Since the oxygen transfer rate to a micro-organism is affected by the presence of salts, proteins, surfactant etc. in the broth, and also by the rate at which the organism can utilise the oxygen, most production fermenters are designed by scaling up from pilot-scale units, based on the existing correlations given below, and modified to reflect the actual experimental behaviour. Little oxygen transfer data is available from full-scale plants, primarily because the information is considered commercially secret.

Several different scale-up methods are available, although design for constant $K_L a$ (volumetric oxygen transfer rate) is most commonly used for yeast and bacterial fermentations. Because aeration and agitation are interdependent, the best scale-up method may depend on factors such as the viscosity of the broth, the resistance to shear of the organism and the importance of constant nutrient levels. In a large fermenter, it is not possible to design for a range of possible agitation levels, so the accuracy of the scale-up method must be confirmed in the progression from pilot plant to semi-scale operation.

Constant power per unit volume. One of the simplest methods of scale-up is to maintain constant power per unit volume, and this is certainly the simplest approximation to use for an initial estimate. For oxygen-limited fermentations, typical power inputs are 2 kW/m^3 for both biomass and penicillin production.

Generally the geometric proportions of the fermenter and impeller are kept constant, in which case the impeller speed is given by:

$$N_1 = N_2 (D_2/D_1)^{2/3}$$

In initial studies, the air rate may be scaled-up based on the flow per unit volume (vvm), and this method is normally used to scale-up oxygen rates from laboratory scale experiments where power requirements cannot be measured. Typical air rates are between 0·5 and 1·0 vvm for a wide range of aerobic fermentations.

Constant K_La. The most commonly used scale-up for fermentations limited by oxygen transfer is on the basis of constant K_La or constant oxygen transfer rate. This will achieve the same growth rate in the scaled-up fermenter and ensure similar kinetics. The method requires that the power required to achieve a given K_La can be predicted for the fermentation system, and many correlations have been proposed. Because of the non-Newtonian nature of most fermentation broths and the effects of surface tension and solution composition on the oxygen transfer rate, it is recommended that the correlations are confirmed or modified by experimental work. Several different methods are available and include sulphite absorption tests, degassing methods, oxygen uptake and measurement of an overall mass balance.[4] If possible, it is best to carry out trials on the organism itself rather than by sulphite absorption since this will give a more accurate estimate.

Typical predictions for K_La are of a similar form, but the exponents vary depending on the scale of operation, the nature of the substrate and its viscosity, i.e.

$$K_La = K(P_g/V)^{0\cdot4}(V_s)^{0\cdot5}N^{0\cdot5} \qquad \text{(Ref. 20)}$$

$$K_La = K(P_g/V)^{y}(V_s)^{x}a^{-0\cdot86} \qquad \text{(Ref. 21)}$$

In a fermenter system the power delivered by an agitator will differ depending on whether the vessel is gassed (or aerated) or ungassed. The ungassed power input is needed to size the agitator motor for start-up conditions such as substrate preparation or sterilisation, and also in many calculations as the basis of calculating the gassed power input.

The ungassed power input for a broth which approaches Newtonian behaviour can be estimated using the equation suggested by Rushton:[22]

$$P_0 = \frac{N_p \rho N^3 D^5}{g}$$

where N_p $(=Q_a/ND^3)$ is the ungassed power number and is approximately 5·0–6·0 for turbulent flow using a six-bladed turbine.

In broths containing mycelia or concentrated yeast the non-Newtonian character has a significant effect on the power input; a modified Reynolds number is proposed by Wang.[6]

Typically, the power input from a gassed impeller is less than 2/3 of the power from an ungassed impeller, and several different correction methods have been proposed.[6,23,24] The equation of Michel and Miller[24] predicts for Newtonian liquids:

$$P_g = K\left(\frac{P_0^2 N D^3}{Q_a^{0\cdot 56}}\right)$$

Alternative fermenter designs, such as air-lift fermenters, require similar methods for scale-up, but there is considerably less data available for mass transfer estimates and experimental confirmation of the scale-up method becomes more important.

Constant shear. If the micro-organism is shear-sensitive, the shear force should be maintained constant during scale-up. The maximum shear force is proportional to the tip speed and so:

$$N_2 = N_1(D_2/D_1)$$

It is possible to maintain constant tip speed and either constant power per volume or constant $K_L a$, although geometric similarity cannot also be maintained for the latter.

If the culture is found to be damaged by shear, it is important to establish whether the average shear force is more important. The average shear is proportional to the agitator speed,[25] and therefore reduces if the agitator is scaled up using constant power per volume or constant $K_L a$.

Alternatively, some antibiotic fermentations require that a minimum shear be exceeded and hence use fast, small-diameter impellers. The correlations for $K_L a$ given above are used to estimate the necessary impeller power input.

5.4. Heat Transfer

The fermenter contents must be cooled to remove both the heat produced by metabolism of the organism (Section 2.3) and also the heat input from agitation. Cooling is usually achieved through a water jacket on the fermenter, and this may be supplemented by internal cooling coils. In general, the mixing provided for aeration will ensure a good heat transfer coefficient, but in viscous or non-Newtonian broths the agitation at the wall may be severely reduced.

The heat transfer coefficient can be estimated using methods available for chemical reactors[26,27] of the form:

$$\mathrm{Nu} \propto \mathrm{Re}^{0.67}\,\mathrm{Pr}\,(\mu/\mu_w)^{0.14}$$

where Nu, Re and Pr are the Nusselt, Reynolds and Prandtl numbers.

As the size of the fermenter is increased, the area available for heat transfer through jacket and coils decreases relative to the fermenter volume and eventually becomes insufficient. For this reason, jacketed fermenters are generally limited to below 150 m^3 and larger fermenters require a heat exchanger operating on an external pumped circuit. This exchanger may be a plate exchanger if only a moderate level of asepsis is needed, but if high sterility is required then exchanger, pump and pipework must all be designed to high standards.

Care is needed in the design of the external cooling system to ensure that the organism is not overcooled, that the organism is not damaged by the shear action of the pump and that its viability is not reduced by removal from the fermenter for too long, due to loss of oxygen, substrate, etc.

6. SAFETY

The introduction of new biochemical processes results in additional safety hazards to those met in the 'conventional' food processing or chemical industries due to the handling of live organisms.[28] However, once these hazards have been understood and overcome, the overall level of safety in biotechnology processes is very likely to be greater than that in the chemical industry. Fermentation processes are carried out in dilute solutions at low temperatures and pressures, and the major hazards associated with the chemical industry (high temperatures and pressures, flammable materials, exothermic or explosive reactions and production of toxic by-products) are usually absent.

The potential hazards that can result from the use of biochemical processes are the risk of exposure to live cells, and the lesser risk associated with exposure to dead organisms or fermentation products. Although dead cellular material and products are not normally toxic, they may present a health hazard to operators if they are not correctly handled. Biologically active molecules such as antibiotics may cause allergic reactions while the health hazards arising from cell debris are similar to those from naturally occurring protein dusts. These hazards normally occur in the separation processes downstream of the fermenter, and the necessary degree of protection corresponds to accepted good manufacturing practice in the food and pharmaceutical industries.[29]

The risks associated with live cells are present in the fermenter and its associated equipment. Unlike the risks caused by exposure to chemicals, there is little quantitative data on the acceptable level of exposure to different micro-organisms and the concept of TLV (threshold limit value) used to determine the protection needed for personnel from chemical processes is not yet applicable. Undoubtedly quantitative data will be developed as more operating experience is gained in fermentation processes but it will take some considerable time before reliable data are available. In the meantime, other methods have been evolved to determine the level of protection that is required for biochemical processes.

There are three broad categories of micro-organism, based on the understanding of the potential risk due to the organism.

(1) A number of micro-organisms are currently in use in conventional food processes. Examples of these are yeasts for beer and baking, and bacteria for dairy products. The hazards associated with these organisms are generally low, and well understood, and acceptable levels of hygiene and protection have been evolved within the industry.

(2) New fermentation processes will make use of genetically engineered organisms which are tailored for efficient production either by classical genetic methods or by the more recent procedures of genetic manipulation. The potential risks associated with these organisms have to be determined by experience, and precautions are required until an organism is proved to be safe. The possible risks may be due to the host organism, modification of the organism by mutation or other unknown characteristics.

(3) Pathogenic organisms have direct toxic effects and are highly unlikely to be used in food manufacturing processes. However, they may be used in laboratory experiments to develop suitable organisms for food-related products.

In the mid-1970s, there was considerable concern both by the public and parts of the scientific community about the dangers of genetic engineering. In the UK, the GMAG (Genetic Manipulation Advisory Group) was formed to report on the risks and to advise on the level of containment and protection required when handling recombinant organisms. A similar body, ACDP (Advisory Committee on Dangerous

Pathogens), had been formed earlier to advise on the risks associated with handling pathogens. Both GMAG[30] and ACDP[31] categorised micro-organisms according to their perceived hazard, and laid down containment levels for handling each category under laboratory conditions. In assessing the risk category, GMAG considered the ability of a recombinant organism to cause damage to the environment (i.e. to humans, other animals and to plant life). The extent of the perceived hazard depends on the organism's ability to infect a host, its capability to produce a toxin and finally whether the organism can survive and produce toxins under the new conditions. The ACDP categorisation relates to individual identified pathogens which are assessed by an expert committee.

Operation of large-scale commercial fermentation processes is recognised to be different from laboratory work, and GMAG[32] published a report advising on the different procedures and precautions required for industrial processes involving recombinant organisms. In the laboratory, the work is experimental and the GMAG report stated: 'Large scale work differs from most laboratory work involving recombinant DNA technology in that organisms used and product (if any) isolated are completely defined prior to the start of development or production'. Although the higher number of cells present in industrial scale fermenters may increase the occurrence of alternate strains by mutation or reversion, the economics of a commercial fermentation usually depends on the use of a stable organism.

Subsequently, the responsibility for much of the work of GMAG (including regulating the safety of industrial fermentation processes) has passed to the H&SE (Health and Safety Executive). H&SE guidelines for industrial scale processes will eventually be published, but their role in regulating the safety of biotechnology processes has been clearly explained.[33] It is likely that three risk categories will be proposed for industrial processes instead of the four categories used in laboratory work. These are shown in Table 5 with the equivalent existing laboratory codes for comparison. In practice, the majority of commercial processes will be included in category 1. The cost of including extensive containment facilities would make the cost of commercial fermentation plant unacceptably high except for high-value products which can be produced in small, laboratory scale equipment.

Where the organism has been identified as presenting a low risk then it is expected that new fermentation processes may proceed without restrictions. As in all other areas of industrial safety, the onus is on the operating company to ensure safe operating conditions 'so far as is

Table 5
Containment Categories

H&SE risk category	*Examples*	*Existing codes*		
		GMAG[30,32]	*ACDP*[31]	*NIH (USA)*[34]
3 (high)	Highly virulent pathogens; rDNA with 'dangerous' vectors/DNA fragments	IV	4	P3
2 (medium)	Most experimental and pilot scale work	III, II	3, 2	P2, P1
1 (low)	'Proven processes'	I	1	—

reasonably practicable', and to show to regulatory authorities that the process and operating procedures are adequate. This is best achieved by consultation with the H&SE from an early stage.

In the USA, similar regulations for laboratory and industrial scale processes have been published by the NIH (National Institutes of Health).[34] However, in the USA an operating company must comply not only with the Federal Guidelines but also with local regulations which may change due to local opinion. This latter complication has affected the choice of sites for both experimental and commercial installations, since local regulations may change retrospectively and such a change could effectively close down a facility.

7. STERILE DESIGN AND CONTAINMENT

7.1. Principles

Both the level of sterility and the degree of containment necessary for a fermentation process vary over a wide range. Strictly speaking, sterility is an absolute term and a sterile environment is one in which no viable organisms exist. However, it is convenient to refer to levels of sterility to represent the probability of organisms passing into the process from the environment.

The reason for maintaining sterility is to protect the fermentation from extraneous organisms which may compete with the preferred micro-organism to reduce the yield, or which may produce unwanted or toxic products. Whereas the containment level is determined by the characteristics of the organism, the level of sterility depends on process economics, the characteristics of the nutrients, the organism and the product.

Continuous processes tend to require a higher level of sterility compared to batch processes, since the process economics usually require operation for a long period between shutdowns. With batch processes, the cost of achieving a high level of sterility may be more than the value of a small proportion of batches lost through contamination. Obviously, many products such as those for the health care and pharmaceutical industries, and foodstuffs such as SCP, must be protected against contamination, and this increases the required level of sterility of the fermenter and its associated equipment. Research and development fermentations must also be carried out under highly sterile conditions in order to obtain

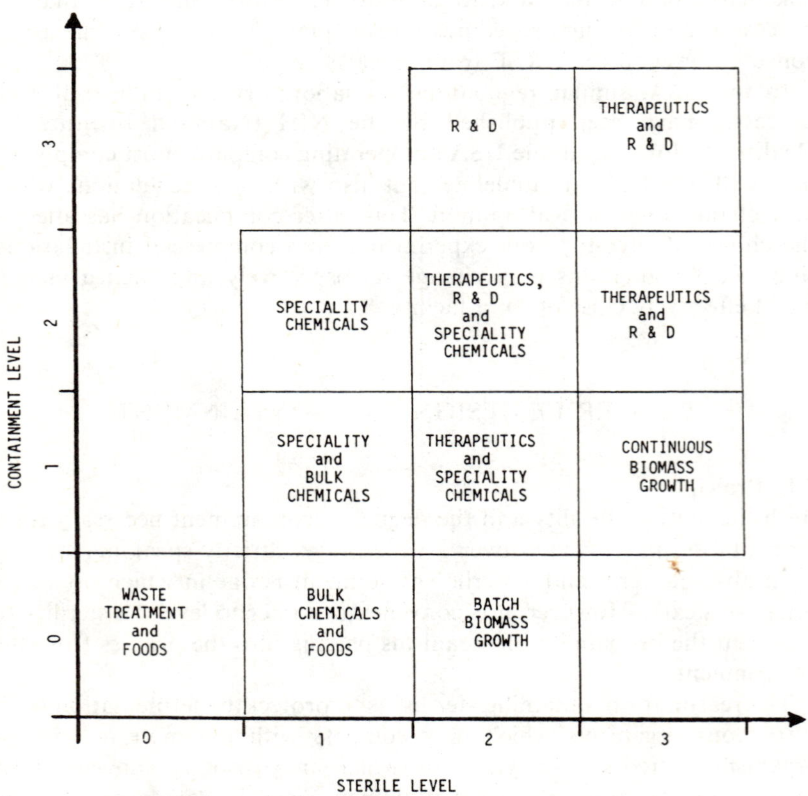

Fig. 8. Levels of sterility and containment.[36]

representative and reproducible data, even if the resulting commercial process is expected to be run under only nominally sterile conditions.

Figure 8 indicates how the requirements of containment and sterility vary between different categories of fermentation process, with the axes showing increasing requirements of each. The longer-established processes can operate under less rigorous standards of process protection, and it is apparent that (apart from R&D operation) the value of the product increases with increasing levels of either sterility or containment. If a new fermentation process is to be competitive with existing comparable products, it must not require a significantly higher level of sterility or containment than its competitors since the cost of achieving increased

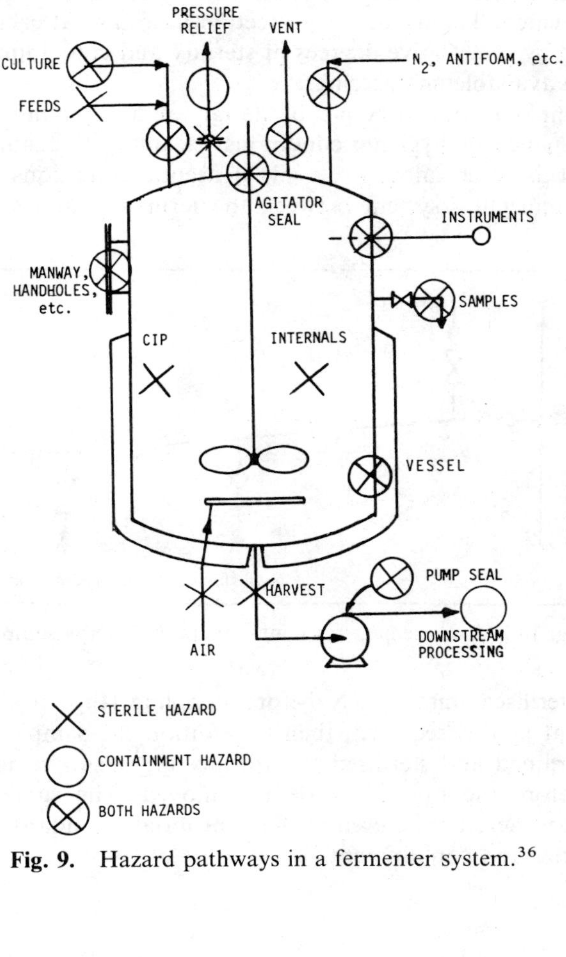

Fig. 9. Hazard pathways in a fermenter system.[36]

levels of either is high. It is therefore essential to define the correct level of barrier between the fermentation process and the environment that is needed to achieve both safe and satisfactory operation before designing the fermentation equipment.

It is also useful to consider a unified approach whereby similar engineering methods can be used to achieve both sterile barriers and containment in order to reduce the overall cost of achieving these aims. A diagrammatic stirred fermenter is used in Fig. 9 to show the 'hazard pathways' typically present in a fermenter. Similar pathways exist in the feed preparation prior to the fermenter and in the downstream separations. The use of different symbols to show where a risk to sterile operation or containment exists highlights the pathways where a common protection can be provided. The degree of protection necessary at each pathway is determined by the relative degrees of sterility and containment required, and by the available engineering.

An example is the principle of design of a sampling system. For the maintenance of hygienic conditions, steaming the sample point to atmosphere is often adequate, while if aseptic conditions are required then the connecting system between the fermenter and sample holder

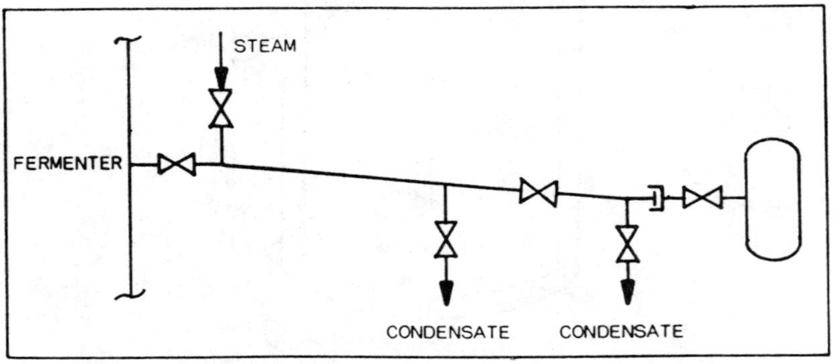

Fig. 10. Arrangement for containment and sterile sampling.

must be sterilised immediately before sampling (Fig. 10). However, if containment is also required, then in addition the sample transfer line must be drained and sterilised (to kill any organisms remaining in the system) before the sample bottle is removed. The same engineering methods can therefore be used at the appropriate standards to achieve a sterile barrier and containment.

Table 6
Application of Engineering Solutions to Hazard Categories[36]

Arbitrary level	*Sterile engineering*	*Containment engineering*
0 (unclassified)	Hygienic practice	As for similar chemical plants
1 (low)	Hygienic practice, but with CIP for vessels Feed sterilisation Clean rotary seals No secondary containment, but fermenter environment cleanable	As GMP CIP for vessels Vents at safe place Aerosol protection No open handling of viable organism Process area containment
2 (moderate)	High sterility achieved by: Sterilisable fermenter Feed and air sterilisation Sterile rotary seals Sterile pipework design Hygienic environment, using building containment Open processing of product in a clean room	High integrity vessel Sterile vent filters Treatment of vents Sterile rotary seals Sterilise before opening Process building containment under negative pressure with restricted access High velocity extraction around leak points
3 (high)	Aseptic operation by rigorously applying (2) above Avoidance of openings including flanged joints and rotary seals Containment around equipment openings if necessary	Vessel design as in (2) Containment around process equipment with personnel access after sterilisation Process building containment with HEPA filtration of exhaust air

7.2. Techniques

In the operation of a sterile fermentation system there are three distinct sets of techniques to be applied. The fermenter and its associated equipment must be pre-sterilised, the feeds to the fermenter must be made sterile, and finally the sterile barrier between the fermenter and the environment must be maintained. It has been stressed that different levels of sterility and containment require differing standards of engineering. Table 6 summarises the engineering solutions necessary for these different hazard levels.

7.2.1. Pre-sterilisation

The most commonly used agent for sterilisation is wet steam, and exposure to wet steam at 120°C (1 bar g) for 30 min is sufficient to ensure asepsis. It is essential that the steam condenses on all parts of the equipment and can drain freely away to avoid non-sterile areas being left in the system. For this reason both the design of the fermentation equipment and the procedures for pre-sterilisation are critical if high levels of sterility are required. Every part of the fermenter system within the sterile barrier must be pre-sterilised to avoid contamination.

All equipment must be free-draining, and this requires careful design not only of fermenter internals such as baffles, supports, agitators etc., but also of piping systems and valve orientation. The surfaces must also be smooth and crevice-free to ensure that deposits can be washed off prior to sterilisation, without leaving pockets of contamination. Although the finish of welds is important to avoid crevices (100% radiography is required for asepsis), it is not necessary to specify highly polished surfaces on vessels since it has been shown[35] that a lesser quality of finish is adequate if spray jets are used for the initial cleaning. Welded joints should be used, as far as is possible, for piping and fittings to avoid crevices.

Not only must equipment be free-draining to ensure asepsis, but every part must be raised to the condensing temperature. Care is needed to ensure that the insulation is adequate to avoid cold spots, especially where heat sinks may be present in the equipment. All the fermenter systems must be designed for full vacuum conditions if steam sterilisation is used, to avoid the possibility of imploding the vessels.

The sequence of pre-cleaning and pre-sterilisation for a complex fermentation system will take a considerable time, and for a batch system this will lead to a high proportion of downtime between batches. Initially the equipment must be cleaned using a dilute caustic detergent to remove deposits, and then rinsed. The steam pre-sterilisation must then be carried out in a predefined sequence to ensure that no unsterilised zones are left between valves or due to blanketing by inert gas or condensate pools. Because of the critical requirements and the complex sequencing required in pre-sterilisation of some systems, microprocessors are necessary to control the operation and provide the correct interlocks.

Lower levels of sterility can be achieved by less rigorous methods. In the brewing industry, the use of cold caustic solutions is adequate to provide the required hygienic standards. This results in significant reductions in the cost of vessels compared to those needed for steam sterilisation since the fermenters are not designed for vacuum duty and lagging is not necessary. It is evident that, with caustic cleaning, the detergent solution must reach all areas and the system must drain freely. For lower sterile levels, hygienic equipment such as dairy couplings for pipes and fittings are appropriate.

7.2.2. Feed Sterilisation

The purpose of sterilising the incoming raw materials is to prevent the ingress of competing organisms. A few fermenter systems are self-sterilising and can operate successfully without the need to sterilise the feeds (e.g. some continuous fuel alcohol processes where ethanol inhibits growth of other organisms), but these are the exception. Sterilisation can be achieved either by moist heat or by filtration; the use of heat is both more economical and more generally effective.

As with pre-sterilisation, the feed must be held at the sterilising temperature for a period of time to ensure that unwanted organisms are killed. Because the destruction rate depends both on the resistance of the infecting organisms and the sterilisation temperature, both the sterilisation temperature and time can be varied to achieve the required degree of sterilisation.[37]

In small batch operated fermenters, the liquid feeds may be sterilised *in situ*. The fermenter is first pre-sterilised and then, after introduction of the feeds, the contents are raised to the required temperature and held before cooling to the fermentation temperature. Because of the time taken for heating and cooling, and the added complexity of the vessel design, this method is normally only used for fermenters below 1000 litres. The feeds to most continuous processes and larger batch fermenters are sterilised continuously before entry using a series of heat exchangers and a holding vessel. The energy needed to sterilise the feeds for a large continuous process is considerable, and significant cost savings can be made by optimising the system and using heat recovery.[37]

Heat sterilisation cannot be used for heat-sensitive feeds or for gaseous feeds such as air, and these feeds are sterilised by filtration to between 0·2 and 0·5 μm depending on the degree of sterility required. Depth filters are relatively low cost items and rely on filtering through a thickness of medium such as glass wool to remove particles. The filters are not absolute devices and a proportion of small particles will pass through the filter. Because particles can both migrate and grow through the filter, the efficiency falls with time and the filters require re-sterilising at intervals. A membrane filter ensures a sharper separation of particles by use of a thin membrane with fine pores. To avoid blinding, a series of filters are employed, making the best use of the filtration area by removing larger particles in the initial stages and, in large installations, optimisation of this arrangement can lead to large reductions in cost. Significant improvements are being made to designs of membrane filters for sterilisation, especially in the ability to steam-sterilise and in reliability; these filters usually offer a more effective and economic design.

Where containment is required, similar methods to those described above can be used. Liquid products may be killed-in-place in the fermenter (using live steam or biocides) before harvesting, or heat-sterilised continuously after leaving the fermenter.

7.2.3. Sterile Barriers

In order to maintain a high level of sterility within a fermenter system, every boundary between the fermenter and the environment must be protected. It is important to minimise the number of entries to the fermenter because of the cost and complexity of providing suitable barriers, and the possibility of their failure.

Valves. For the highest sterility and containment levels every pipework connection to the fermenter must be protected by a combination of valves incorporating a live-steam barrier. The overriding consideration when specifying valves for the most stringent sterile levels is the absence of non-draining pockets, crevices and voids which cannot be steam-sterilised. Valves for high levels of sterility employ either steam-purged glands or are totally enclosed and crevice-free (e.g. special diaphragm valve designs). For lower levels of sterility and containment engineering, special designs of ball or diaphragm valves suffice.

Sterile barriers for out-flowing streams can take several forms but are mainly based on forward velocity or a combination of forward velocity and a valve combination. A system based solely on velocity is only permissible for the lower levels of hazard, since the barrier does not work when flow ceases, and unforeseen flow upsets can allow contamination.

The requirements for sampling are more critical, since not only must the sterile barrier be effective, but a representative and viable sample must be obtained. The principles have been described in Section 7.1, and proprietary systems are available to ensure sterile sampling.

Safety relief valves, which do not provide an effective biological barrier, are not allowed for sterile applications, and bursting discs incorporating stainless steel diaphragms should be used. In containment cases, the relief must be led to a secondary vessel, designed to contain and ultimately process the vented material.

Rotary seals. Sterile rotating seals are needed both for agitator shafts and pumps within the sterile environment, and either double mechanical or labyrinth seals are suitable. Moist steam is used as the sterilising medium between the seals for high-risk applications, but hot water or biocide is acceptable for lower risks. If possible, the use of seals should be avoided where the risk is high, and magnetic couplings or alternative designs should be used.

Vent gases. For sterile design the concern is to stop contamination. Depending on the scale and aeration rate, either a sub-micron filter or a sterile flow barrier in the vent line is used. One particular problem with vent filters on multipurpose fermenters is the risk of batch-to-batch contamination, and a policy of changing or steam-sterilising the vent filter has to be adopted in these cases. Where a lower sterile level is needed, adequate protection can be provided by a 'Pasteur bend' or similar vent pipe. This relies on a constricted 180° bend to provide both a gravity and a velocity barrier to organisms entering via the vent line.

For vent gas containment, the highest standard is to pre-filter the gases to knock out spray and then use two 0·5 μm filters in series or to pass the gas stream to an incinerator.

On large-scale units, the size and cost of adequate filters and fans (to overcome the pressure drop) may make filtration uneconomic and an incinerator the only suitable means. At lower hazard levels only one filter is needed, or caustic scrubbing may be suitable.

Enclosures. The barriers discussed so far have all been primary barriers to separate the fermenter from its surrounding environment. The use of secondary enclosures (providing a high degree of ventilation to local areas of risk) and tertiary enclosures (enclosing the process area in a specially ventilated building or room) is well understood for small-scale sterile operations in the pharmaceutical industry. In this context, sterile cabinets are used to provide secondary enclosure, and these are sited within 'clean rooms', to provide air at a high level of purity around critical sterile operations.[29,38,39] Within fermentation halls, special building finishes and standards are used to discourage the growth of unwanted organisms.

The critical difference between the types of enclosure for sterility and containment is the air flow pattern. For sterile engineering the air flow is outwards, in order to protect the contents by an elevated pressure in the enclosure, and the inlet air is passed through sterile filters. The converse is used for containment. If a fermentation requires both precautions, containment will take precedence and the enclosure will be at negative pressure, but include sub-micron HEPA filtration of both inlet and outlet air. Also in this case the secondary enclosure would be enclosed in a 'clean' tertiary enclosure maintained at a positive pressure with respect to atmosphere. The cost of providing sterile or containment enclosures for other than high-value products would be prohibitive.

8. FUTURE TRENDS

It is probable that over the next decade at least, a high proportion of commercial fermentations will still be carried out in conventional, stirred tank fermenters. The reasons have been discussed earlier and are both practical and economic. The equipment will allow most fermentations to be carried out economically and the fermenter system can be designed using existing scale-up techniques, allowing a new process to be developed in the shortest time and at minimum cost. Improvements are likely in such aspects of the detailed design as impeller design (to achieve better oxygen transfer and mixing), arrangements for sterile and containment barriers, and in the development of process control.

Alternative fermenter designs will tend to be used in processes for bulk products, where the product cost is crucial to the competitiveness of the product and it is necessary for the fermenter to operate at the best possible efficiency. This incentive is not normally present for high value products where the overriding consideration is to get the product on the market as quickly as possible. For these products, alternative designs will only be used where necessary to satisfy the demands of the organism.

The airlift fermenter is likely to find increased applications for production of bulk products, since it has several advantages for large scale, aerated fermentations. The construction is suitable for large volume fermenters and the oxygen transfer efficiency is high. In addition, because the power for agitation and aeration is supplied to the air compressor, it is possible to use alternative sources of power (other than direct electric drive) depending on local availability and costs.

Potentially important advances could be made in the use of immobilised fermenter systems. These offer the potential both to decouple cellular growth from production and expression of the desired product, and the means to retain a high cellular concentration in the fermenter. The conversion efficiency will therefore increase (since substrate is not 'wasted' on cell growth) and the cost of downstream separation equipment will be reduced, both of which have a significant effect on the economics of most processes.

Finally, there is considerable scope for improving the characteristics of fermentation micro-organisations by genetic engineering, and it is certain that improved strains will continue to be developed to make many fermentation processes more economic. The most significant improvements would be the development of thermophilic organisms, and those having high tolerance of feed and product concentrations.

REFERENCES

1. Levenspiel, O., *Chemical Reaction Engineering*, John Wiley and Sons, Chichester and New York, 1962.
2. Monod, J., *Ann. Rev. Microbiol.*, **3**, 371 (1949).
3. Mateles, R. I., *Biotechnol. Bioeng.*, **13**, 581–2 (1971).
4. Finn, R. K., Agitation and aeration, in *Biochemical and Biological Engineering Science*, Vol. 1, Blakeborough, N. (Ed.), Academic Press, London and New York, 1967.
5. Atkinson, B. and Mavituma, F., *Biochemical Engineering and Biotechnology Handbook*, The Nature Press, New York, 1983.
6. Wang, D. I. C., Cooney, C. L., Demain, A. L., Dunnill, P., Humphrey, A. E. and Lilly, M. D., *Fermentation and Enzyme Technology*, John Wiley and Sons, Chichester and New York, 1979.
7. Cooney, C. L., Wang, D. I. C. and Mateles, R. I., *Biotechnol. Bioeng.*, **11**, 269–81 (1968).
8. Zanetti, R., *Chem. Engng*, **91**, 18–21 (1984).
9. Kolot, F. B., *Process Biochem.*, 7–13, February (1984).
10. Klein, J. and Wagner, F., *Immobilised Whole Cells*, Dechema Monograph No. 84, Frankfurt, 1979.
11. *Process Engineering Aspects of Immobilised Cell Systems*, Proc. Symp. at UMIST, March 1984, Inst. Chem. Eng., 1985.
12. Bucke, C., *Phil. Trans. Roy. Soc.*, **B300**, 369–89 (1983).
13. Anon., *Biotech. Newswatch*, **3**(19), 8 (1983).
14. Margaritis, A. and Wallace, J. B., *Biotechnology*, **2**(5), 447–53 (1984).
15. Schügerl, K., *Int. Chem. Engng*, **22**(4), 591–610 (1982).
16. Guideboni, G. E., *Enzyme Microb. Technol.*, **6**(5), 193–200 (1984).
17. Jones, R. T. and Eliezer, E. D., A novel bioreactor, *Proc. Biotech. '84*, USA (1984).
18. Atkinson, B., *Biochemical Reactors*, Pion Limited, London, 1974.

19. Aiba, S., Humphrey, A. E. and Millis, N. F., *Biochemical Engineering*, Academic Press, London and New York, 1973.
20. Richards, J. W., *Progr. Ind. Microbiol.*, **3**, 141 (1961).
21. Ryu, D. D. Y. and Humphrey, A. E. J., *Ferm. Tech.*, **50**, 424 (1972).
22. Rushton, J. H., Costich, E. W. and Everett, H. J., *Chem. Eng. Progr.*, **46**, 395 (1950).
23. Calderbank, P. H., *Trans. Inst. Chem. Eng.*, **36**, 443 (1958).
24. Michel, B. J. and Miller, S. A., *Am. Inst. Chem. Eng. J.*, **8**, 262 (1962).
25. Midler, M. and Finn, R. K., *Biotechnol. Bioeng.*, **8**, 71–84 (1966).
26. Edwards, M. F. and Wilkinson, W. L., *Chem. Engr*, 310–19, 328–35 (1972).
27. Kern, D. Q., *Process Heat Transfer*, McGraw-Hill, New York, 1950.
28. Russell, P. D., *Process Engng*, **64**(8), 18–21 (1983).
29. *Guide to Good Pharmaceutical Manufacturing Practice*, HMSO, London, 1983.
30. Genetic Manipulation Advisory Group, GMAG Note No. 14, Medical Research Council, London, 1982.
31. Advisory Committee on Dangerous Pathogens, Report No. 1, *Categorisation of Pathogens according to Risk and Categories*, HSE/DHSS, March 1983.
32. Genetic Manipulation Advisory Group, GMAG Note No. 12 (revised), Medical Research Council, London, February 1982.
33. Norish, R., Safety in biotechnology, *Proc. Biotech. '83*, London (1983).
34. National Institutes of Health, *Guidelines for Research involving Recombinant DNA Molecules*, US Federal Register 48, No. 106, June 1st, 1983, p. 24577.
35. Milledge, J. J. and Jowitt, R., *Proc. IFST*, **13**(1), 57–62 (1980).
36. Brooks, C. H. and Russell, P. D., *Process Biochem.*, **21**(3), 77 (1986).
37. Ashley, M. H. J., *Chem. Engr*, 54–8, February (1982).
38. BS 5295, *Environmental Cleanliness in Enclosed Spaces*, Parts 1 and 2, British Standards Institution, 1976.
39. BS 5726, *Specification for Microbiological Safety Cabinets*, British Standards Institution, 1979.

2

Improvement of the Processing Properties of British Crops by Genetic Engineering

P. R. SHEWRY, M. KREIS, M. M. BURRELL and B. J. MIFLIN

Biochemistry Department, Rothamsted Experimental Station, Harpenden, UK

1. INTRODUCTION: AGRICULTURAL PRODUCTION IN THE UK

Two events have resulted in an unprecedented change in the face of UK agriculture over the past 20 years (see Table 1). The first was the passing of the Plant Varieties Act in 1964. This enabled plant breeders to collect royalties on the sales of seed of their cultivars, and provided a stimulus for the production of high-yielding cultivars. The second was the entry of the UK into the EEC in 1973, which resulted in political control of agricultural production and use by combining financial incentives (or disincentives) to grow certain crops with high tariffs on the import of crops from outside the EEC.

Improved cultivars and agronomy have resulted in substantial increases in the yield/ha of all the major crops (Table 1). The most noticeable changes since 1973 are in the growth of oilseed rape and wheat, and both result from direct financial incentives under the Common Agricultural Policy. The area under oilseed rape has increased over 15-fold since 1973 to become the third most important crop in 1983. Increases of about 50% in the area under wheat and 40% in the yield resulted in an almost two-fold increase in the total production between 1973 and 1983. It is now the major UK crop in terms of total production, although second to barley in area. Even allowing for increased utilization of home-grown wheat in baking (see below), it is now in surplus.

Potatoes and sugar beet are traditionally the two major non-cereal crops. Whereas the area under sugar beet increased in both decades, the

Table 1
Changes in UK Agricultural Production between 1963 and 1983

	Barley	*Wheat*	*Potatoes*	*Sugar beet*	*Oilseed rape*[a]
1963–4[b]					
area (1 000 ha)	1 907	780	311	170	—
yield (t/ha)	3·5	3·9	21·6	31·4	—
production (1 000 t)	6 705	3 046	6 682	5 338	—
1973					
area	2 268	1 146	225	193	14
yield	4·0	4·4	30·4	38·4	—
production	9 007	5 002	6 808	7 427	—
1983					
area	2 144	1 695	195	199	222
yield	4·7	6·4	29·9	37·7	—
production	9 985	10 801	5 814	7 494	—

Data taken from *Agricultural Statistics, United Kingdom*, HMSO, London.
[a] Figures only available for areas in 1973 and 1983.
[b] Year from 1963 to 1964.

area of potatoes declined until the late 1970s and it is currently only the fifth most important crop.

Unlike soybeans in the USA, legumes have never achieved the status of a major crop in the UK. Although peas, lupins and field beans have been evaluated, only the latter is grown sufficiently widely (33 902 ha in 1983) to be listed in UK agricultural statistics.

The changes in production levels of UK crops have been a major factor in determining changes in their pattern of utilization. These will now be discussed.

2. UTILIZATION AND QUALITY CONSIDERATIONS OF THE MAJOR UK CROPS

Quality is defined by use. It is therefore necessary to discuss these together. The traditional and major use of wheat is in baking. British wheats have always been soft with weak gluten, more suitable for cakes and biscuits than bread. It was therefore usual, before our entry into the EEC, to fortify home-grown wheat with strong breadmaking wheats, mainly from North America. The introduction of a high import tariff by the EEC has provided a considerable economic stimulus to increase the proportion of home-grown wheat in British bread, whether by technological innovation or improving the breadmaking quality of UK cultivars. As a result the imports of wheat have fallen from over 3×10^6 tonnes during the period from August 1973 to July 1974, to less than $1{\cdot}2 \times 10^6$ tonnes (provisional estimate) a decade later. Home-grown wheats now account for about 80% of the breadmaking grist compared to less than 50% in 1973–4. Although technological innovation has contributed, a major factor in the increased utilization of home-grown wheats has been the development of improved varieties at the Plant Breeding Institute, Cambridge. It is still, however, usual to fortify home-grown grain with imported grain and/or isolated gluten (see below).

An increasing quantity of wheat is also used for animal feed, an estimated $5{\cdot}1 \times 10^6$ tonnes in 1983–4 compared to $2{\cdot}6 \times 10^6$ tonnes in 1973–4. Almost 100% of this is now home-grown compared to about 90% in 1973–4. The major quality consideration for feed, the amino acid composition, is outside the subject of this review. Wheat is also used to produce starch and gluten by industrial separation processes. Although figures for the UK are not readily available, it is probable that much of the gluten produced is currently used to fortify wheat for breadmaking. Schofield and Booth[1] have compiled a detailed survey of wheat gluten uses and note that gluten fortification is particularly important for the rapidly growing production of high-fibre bread. They also list a wide range of uses in other countries including breakfast cereals, pasta, breading and batter mixes, processed and synthetic meats and pet foods. These all make use of the unusual physical properties of unmodified gluten.

At present the gluten produced industrially is generally of poor quality, probably due to the effects of high-temperature drying. This problem would be most readily solved by more careful control of the processing conditions, rather than genetic engineering of wheat.

Several factors will almost certainly result in an increase in the availability of gluten for other uses in the UK. There are further increases in the production and baking quality of UK wheat, advances in baking technology and an increase in the number of gluten plants in Western Europe. Apart from the food uses listed above, it is probable that gluten will be modified for other uses, for example as an emulsifier or foaming agent. These will require extensive structural changes which would be most readily achieved by chemical modifications of the gluten rather than genetic engineering of the wheat plant.

The other product from the industrial separation of wheat is starch.

The utilization of starch from various sources is considered in a later section.

The major use of barley is as feed for livestock, although the total amount used has decreased ($5{\cdot}3 \times 10^6$ tonnes in 1983–4 compared to $7{\cdot}3 \times 10^6$ tonnes in 1973–4) with the increasing use of wheat. The major quality consideration is again the amino acid composition for monogastric animals.

The second major use of barley is termed 'human and industrial', which is predominantly malting for brewing and distilling. There have only been minor fluctuations in the total consumption (about $1{\cdot}8$–$1{\cdot}9 \times 10^6$ tonnes) over the last decade, and the proportion of home-grown grain has remained at over 95%. The requirements for malting and brewing are complex, and include low grain N, low β-glucans and a high level of starch-degrading enzymes. Limited amounts of malt and barley are also used in the food industry, for example in health foods and malted beverages.

The increases in production of wheat and barley are also reflected in the exports, which are expected to amount to about 5×10^6 tonnes in total in 1983–4 compared to 3×10^5 tonnes in 1973–4.

Despite the increases in utilization and exports, surplus grains of wheat and barley are still sold into intervention. Although the extent of this and the total intervention stocks vary, it can again be expected to increase in the future. It is therefore increasingly important not only to optimize the quality of cereal grain for the traditional end uses, but also to consider other uses, both in the food industry and as industrial raw material.

Other cereals are only grown to a limited extent in the UK. Oats are currently in decline, which is perhaps surprising in view of their high feed quality and yields which are often equal to those of barley. Rye and durum wheat are only grown to contract, for use in crispbread and pasta manufacture respectively.

Although a large amount of legume protein is used in the food industry, this is all imported and derived from soybean. Home-grown field beans are used almost entirely for animal feed, and other legumes as fresh and preserved vegetables. The lack of a high-yielding legume suited to the British climate has long been recognized as a major problem. Until such a crop is developed the processing properties of UK legumes are of purely academic interest.

The processing of potatoes for snack and convenience foods increased from about 16% of the total UK production in 1981 to 24% in 1984. It is certain to increase even more in the future. The main products are crisps and other snack foods, French fries, dehydrated mashed potatoes and canned whole and diced potatoes. The precise quality requirements for these products are listed in Table 2. In addition, the tubers need to be free from disease, of uniform size with shallow eyes and smooth uncracked skin, and with pigments restricted to the surface. These characteristics can, to some extent, be controlled by cultural practice, for example tuber size by planting density. In general, however, it is necessary to grow different cultivars for specific processing applications. For a more detailed discussion than given below see Talburt and Smith,[2] Harris[3] and Young.[4]

Dry matter content is one of the most important quality factors and the precise value varies with the product. Thus crisps need a dry matter content above 21%, and some canned products only about 10%. Although about 60–80% of the dry matter is starch, this is not necessarily the most important. For example pectin content, types of pectins, cell size and ion content may all affect whether a tuber disintegrates on cooking. Although proteins are also an important component of dry matter and have high nutritional quality, the protein content is not usually considered to be important in processing. It is important to define more precisely the chemical components of dry matter which are important for specific processing applications so that varieties can be improved in a more predictable and rational way.

Internal bruising of potatoes can cause problems for all types of processing, and varieties vary considerably in their tendency to bruise. It is not clear whether this is due mainly to biochemical factors or the mechanical strength of the tissues. The former would be factors such as the amounts of tyrosine and enzymes that produce the brown and black pigments. The latter would be the structure of the cell walls and the osmotic properties of the cells. Since we have some knowledge of the pathways of synthesis of quinones and melanins (the brown and black pigments) this aspect should be amenable to genetic manipulation.

Since the potato crop is seasonal, potatoes often have to be stored for up to several months. It is therefore important that the quality of the stored crop is maintained. The quality during storage may deteriorate due to water loss, or to an increase in sugar content. A high content of sugar leads to the production of unacceptable coloured products during the production of crisps, French fries and dehydrated potatoes. Despite a reasonable knowledge of the enzymes of starch and sucrose synthesis and degradation, the exact pathways and their regulation are still not known.

Table 2
Quality Requirements for Potatoes for Processing

	Shape and surface	*Dry matter content*	*Reducing sugar*	*Flesh colour*
Dehydrated for mash	Large; minimum peeling loss	>20% S.G. 1080	<0·5%	Creamy
French fries	Medium to large, >45 mm; regular shape; shallow eyes	>20% S.G. 1080	<0·3%	White to cream
Crisps	Medium, >40 mm; uniform; shallow eyes	>21% S.G. 1085	<0·25%, preferably <0·1%	Cream
Canned	Small, 18–38 mm; uniform shape	<19%		Cream
Canned diced	Large, >50 mm	<10%		White

Source: Potato Marketing Board, UK.

The potential for modifying the composition and properties of two major groups of plant biopolymers (cereal and potato starches and wheat gluten proteins) will be discussed after a brief consideration of the methods of genetic engineering.

3. GENETIC MANIPULATION OF PLANTS: PROGRESS AND LIMITATIONS

Although the quality of UK crops may be affected by the environment in which the crop is grown and by post-harvest processing, the genetic make-up of the plant is of overwhelming importance. The improvement of quality by manipulation of the genetic make-up is now becoming possible for some crop plants.

In the present section we describe the recent advances in molecular and cell biology which make these manipulations possible. These involve the identification and isolation of the genes of interest, their modification *in vitro* and re-introduction into the crop plant. These advances in recombinant DNA technology also allow us to obtain more detailed information about the components which determine quality in food crops, which in itself is of value to industry.

3.1. Gene Isolation

A number of approaches have been used to isolate genes. That most widely used involves first making a double stranded DNA copy (called complementary DNA or cDNA) of the messenger RNA (mRNA) transcribed from the gene. This is then inserted into the DNA of a phage or plasmid which is grown in the bacterium *E. coli*. The steps in this procedure have been described in detail in a number of textbooks (for example, Old and Primrose[5]). An essential point is that the population of cloned DNAs (termed a library) must be sufficiently large to contain one or more cDNAs related to the gene of interest. It is therefore necessary to construct a larger library when mRNA transcripts from the gene are rare than when they are abundant. Because only one cDNA sequence is inserted into each phage or plasmid and only one phage or plasmid infects each *E. coli* cell, colonies of bacteria grown from single cells of *E. coli* contain only one type of cDNA sequence. Molecular cloning is therefore a powerful purification of individual cDNAs. Because DNA can be sequenced more readily than protein, it is now often more practicable to obtain protein sequences by deduction from the nucleotide sequences of cDNAs than by direct sequencing of the protein itself. Advances in the techniques of molecular cloning now allow large libraries of cDNAs to be routinely constructed, with a high probability of obtaining copies of even rare mRNA transcripts. In addition, many of the cDNAs will represent near or full length copies of even long mRNAs.

It is next necessary to identify the cDNAs of interest. Where the mRNA transcribed by the gene of interest forms a significant proportion of the total population and is only found in one tissue or organ (e.g. the developing seed), this is relatively easy. Clones which are tissue-specific can be identified by comparison of the patterns of hybridization to mRNA populations purified from the organ of interest (i.e. the seed) and from green leaves or shoots (see Forde *et al.*[6] for an example). Their specific identity can then be determined in two ways.

(1) Comparison of the amino acid sequences predicted from their nucleotide sequences with that determined by direct sequencing of the protein. In many cases such an identification can be made even if only short sequences (less than 10 amino acids) of the proteins are known.
(2) Identification of the proteins encoded by the cDNAs by hybrid selection translation. In this procedure the mRNA corresponding to the cDNA is isolated from a mixed population of mRNA by hybridization. The mRNA/cDNA hybrid is then 'melted' by raising the temperature and the mRNA translated *in vitro*. The polypeptide product and authentic proteins can then be compared by electrophoresis or isoelectric focusing. It is also possible to identify the polypeptide synthesized *in vitro* by reaction with specific antibodies.

A more direct approach to identify cDNA clones of interest is to use specific oligonucleotide probes. Where part of the amino acid sequence of the protein is known it is possible to deduce all the possible nucleotide sequences that might be present in the mRNA. A series of DNA oligomers containing all these sequences can then be synthesized (typically this may consist of 32 or 64 different oligomers, each 14–17 bases long), and used to identify related cDNA sequences by hybridization.

Where antibodies to the protein coded by the gene are available it is possible to isolate mRNA for a probe by immunoprecipitation of polysomes. The antibodies react with the nascent polypeptide chain being synthesized on the polysomes, and the polypeptide/polysome/antibody complex can then be isolated. The purified mRNA is then used to recognize its complement in the cDNA library.

Antisera can also be used more directly if the cDNA is cloned into an expression vector. In this case the cDNA is inserted into a plasmid or phage downstream from a promoter that initiates transcription, and a codon that initiates translation. One in every six insertions of a given cDNA sequence may be expected to be in reading frame and in the correct orientation, and so can be transcribed and translated in *E. coli* to give a protein which reacts with antiserum raised against the gene product. The individual colonies of *E. coli* are therefore transferred to a filter, lysed and reacted with radioactively labelled antiserum. Colonies producing the protein are detected by autoradiography and rescued from a duplicate filter.

A final approach is to use a heterologous cDNA or genomic DNA probe from another species. This procedure relies on the presence of sufficient homology in the nucleotide sequences of the genes encoding related proteins from different species to allow hybridization. It has proved possible to use cDNA or genomic DNA probes from other plant species or even microorganisms to identify cDNAs for specific plant proteins.

Although much information on the sequence of the encoded protein can be obtained from analysis of cDNA clones, their main application is often to facilitate the identification of genes from a library of cloned genomic DNA. The direct identification of genes in a higher plant genomic library without using a cDNA clone is fraught with difficulties. This is because the genomic DNA of most plants consists largely of repeated sequences (over 90% of the total in cereals) which do not encode functional proteins. In addition, even the remaining small proportion of non-repetitive DNA may contain non-functional copies of genes (pseudogenes). A cDNA clone is not only valuable as a probe to identify genes in the library, but also provides a 'reference' nucleotide sequence for identification of the true expressed gene. Although other probes (oligonucleotides, mRNA or heterologous probes) can be used to identify genes it is not possible to conclude whether the genes identified in this fashion are expressed.

The genomic library is constructed by cutting the total plant genomic DNA into small (typically about 15–20 kilobase pairs) sections, inserting these into phage vectors and growing the phages in *E. coli*. The total haploid genome of a cereal such as barley is 5·5 picogrammes (about $5{\cdot}4 \times 10^6$ kilobase pairs) of DNA. It is therefore necessary to construct a library of about 10^6 recombinant plasmids with 17 kilobase inserts to have a 99% probability of the required gene being present. If the construction of a library with larger inserts is required (for example for the analysis of the structure and organization of multigene families), it is usual to insert the genomic DNA into a cosmid, a man-made 'hybrid' containing phage and plasmid DNA which will accept inserts of up to about 50 kilobase pairs.

In summary, therefore, the current state of the art is such that there are no major difficulties in isolating cDNA and genomic DNA clones for any given plant protein once a basis for identification is available.

3.2. Protein Engineering

This is the term applied to the alteration of the amino acid sequence of a protein by changing the nucleotide sequence of the gene encoding it. The changes may be minor, such as the deletion, insertion or substitution of single residues, or more extensive. It provides a direct means to study the relationships between protein sequence, conformation and function. It also has great potential for producing proteins specifically designed for use in the food industry.

To introduce the changes it is first necessary to isolate, clone and characterize the gene. The methods for this are now well established (see above), and it usually poses no problem. It is then possible to use a number of different strategies to introduce specific types of changes.

Substitution, insertion or deletion of single or small numbers of bases is most readily achieved by oligonucleotide-directed mutagenesis[7] (Fig. 1). The DNA is initially inserted into a vector, preferably single stranded such as the M13, ϕX174 and fd phages, or a double stranded plasmid such as pBR322. An oligonucleotide is then synthesized to direct the mutagenesis. This is short (usually 8–21 residues), and is completely complementary to a region of the cloned (or template) DNA, except for a single nucleotide change (or mismatch). The oligonucleotide is hybridized to the template DNA and extended by the enzyme DNA polymerase from *E. coli* to give a perfect, but linear, complementary copy of the template DNA with the exception of the single nucleotide mismatch. The two ends of the new strand are then joined (or ligated) using a second enzyme, T4 DNA ligase, to form a cloned circular double stranded molecule. When this is used to transform cells the mutant and wild type strands give rise to a mixed population of cells with the potential to synthesize normal or modified proteins. It is therefore necessary to select for the mutant phenotype.

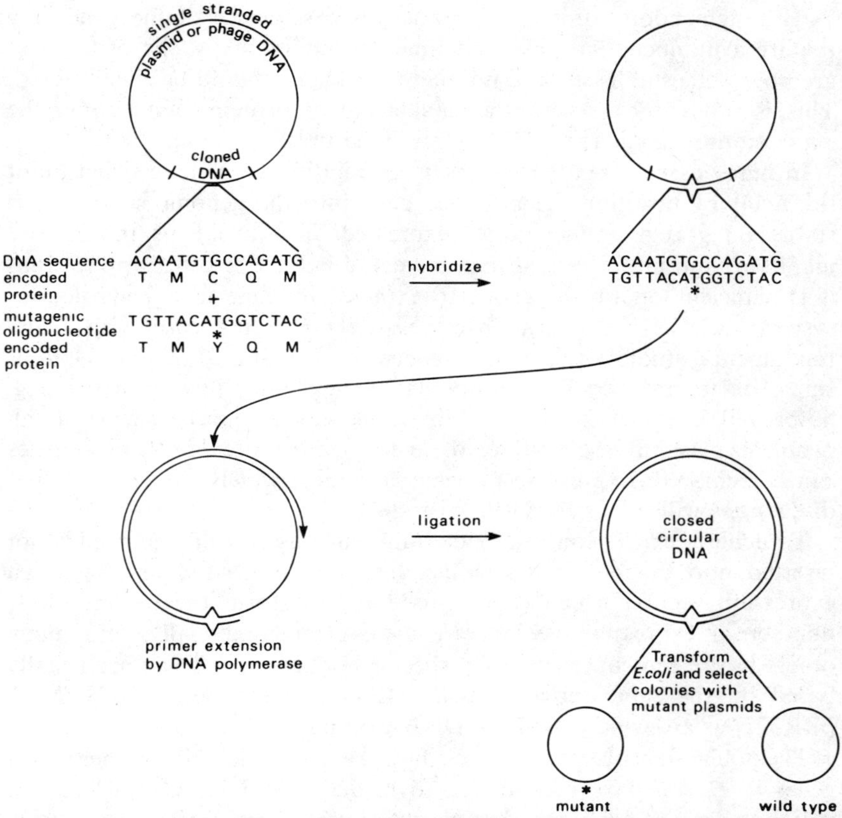

Fig. 1. Steps in the oligonucleotide-directed site-specific mutagenesis of cloned genes. The experiment illustrated is the substitution of a tyrosine residue (Y) for a cysteine (C) in an HMW subunit of glutenin (see Fig. 6). See the text for a detailed discussion.

Other strategies can be used to effect more drastic changes. Whole sequences or domains can be readily deleted from proteins by specific digestion with restriction endonucleases, followed by re-ligation of the cut ends. In many cases, however, it is necessary to insert specific restriction sites at the correct positions using oligonucleotide-directed mutagenesis. It is also possible to construct chimeric genes by ligation of the cut ends to synthetic oligonucleotides, or to DNA fragments prepared from other genes. Again it is often necessary to insert specific restriction sites.

Although modifications of the nucleotide sequence of the gene may require a number of precise and time-consuming steps, the procedures are now well established and no major problems should be encountered. This is not the case with the final stage in protein engineering, the expression of the gene to give the modified protein.

In many cases it may be necessary to modify the genetic structure of the plant by inserting the modified gene into the genome so that it is stably integrated and correctly expressed. The problems in this are manifold, and are discussed in some detail later. For other applications it is sufficient or advantageous to express the gene in a heterologous system such as a bacterium or yeast. This is the case with many fundamental studies of protein sequence/conformation/function relationships, or to produce large amounts of single modified proteins (e.g. 'improved' enzymes for industry) by large-scale fermentation. Although problems may still be encountered, heterologous expression of most genes can be achieved. Expression in bacteria is most widely used and further discussions will be restricted to this.

To achieve expression in a bacterium (such as *E. coli*) the gene is not inserted into the bacterial genome, but into a plasmid or phage (an expression vector) which is then used to transform the bacterium. A number of expression vectors are available commercially, and many others have been constructed for specific applications. These are usually based on the same series of plasmids and phages (e.g. pUC8, M13, pBR322) as are widely used for DNA cloning.

The potential problems in expressing eukaryotic (including higher plant) genes in *E. coli* have been discussed in detail by Harris.[8] He lists five major sources of problems, and ways to avoid them.

(1) Prokaryotic genes do not contain introns. If these are present in the plant gene, it is not possible to achieve expression in *E. coli*. The easiest solution is to use a cDNA copy of the mRNA, from which the introns have been removed by splicing.

(2) The transcription signals of prokaryotes are different from those in eukaryotes. It is therefore usual to express eukaryotic genes under the control of a strong prokaryotic promoter, such as the *lac* and *trp* promoters from the *lac* and *trp* operons respectively of *E. coli*, or other promoters derived from phages or plasmids.

(3) Differences in the structures of prokaryotic and eukaryotic RNAs may lead to problems in translation, due to effects on mRNA stability and ribosome binding. This may be overcome by fusing the gene to a prokaryotic gene so as to synthesize a fusion protein (most commonly

with β-galactosidase from *E. coli*), or by placing it adjacent to an *E. coli* ribosome binding site.

(4) Codon usage may differ between prokaryotic and eukaryotic genes. This preference is correlated with the populations of tRNAs, so it is possible that the translation of some eukaryotic genes will be limited by availability of tRNAs. This is difficult to overcome experimentally.

(5) Prokaryotic cells do not carry out many of the post-translational modifications which occur in eukaryotes. These may include glycosylation, phosphorylation and the introduction of disulphide bonds, and may affect the activity and/or stability of the proteins.

To these may be added a sixth possible problem, the toxicity of some plant proteins to prokaryotic cells. This may be at least partially overcome by using inducible promoters. These allow the gene to be switched on only after a high cell density is reached. Thus, although the product may be toxic, the total yield may be greater than with constitutive expression.

The present brief discussion shows that it is now realistic to isolate plant genes, make specific changes to their nucleotide sequences and then express them in a prokaryotic cell. The major barrier is how to re-insert such modified genes into crop plants.

3.3. Modification of Gene Expression

In some cases it may be necessary to increase or decrease the production of a protein by altering the regulation of its structural gene, rather than to change the structure of the protein itself. The techniques for this are the same as those described above. However, it is first necessary to identify the sequences which regulate gene expression and to establish which changes in nucleotide sequences are required to increase or decrease the level of expression. It is now generally accepted that sequences which regulate the expression of genes are present in the DNA upstream (5′) to the coding region. However, we are still a long way from identifying these sequences for most plant genes, and from understanding the molecular basis of their action.

It is probable that the regulatory 5′ upstream sequences are binding sites for proteins encoded by specific regulatory genes. It may be possible, therefore, to affect the level of expression of whole families of structural genes which are expressed co-ordinately by manipulation of the amounts or amino acid sequences of the regulatory DNA-binding proteins. This may be particularly important if we wish to change the expression of multigene families such as those encoding cereal seed proteins.

The study of plant gene expression and particularly the isolation and characterization of regulatory genes is clearly a major priority if we wish to manipulate crop quality.

3.4. Transformation of Crop Plants

A major limitation to the genetic engineering of crop plants is the re-insertion of the modified genes into the genome so that they are stably integrated and expressed correctly.

Two types of approach have been used to transfer genes into plants.[9,10] Modified tumour-inducing (Ti) plasmids from the crown gall bacterium *Agrobacterium tumefaciens* have been used to introduce genes into a number of dicotyledonous plants including tobacco, petunia, potato, carrot, *Arabidopsis* and oilseed rape. The transformation has been achieved by inoculation of either roots or cultured shoots, or by co-cultivation of protoplasts with *Agrobacterium*. The latter approach depends on the ability to regenerate whole plants from transformed protoplasts, which is not possible for all species. The genes introduced using *Agrobacterium*-based vectors are stably integrated into the genome, and have been followed in the offspring of sexual crosses.

A second approach is direct gene transfer. Although it is possible to introduce DNA into protoplasts or other morphogenetic cells by microinjection, this is too time-consuming for routine use. More promising is the reported uptake of DNA and transformation after treatment of protoplasts with chemicals, heat shock or electric shock. The uptake may result from the transient presence of pores in the plasmalemma. Genes introduced in this way into tobacco have been shown to be stably integrated into the nuclear genome, and transmitted in a Mendelian fashion to the progeny of crosses.[11]

Transformation of cereals has been limited by two major problems: the inability to use *Agrobacterium*-based vectors, and to regenerate whole plants from protoplasts. However, the prospects for the future are promising. On one hand, successful infection of several monocotyledonous species, but not cereals, has been achieved using *Agrobacterium tumefaciens*. These include *Narcissus* and *Asparagus*.[12] On the other hand, successful transformation of protoplasts of diploid wheat (*Triticum monococcum*)[13] and Italian rye-grass (*Lolium multiflorum*)[14] has been reported by direct gene transfer. In the latter case the gene was shown to be integrated into the nuclear genome. It appears, therefore, that transformation of the major cereals will soon be possible. However, there is a problem in that the major cereals (with the exception of pearl millet) cannot be regenerated from protoplasts. Even when this barrier is overcome, it is still necessary to ensure that the transferred genes are correctly expressed, for example in the developing endosperms of cereals and tubers of potato. Research on the tissue-specific expression of plant genes is currently in progress in a number of laboratories.

4. STRUCTURE AND GENETIC ENGINEERING OF BIOPOLYMERS

4.1. Starch

World starch production in 1982 was about 16 million tonnes, with 3·5 million tonnes produced in the EEC.[15] The largest part is used in the food industry, while the rest finds applications in the chemical, paper and textile industries. The most important crops for starch production are maize, potatoes, wheat, and, to a lesser extent, sorghum, rice and barley. The raw material used for processing is mainly determined by the properties of the starch, the availability of the crop and market prices. We will only discuss starches produced from maize, potatoes, wheat and barley. The production of starch is likely to increase considerably in the next decade due to a greater use in the production of sweeteners and ethanol (the latter due in part to higher oil prices).[15,16] Increased use in other industries will depend on the production of starches with a broader spectrum of qualities[15] (e.g. uniform granule size, different proportions of amylose and amylopectin).

Starch is the major form of carbohydrate deposited in all higher plants. In the cereal endosperm, potato tuber and other reserve organs it is packaged into starch granules during one phase of the life cycle and utilized during a later phase. Starch owes much of its functionality (e.g. as a thickening agent in the food industries or plasticizer in the film industries) to its two major components, amylose and amylopectin (Table 3), and to the organization of these two biopolymers into a granular structure. The fine structure of amylose and amylopectin has recently been reviewed by Manners[19,20] and Whistler *et al.*[16] and will now be briefly described.

4.1.1. Structure of Amylose

Amylose is essentially a linear chain of 1000 to 10 000 D-glucopyranosyl units. Studies with the pure enzyme β-amylase showed that most samples of amylose gave about 70% conversion to maltose.[21,22] The limit to complete hydrolysis by β-amylase is probably the presence of some (1,6)

Table 3
Properties of Starches from Different Sources

	Maize	*Wheat*	*Rice*	*Potatoes*
Amylose (%)	23–27	19–28	14–25	20–24
Amylopectin (%)	73–77	72–81	75–86	76–80
Gelatinization (°C)	64–71	53–64	61–78	56–68
Density (g/cm³)	1·514	1·515	1·518	1·519
Crystallinity index	0·39	0·36	0·38	0·25
Granule size (μm)	10–15	20–40		20–185

From Woelk;[17] Skrede.[18]

α-D glucosidic interchain linkages. The molecular size and degree of branching of amylose is heterogeneous,[23] and Hizukuri *et al.*[24] showed about 9–20 branch points per molecule in different samples of amylose. The branch points are the result of a limited action of branching enzyme but the factors which control the action of this enzyme *in vivo* are not known. In solution amylose forms a rigid and opaque gel stabilized by hydrogen bonds. These properties are useful in the confectionery industry where candy pieces require a stabilizer to give shape and integrity.

4.1.2. Structure of Amylopectin

Amylopectin is a highly branched molecule of units with α1–4 and α1–6 linkages. It does not give a starch–iodine blue colour like amylose, but a purple or sometimes a reddish-brown colour depending on the source of starch.[25] Amylopectin has an average chain length of 20–25 and a degree of polymerization (number of glucose units per reducing group) of approximately 1000. Because of the highly branched structure, fewer hydrogen bonds are formed in solutions which remain fluid and clear. These properties are exploited in the food and other industries. The two polymers, amylose and amylopectin, are hydrophobic, and upon heating together form a paste.

4.1.3. The Starch Granule

The starch granules of cereals are present in amyloplasts and vary considerably in size. In wheat and maize the endosperm starch granules vary from about 5 μm to 30 μm. In barley two distinct populations with mean diameters of 5 μm and 25 μm are present, and methods have been described for their separation.[26,27] At maturity the larger starch granules

Table 4
Summary of Mutant Effects on the Contents of Amylose and Amylopectin and the Activities of Synthetic Enzymes

Species	*Biochemical change*	*Enzyme affected*
Maize		
waxy (*wx*)	100% amylopectin	Starch granule synthase
amylose-extender (*ae*)	45–70% amylose	Branching enzyme IIb
dull (*du*)	55% amylose	Branching enzyme IIa Soluble starch synthase II
sugary-1 (*su*)	Phytoglycogen	Phytoglycogen branching enzyme
Barley		
waxy (*wx*)	100% amylopectin	
amylose-extender	43% amylose	
Pea		
amylose-extender (Progress No. 9)	70% amylose	Branching enzymes I and II
Amaranthus		
waxy (*wx*)	100% amylopectin	Starch granule bound starch synthase

Data from Matters and Boyer;[29] Walker and Merritt;[30] Tsai;[31] Nelson;[32] Preiss.[33]

contained a greater proportion of amylopectin. The small granules represent about 90% of the total by number, but 11% by weight.[25] In barley and wheat the smaller starch granules have a higher gelatinization temperature than the large granules, and are usually associated with more protein. The latter may cause effluent problems in wheat starch processing.[28] In the high amylose (amylose extender) barley line (Table 4) the small starch granules are absent and the mean starch granule size is about 15 μm.[27,34] Bathgate and Palmer[27] discuss the possibility that the small starch granules of barley may be under separate genetic control. In potatoes the starch granules vary in size from 10 μm to 100 μm. As the potato tuber grows the average granule size and its proportion of amylose increase. However, the large granules retain the same fibrillar structure as the small granules.

4.1.4. Starch Synthesis

Two families of enzymes are required to synthesize the polymers amylose and amylopectin from the monomeric substrate. These are the starch synthases which synthesize chains of α-1,4-glucan and branching enzymes which synthesize the α-1,6 linkages or branches.

Starch synthase exists in multiple forms which can be divided into two classes depending on whether the enzymes are soluble or bound to starch granules.[35] It remains to be established whether the soluble forms are

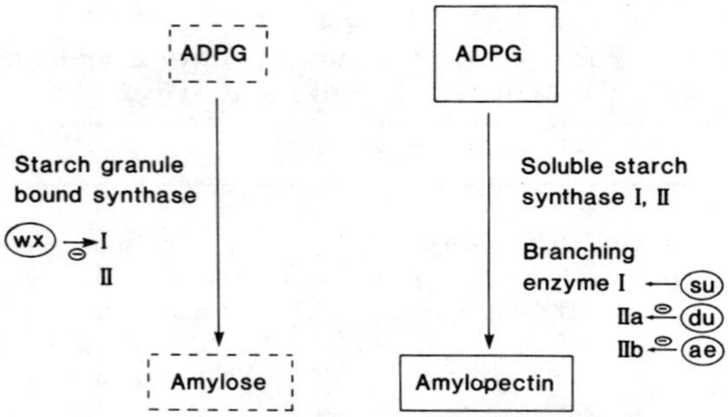

Fig. 2. Possible pathways for the biosynthesis of starch polymers (amylose and amylopectin) in cereals and potatoes.

truly soluble *in vivo*. However, it is clear from work on maize that the amounts of soluble and bound enzymes affect the proportions of amylose and amylopectin; for example the waxy mutant has little bound enzyme and a small proportion of amylose.[33] It has been suggested that the bound enzyme is responsible for the synthesis of amylose while the soluble enzyme synthesizes amylopectin. Whereas the soluble enzyme almost certainly uses ADP glucose as a substrate, there is uncertainty whether the bound enzyme uses only ADP glucose (as shown in Fig. 2) or UDP glucose as well.

The soluble and bound forms of starch synthase can each be divided into at least two types (I and II), which differ in their abilities to catalyse unprimed glucan synthesis, and to use glycogen as a primer.[35] The type I enzymes appear to have lower molecular weights than the type II enzymes (58 000–70 000 compared to 92 000–95 000), and there is also evidence that the type I insoluble enzyme has a lower molecular weight than the soluble form. The relationships of these multiple forms of starch synthase are not known, but immunological studies indicate that they are distinct.

The synthesis of the branches in amylopectin is also complex and incompletely understood. In maize three branching enzymes are present, which show some immunological cross-reactivity and similarities in amino acid composition.

The ADP glucose used by the soluble starch synthases is produced from glucose-1-phosphate and ATP by ADP glucose pyrophosphorylase in a reaction which also produces pyrophosphate and AMP.[36] The enzyme extracted from leaf tissue, maize endosperm and potato tubers is activated by 3-phosphoglycerate and the activation is inhibited by phosphate. This implies that, whether starch is produced from photosynthetic intermediates in the leaf or from sucrose supplied by the translocation stream from distant leaves, the level of 3-phosphoglycerate is crucial in regulating active starch synthesis.

4.1.5. The Genetics of Starch Synthesis

Although most starches from non-mutant lines usually contain about 20–30% amylose, the precise proportion varies between species and cultivars (Table 3). This variation has been shown to be genetically controlled.[30,32] Other properties also vary including the physical properties of the whole starch (Table 3) and of the purified amylose and amylopectin. The latter include β-amylolysis, average degree of branching, average length of unit chain and viscosity.[20,25,37]

Even more variation is present in mutant lines (Table 4) and analysis of these has contributed much to our knowledge of the pathways of starch synthesis and its regulation (Fig. 2). Such mutant lines have been identified and analysed in barley, pea and Amaranthus, but most work has been on the extensive series of maize mutants. The potential uses of these mutants in industry have been discussed.[38,39]

Tsai[31] showed that the waxy allele of maize was a mutation in the structural gene coding for the bound form of starch synthase. The absence of this enzyme results in starch consisting of almost 100% amylopectin.[40,41] More recently Nelson *et al.*[42] reported the presence of a second form of the bound starch synthase in waxy endosperms. This type II enzyme has a lower molecular weight than the normal form of the enzyme, and different kinetic properties including a much lower ($\times$ 100) K_m for ADP glucose.[35,36,42] The presence of this second enzyme is only detected when the more active form is eliminated by the waxy mutation. The waxy gene of maize has recently been cloned and characterized.[43] A waxy mutant has also been identified in barley,[44] and may affect a homologous locus.

Amylose-extender mutations are known which result in increased proportions of amylose in maize, barley and peas. Boyer and Preiss[45] purified three branching enzymes (I, IIa, IIIb) from maize and showed that the amylose-extender mutant lacks branching enzyme IIb (Table 4). The three branching enzymes appear to be under separate genetic control.[46] Matters and Boyer[29] have reported a comparable enzymic deficiency in the amylose-extender line of pea (Progress No. 9), which has only 10% of the total branching activity present in the normal line (Alaska). An amylose-extender phenotype is also present in the barley cultivar Glacier (Ac38), which has a normal amount of starch but 43% amylose[30] (Table 4). The enzymic deficiency in this line is not known. Maize varieties containing up to 70% amylose have now been produced using the amylose-extender mutant and minor modifying genes.

A modest increase (about 10%) in the proportion of amylose is also present in the dull mutant of maize (Table 4). Preiss and Boyer[47] showed that this line had lower activities of soluble starch synthase II and branching enzyme IIa, and proposed that the mutation was at a 'regulatory' locus.

Sweet maize varieties which are grown commercially contain high levels of phytoglycogen (30% of endosperm dry weight) and sucrose, and are based on the homozygous sugary (*su*) genotype. Phytoglycogen is a water-soluble glucan that is more branched than amylopectin[25] and gives the starch a creamy consistency. Black *et al.*[48] showed that maize endosperms which contain phytoglycogen also show phytoglycogen branching activity, and several workers[45,49,50] showed that *su* kernels contained a branching enzyme capable of forming phytoglycogen *in vitro*. Boyer and Preiss[45] showed that branching enzyme I in *su* endosperms had an increased branching activity compared to the normal enzyme, and suggested that modified enzyme was a major cause of increased phytoglycogen synthesis. The *su* locus is not, however, the structural gene for branching enzyme I.[32,48] Other genes also affect kernel sugar levels (e.g. *sugary-2*)[32] and sugary enhancer (*se*). The double mutants *se su* have sucrose levels of more than 20% the dry weight at 21 days after pollination.[51]

4.1.6. Limits to Our Knowledge of Starch Synthesis

Although analysis of the numerous maize mutants has contributed to our knowledge of starch synthesis in the plant, it is still incomplete. This is perhaps surprising in view of the important roles of starch in human and animal nutrition and as industrial raw material. Although starch phosphorylase is probably degradative, this is still not certain and interpretation of its role would be facilitated by the identification of a phosphorylase mutation. Little is known of the mechanisms by which starch molecules start to grow. Lavintman *et al.*[52] suggested that a glycoprotein may be involved in the initial stages, and Krisman and Barengo[53] suggested a similar mechanism for glycogen synthesis. The exact roles of the various glucosyl-transferases are also unknown. In addition, further information is still required on the roles and mechanisms of the various forms of starch synthase and the three branching enzymes discussed above. Even when the enzymology of starch synthesis and its regulation is fully understood, we still need to know how amylose and amylopectin are packaged into starch granules, and how this relates to differences in starch granules between species and genotypes (including mutants).

Only when this information is available will it be possible to manipulate the composition and properties of cereal and potato starches by genetic engineering. In the meantime molecular biology provides a very powerful tool to obtain this information.

4.1.7. Present Sources and Uses of Starch

Maize is currently the major source of starch, accounting for about 75% of total production within the EEC (Table 5). The major uses of starch are in the food, chemical, paper and corrugated product industries (Table 6).

There are several reasons for the predominant use of maize, including its lower market price, ease of production and the development of highly efficient modern processing plants. Maize starch, in its normal and mutant

Table 5
Raw Materials used for the Production of Starch in the EEC

Raw material	*Consumption (tonnes/year)*	*Production of starch (tonnes/year)*
Maize	4 066 000	2 505 000
Potato	4 612 000	784 000
Wheat	350 000	184 000
Total		3 473 000

Data for 1981–2, taken from Rexen and Munck.[15]

forms, is also well suited to a range of food and industrial uses. Cereals in general are also preferred to potatoes because they can be stored for long periods without deterioration of the starch granule.[15] However, there are now factories built to use both maize and potatoes for starch production, and if future economics were favourable the waste from chip and crisp production could be used.

The uses of starch fall into three categories: unmodified, modified and hydrolysed.[55] Unmodified starch is used in the food industry as a thickener and bulking agent, and chemically modified starches for special uses in canned and frozen products. However, these together only account for about 15% of the total starch used by the food and drink industry in the UK. The bulk is hydrolysed, with controlled enzymic hydrolyses giving a range of syrups for confectionery, soft drinks, other food uses and as substrates for fermentation.[55]

The same three forms of starch are also used in the non-food industries, although the proportions differ with unmodified starch accounting for over 50% of the total used in the UK.[55]

Table 6
Utilization of Starch in the EEC

	$\times 10^6$ *tonnes*	%
Total produced	3·6	—
Starch used in		
food industry	2·3	64
paper industry	0·55	15
chemical industry	0·36	10
corrugated cardboard industry	0·21	6
other industries	0·18	5

Data for 1981, taken from Budzinski.[54]

Potato starch has a number of specific applications in the non-food industries in the manufacture of paper, plastics and adhesives and in the oil industry as a viscosity and fluid-loss control agent.[4,15,16] It is preferred to cereal starches in the paper industry because the larger granules act as bridges between fibres, resulting in increased strength. Modified (cationic) potato starch is also used to improve the quality of paper and dextrinized starch for the manufacture of adhesives.

Wheat starch is only produced on a small scale in Europe (Table 5); its utilization has been reviewed.[55,56] The potential for increasing the utilization of wheat and barley is discussed below.

4.1.8. Storage of Potatoes and Starch Breakdown

Unlike cereals, storage of potatoes requires carefully controlled conditions depending on the final use. During storage the amount of starch decreases due to respiration and slow conversion to sugars. Assuming that starch is the primary respiratory substrate, it can be estimated that 1·5% of the starch (about 0·3% of the total fresh weight) is lost after 1 month. The extent of the conversion of starch to sugars depends on the variety of potato, the physiological maturity of the tuber when harvested and the temperature of storage.[3] Storage at 15°C will minimize sugar accumulation but losses due to respiration, decay and sprouting will be greater. Storage at lower temperatures will increase sugar accumulation and values can exceed 1% of the fresh weight in 2 months.[57] In extreme cases 25% of the starch can be degraded before tubers are used.

The pathway of starch breakdown in potato tubers is not clearly understood (for reviews see references 58 and 59). A possible pathway is presented in Fig. 3. It should be emphasized that this pathway is hypothetical and is not likely to be identical in maize, wheat, barley and potato. It is not clear what type of amylase activity is present in potato tubers since the assays used in the past would not have clearly distinguished the enzymes recognized classically as α and β amylases (for a discussion see reference 59). However, it is probable that the initial degradation of the potato starch granule involves some type of amylolytic attack which releases soluble dextrins. These are then degraded further by starch phosphorylase although the activity of phosphorylase has not been correlated with the observed increase in sucrose in stored potatoes, and it does not appear that the increased breakdown results from active enzyme synthesis.

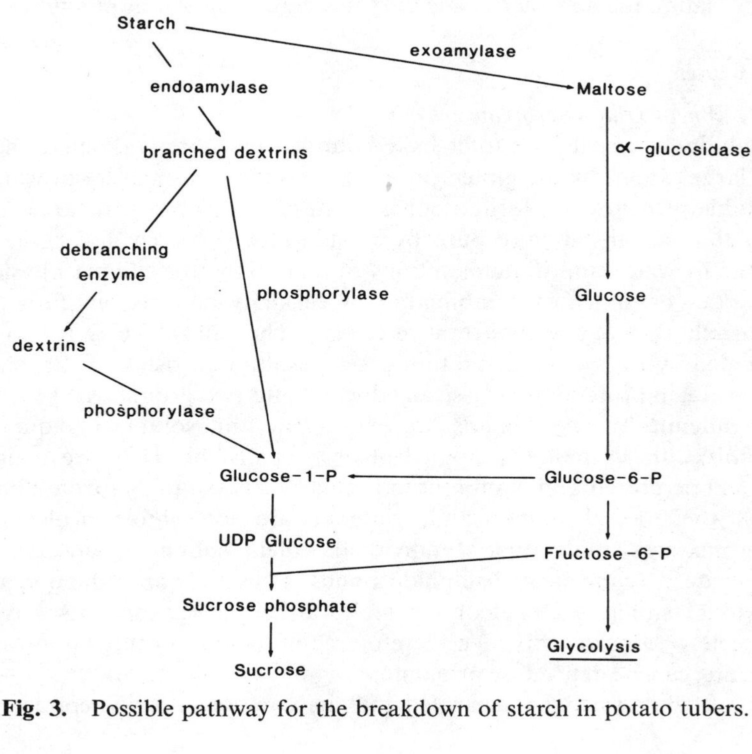

Fig. 3. Possible pathway for the breakdown of starch in potato tubers.

4.1.9. Prospects for the Manipulation of the Synthesis, Properties and Storage of Starch

It is anticipated that genetic engineering will be used to modify three aspects of cereal and potato starches.

(1) To increase and unify the size of cereal starch granules, to make them easier to process, and better suited for use in the paper industry. This may be achieved by identifying and manipulating the genes regulating amyloplast development.
(2) To manipulate the chain lengths and degrees of branching of amylose and amylopectin, and the ratios of the two polymers in cereal and potato starches. This will enable us to optimize the suitability of the starch for different uses in the food industry. These changes may be achieved by altering the relative activities of the free and bound forms of starch synthetase, and the activities of the branching enzymes.
(3) To improve the storage properties of potato tubers, by manipulating the activities of the enzymes regulating starch mobilization.

4.2. Gluten

4.2.1. Wheat Gluten Proteins

The ability of wheat flour to be baked into leavened bread is determined, to a large extent, by the gluten proteins.[60] Gluten is a mixture of water-insoluble proteins that form a cohesive mass, and can be prepared in a form that is substantially pure by washing away starch and proteins soluble in water and dilute solutions of salts. It is the unique physical properties of gluten, a combination of elasticity and viscous flow (or extensibility), that are important in baking. They allow the gluten to be expanded by trapped fermentation gases, resulting in rising of the loaf.

The gluten proteins are classically divided into two groups: the gliadins and glutenins.[61] The gliadins are extractable and soluble in aqueous alcohols, either ethanol, propan-1-ol or propan-2-ol. They are mainly monomeric proteins with subunit molecular weights ranging from about 30 000 to 75 000. In contrast the glutenins are not soluble in alcohol–water mixtures, and consist of individual protein ‘subunits’ associated in ‘aggregates’ by covalent disulphide bonds. However, on reduction the individual subunits are alcohol-soluble, although in some cases only completely so at low pH. We therefore consider that both gliadins and glutenins can be defined as prolamins.

Whole gluten can be separated into its component polypeptides by

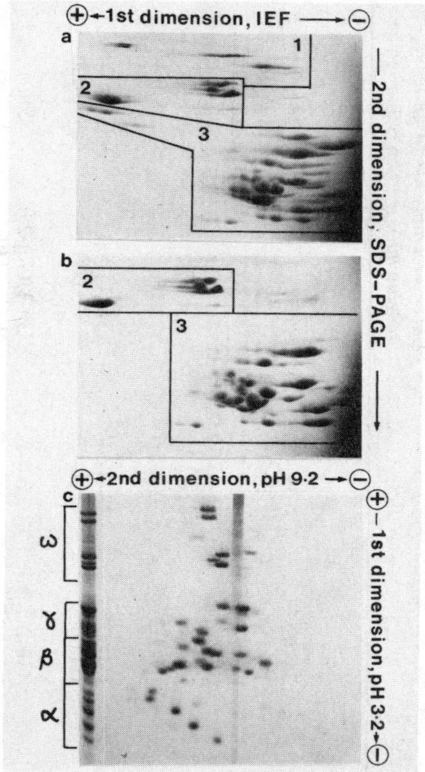

Fig. 4. Two-dimensional electrophoresis of wheat gluten proteins. (a) Total reduced prolamins (gliadins and glutenins) separated by IEF (pH range 3·5–10)/SDS-PAGE. (b) Monomeric prolamins (gliadins) separated by IEF (pH range 3·5–10)/SDS-PAGE. (c) Monomeric prolamins (gliadins) separated by the two pH system of Lafiandra and Kasarda.[62] Areas 1, 2 and 3 in (a) and (b) are HMW prolamins, S-poor prolamins (ω-gliadins) and S-rich prolamins respectively. (Rearranged from Shewry *et al.*[63])

electrophoresis under dissociating conditions and after reduction of intermolecular disulphide bonds. The most widely used procedure is polyacrylamide gel electrophoresis (PAGE) in the presence of the detergent sodium dodecyl sulphate (SDS). Further resolution is given by combining SDS-PAGE in a two-dimensional system with isoelectric focusing (IEF), a procedure which separates on isoelectric point (Fig. 4a). This shows a large number of individual polypeptides.

Table 7
Characteristics of Wheat Gluten Proteins

Prolamin group	*S-poor*	*S-rich*			*HMW*
Classical nomenclature	*ω-gliadins*	*α-type gliadins*	*γ-type gliadins*	*aggregated gliadin (LMW subunits of glutenin)*	*HMW subunits of glutenin*
Partial amino acid composition (mol%)					
Glx	41–53	36–42	39–40	38	34–39
Pro	20–30	15–16	18–19	15	13–16
Gly	0·9–1·4	1·9–2·7	2·7	3·3	14–20
Phe	8·1–9·0	3·7–3·9	1·4–1·7	4·7	0·3–1·1
Cys	0	1·8–1·9	1·9–2·0	2·7	0·4–1·5
Met	0–0·1	0·9–1·2	0·9–1·7	0·6	t[c]–0·4
Molecular weights by					
SDS-PAGE	44–74 000	32 000[a]	38–42 000	36–44 000	95–136 000
sequencing	—	31 000[a]	—	33 000	64–70 000
Structural locus					
designation	Gli-1	Gli-2	Gli-1	Gli-1/Glu-2	Glu-1
location	1AS, 1BS 1DS	6AS, 6BS 6DS	1AS, 1BS 1DS	1AS,[b] 1BS 1DS	1AL, 1BL 1DL

Data taken from Bietz and Wall;[64] Bietz *et al.*;[65] Ewart;[66] Forde *et al.*;[67] Jackson *et al.*;[68] Kasarda *et al.*;[69,70] Payne and Corfield;[71] Payne *et al.*;[72] Platt and Kasarda;[73] Shewry *et al.*;[74,75] Thompson *et al.*[76]

[a] Values for the A group of α-gliadins.
[b] Glu-2 is not present on chromosome 1A.
[c] t = trace.

Unreduced glutenins cannot be separated by gel electrophoresis, probably because the aggregates are too large to enter the gel matrix, and do not exist in discrete size classes. In contrast the monomeric gliadins can be separated without reduction, either by using one-dimensional systems (e.g. SDS-PAGE or electrophoresis at low pH in starch or polyacrylamide gels), or two-dimensional systems. Two different two-dimensional systems, IEF/SDS and electrophoresis at pH 3·2 followed by electrophoresis at pH 9·2, are compared in Fig. 4b,c.

Although the multitude of gluten components revealed by electrophoresis is initially confusing, analyses of purified sub-fractions and individual components show that they can be classified into a small number of groups. We have defined three such groups on the basis of their amino acid sequences and the chromosomal locations of their structural genes. These are called the high molecular weight (HMW), sulphur-poor (S-poor) and sulphur-rich (S-rich) prolamins.[74] Their characteristics are summarized in Table 7.

The HMW prolamins (HMW subunits of glutenin) and S-poor prolamins (ω-gliadins) are present only as aggregates and monomers respectively. In contrast the S-rich prolamins contain aggregative (LMW subunits of glutenin) and monomeric (α-, β- and γ-gliadins) components. Although the three groups differ in a number of their characteristics, they do have limited homology in their amino acid sequences, indicating a common evolutionary origin. The structural and evolutionary relationships of wheat gluten proteins to each other and to other seed proteins have been discussed in detail by Shewry and Miflin[77] and Kreis *et al.*[78] The gliadins and glutenins also have similar biological properties, being deposited in protein bodies in the developing endosperm, and having no apparent function except storage.

Although the gliadins and glutenins may not be valid groups on genetic, biological or chemical grounds, they do have technological significance. Whereas glutenins are considered to be the major determinants of gluten elasticity, the gliadins are considered to contribute to extensibility and viscous flow.[60,79]

It is possible to estimate the numbers of genes related to different prolamin groups in the genome of wheat by molecular analysis. This shows that the HMW subunits of glutenins are encoded by a small multigene family, with one or two copies each of two genes per genome.[67,80] In contrast the S-rich gliadins are encoded by larger families of genes; for example 15–20 copies of a single A(α)-type gliadin gene appear to be present in hexaploid bread wheat cultivar Cheyenne.[81]

4.2.2. Factors Affecting Baking Quality

The baking quality of wheat is affected by environment and genotype. The total amount of gluten proteins present in the grain is directly correlated with baking quality. Since these proteins act as a store of N, their amounts increase disproportionately when wheat is grown under high levels of N availability. It is therefore usual for farmers to grow wheat for baking under high levels of N fertilization. There are also more subtle and less readily manipulated environmental effects, which result in a climatically determined year-to-year variation in the baking quality of home-grown wheats.

Of more interest to us here is the presence of extensive genetic variation in baking quality. A considerable amount of research has been directed towards understanding this variation in terms of gluten structure. A major problem with the earlier studies was the lack of solvents capable of dissolving all the un-reduced gluten proteins. As a result negative correlations were reported between baking quality and the amounts of proteins soluble in 3M urea,[82] dilute acetic acid[83,84] and several other solvents. Further studies with more effective solvents showed a positive correlation between the amount of high molecular weight aggregated glutenin and baking quality.[85,86] This is illustrated by Fig. 5,[86] which shows a strong correlation between the proportion of gluten excluded from a column of controlled pore glass (exclusion limit of about 1×10^6) and the NIAB baking score. Electrophoretic analysis of the excluded fraction (Fig. 5) shows that it is enriched in the HMW subunits of glutenin.

Further evidence for a role of the HMW subunits in determining gluten quality comes from the studies of Payne and co-workers at the Plant Breeding Institute, Cambridge. They showed that the presence of specific HMW subunits was associated with quality, and that this correlation held for the progeny of crosses between good and poor quality cultivars.[87,88] Similar results were reported independently.[89] Although these results indicate a role of the HMW subunits in determining gluten quality, this is clearly not the complete story. Other workers have failed to find similar correlations in North American[90] and Australian[91] cultivars, while some UK cultivars have good baking quality without 'quality'-associated subunits, and vice versa. Furthermore, associations between gliadins and quality have been reported by workers in Australia[91] and France.[92]

4.2.3. What is the Molecular Basis for Gluten Properties?

A number of models for the structure of gluten have been proposed. With

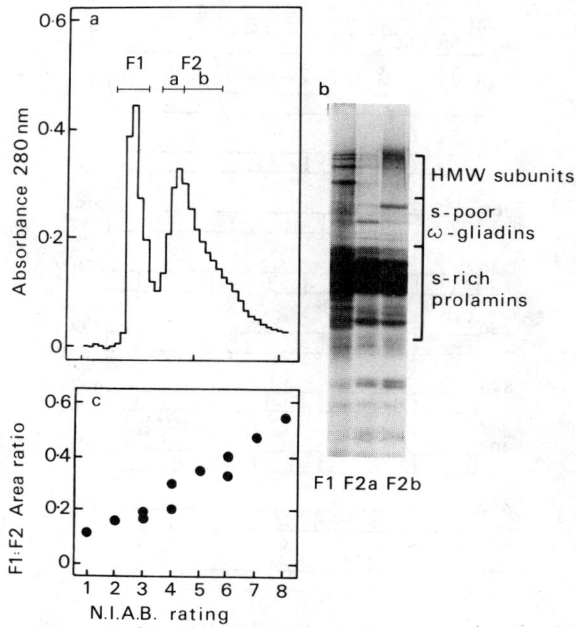

Fig. 5. Association of the amount of high molecular weight aggregated gluten proteins and baking quality in British wheats. (a) Fractionation of total gluten proteins of the cv. Copain on a column of controlled pore glass; fraction F1 consists of high molecular weight (probably over 1×10^6) aggregated proteins. (b) SDS-PAGE, after reduction, of fractions F1, F2a and F2b from (a); note that the HMW subunits of glutenin are only present in the F1 fraction. (c) The relationship of the ratio of the F1 and F2 (a + b) fractions and the baking quality (taken as the NIAB scores for the year of harvest) of eleven cultivars of winter and spring wheat. (Taken from Field *et al.*[86])

one exception,[93] these all consider that intermolecular disulphide bonds, and non-covalent interactions (notably hydrogen bonding and hydrophobic interactions) are important in determining the unique combination of viscosity and elasticity.[60,94–97] However, these models are based on little experimental evidence, and none provides an explanation of gluten properties in molecular terms.

We have adopted a different approach, by making detailed studies of the amino acid sequences, conformations and stabilities of individual gluten polypeptides, and then considering how these may interact to determine the properties of whole gluten. Although we have studied a number of different groups of gluten polypeptides, emphasis has been

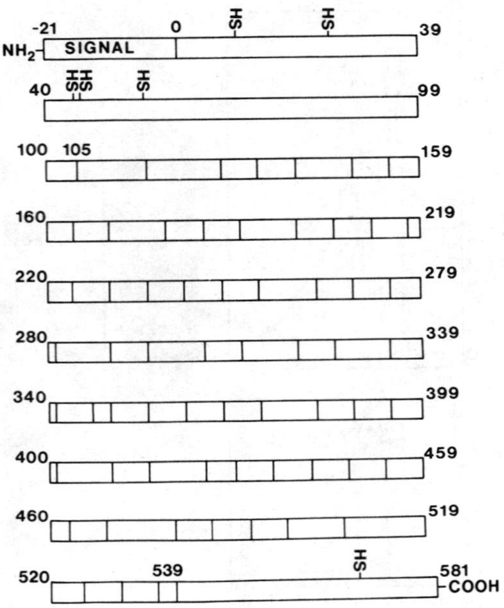

Fig. 6. The structural organization of an HMW subunit of wheat gluten. The signal sequence (−21 to 0) is cleaved co-translationally and is not part of the mature protein. A central domain (106 to 539) of repeated blocks of six and nine residues (boxed) is flanked by non-repetitive N-terminal (1–105) and C-terminal (540–581) domains containing cysteine residues. The figure is based on the amino acid sequence deduced from the nucleotide sequence of an HMW subunit gene by Forde *et al.*[67]

placed on the HMW subunits of gluten. This is because of their implied involvement in determining baking quality, and discussion here will be restricted to this group.

Two complete genes and several partial cDNAs related to HMW subunits have been sequenced at Rothamsted and at the Plant Breeding Institute, Cambridge.[67,76,80,98] The deduced protein sequences show a clear domain structure. Although the gene sequenced by Forde *et al.*[67] is probably not expressed, the structural organization of the encoded protein (Fig. 6) is almost certainly representative of HMW subunits in general. The most striking feature is a central domain (434 residues in total) consisting of tandem and interspersed repeated blocks of six and nine residues, with consensus sequences of Pro-Gly-Gln-Gly-Gln-Gln and Gly-Tyr-Tyr-Pro-Thr-Ser-Leu-Gln-Gln. This is flanked by non-repetitive N-

and C-terminal domains of 105 and 42 residues respectively. The subunit represented in Fig. 6 has six cysteine residues, but these are restricted to the N-terminal (5 cysteines) and C-terminal (1 cysteine) domains. The sequence reported by Thompson *et al.*[76] is essentially similar, the only notable difference being the presence of an additional cysteine residue within the repetitive domain. Chemical analyses of purified subunits have also shown that, with few exceptions, the cysteine residues are close to the N- and C-termini.[75,99,100]

Secondary structure determinations by circular dichroism spectroscopy and computer predictions based on amino acid sequences indicate the presence of little or no β-sheet, but some α-helix in the N- and C-terminal regions.[101,102] The central domain appears to have an unusual secondary structure with regularly repeated β-turns reflecting the repetitive primary structure.

We can now attempt to answer the questions 'how do the HMW subunits interact to form an elastic glutenin aggregate?', and 'how does this relate to baking quality?'. The distribution of cysteine residues is clearly significant, as these are potential sites for cross-linking. Their sub-terminal location could lead to the formation of predominantly linear polymers, although the presence of additional cysteine residues in the N-terminal domain and, in at least one subunit, also in the repetitive domain, will allow appreciable branching and cross-linking of polymers and interaction with LMW subunits of glutenin. Given that such a structure is formed it is now necessary to explain how this confers elasticity.

In order to understand the elastic mechanism we have made comparisons of the structures of the HMW subunits with those of other elastic proteins such as abductin, resilin, connectin and elastin. Of these only extensive amino acid sequences are available for elastin, and this shows a remarkable similarity in molecular organization to the HMW subunits of gluten. Thus elastin also has α-helical regions which are covalently cross-linked, although in this case the cross-links are via demosine residues, formed from the ε-amino groups of lysine. Between these are repetitive regions based on tandemly repeated tetrapeptides, pentapeptides and hexapeptides. All three repeated peptide regions are rich in β-turns, and in the pentapeptide regions these turns probably form a regular helical structure called a β-spiral.[103] Despite a considerable amount of research, there is still controversy about the molecular basis for the elasticity of elastin. It has been suggested that the β-spiral conformation is intrinsically elastic, but this is disputed by some workers.

In view of the long-running controversy about the molecular basis of the elasticity of elastin, it would be unrealistic to draw firm conclusions about the elasticity of gluten on the basis of the data available at present. The distribution of β-turns in the repetitive domains of the HMW subunits indicates that these could form a β-spiral and, as in elastin, this could be intrinsically elastic. However, the number and distribution of disulphide cross-links will almost certainly also affect the properties of the polymer, notably its rigidity. A further factor, not present in elastin, is the possibility of disulphide cleavage and exchange under stress. Variation in baking quality could result from differences in the abilities of HMW subunits to form aggregates, either with themselves or with other gluten proteins.

4.2.4. What Types of Modifications are Required?

We are clearly still a long way from explaining the properties of wheat gluten, including subtle differences in baking quality between cultivars, in molecular terms. We are, however, in a position to propose models, and to test these by protein engineering. Such experiments will give a clear indication of the importance of such factors as aggregate size, the degree of branching and cross-linking of aggregates, the types of cross-links, and the importance of intrinsic polypeptide elasticity.

These should allow us to make specific modifications to the HMW subunit genes, with confidence that they will affect the physical properties of the whole gluten. These modifications can be designed to affect not only the baking properties of wheat, but also the properties of gluten for other uses in the food industry and elsewhere.

5. CONCLUSIONS

We have discussed how a number of factors have combined to result not only in increased utilization of home-grown cereals for the traditional end uses, but also increased availability for other industrial uses. It is therefore important to think of the quality of cereals as industrial raw material as well as for food uses. Also, the production of other crops such as potatoes could be increased if their suitability for industry was improved.

There are currently some technical problems in the improvement of crop plants by genetic engineering. These include the development of transformation systems and of methods to regenerate the transformed plants from protoplasts and tissue cultures. At present, these problems are more severe for monocotyledonous plants (cereals) than for dicotyledonous plants. However, it is anticipated that the major barriers will be overcome in the next few years, certainly by the end of the century.

A more serious limitation to improvement, exemplified by the detailed discussions of starch and wheat gluten, is a lack of understanding of many aspects of crop quality at the molecular level. Obtaining this information must be a major priority, and will be facilitated by application of the powerful techniques of molecular biology, notably protein engineering.

In addition, even when genetically engineered lines are available they must still be incorporated into conventional plant breeding programmes to ensure that the improved varieties are able to compete with the most successful conventional varieties in terms of yield, disease resistance and other characteristics of agronomic importance. This will inevitably result in a delay in their impact on the processing industries.

REFERENCES

1. Schofield, J. D. and Booth, M. R., in *Developments in Food Proteins—2*, Hudson, B. J. F. (Ed.), Applied Science Publishers, London, 1983, pp 1–65.
2. Talburt, W. F. and Smith, O., *Potato Processing*, Avi Publishing Co., Connecticut, 1975.
3. Harris, P. M., *The Potato Crop*, Chapman and Hall, 1978.
4. Young, N. A., *Potato Products: Production and Markets in the European Communities*, Information on Agriculture, No. 75, EEC, 1981.
5. Old, R. W. and Primrose, S. B., *Principles of Gene Manipulation*, 3rd Edn, Blackwell Sci. Publ., Oxford, 1985, 409 pp.
6. Forde, B. G., Kreis, M., Bahramian, B., Matthews, J., Miflin, B. J., Thompson, R., Bartels, D. and Flavell, R., *Nucl. Acids Res.*, **92**, 6689–707 (1981).
7. Zoller, M. J. and Smith, M., *Methods Enzymol.*, **100**, 468–500 (1983).
8. Harris, T. J. R., in *Genetic Engineering*, Vol. 4, Williamson, R. (Ed.), Academic Press, London, 1983, pp 127–85.
9. Jones, M. G. K., *Nature*, **317**, 579–80 (1985).
10. Jones, M. G. K. and Karp, A., in *Advances in Biotechnological Processes*, Vol. 5, Alan R. Liss Inc., 1985, pp 91–121.
11. Hain, R., Stobel, P., Czerniloisky, A. P., Steinbib, H. H., Herrera-Estrella, L. and Schell, J., *Mol. Gen. Genet.*, **199**, 161–8 (1985).
12. Hooykaas-Van Slogteren, G. M. S., Hooykaas, P. J. J. and Schilperoort, R. A., *Nature*, **311**, 763–4 (1984).
13. Lorz, H., Baker, B. and Schell, J., *Mol. Gen. Genet.*, **199**, 178–82 (1985).
14. Potrykus, I., Saul, M. W., Petruska, J., Poszkowski, J. and Shillito, R. D., *Mol. Gen. Genet.*, **199**, 183–8 (1985).

15. Rexen, F. and Munck, L., *Cereal Crops for Industrial Use in Europe*, report prepared for the Commission of the European Communities, 1984.
16. Whistler, R. L., Bemiller, J. N. and Paschall, E. F., *Starch*, Academic Press, New York, 1984.
17. Woelk, H. U., *Stärke*, **33**, 397–408 (1981).
18. Skrede, G., *Informasjon*, Vol. 6, NINF, Oslo, Norway, 1982, pp 77–86.
19. Manners, D. J., *Cereal Food World*, **30**, 461–7 (1985).
20. Manners, D. J., in *Biochemistry of Storage Carbohydrates in Green Plants*, Dey, P. M. and Dixon, R. A. (Eds), Academic Press, New York, 1985, pp 149–203.
21. Peat, S., Pirt, S. J. and Whelan, W. J., *J. Chem. Soc.*, 705–13 (1952).
22. Peat, S., Thomas, G. J. and Whelan, W. J., *J. Chem. Soc.*, 722–33 (1952).
23. Cowie, J. M., Fleming, I. D., Greenwood, C. T. and Manners, D. J., *J. Chem. Soc.*, 4430–7 (1957).
24. Hizukuri, S., Takeda, Y. and Yasuda, M., *Carb. Res.*, **94**, 205–13 (1981).
25. Banks, W. and Muir, D. D., in *The Biochemistry of Plants*, Vol. 3, Academic Press, 1980, pp 321–69.
26. Williams, J. M. and Duffus, C. M., *Plant Physiol.*, **59**, 189–92 (1977).
27. Bathgate, G. N. and Palmer, G. H., *Stärke*, **24**, 336–41 (1972).
28. Edelman, J. and Fewell, A., *Phil. Trans. Roy. Soc.*, **B31**, 317–25 (1985).
29. Matters, G. L. and Boyer, C. D., *Biochem. Genet.*, **20**, 833–48 (1982).
30. Walker, J. T. and Merritt, N. R., *Nature*, **221**, 482–3 (1969).
31. Tsai, C.-Y., *Biochem. Genet.*, **11**, 83–96 (1973).
32. Nelson, O. E., in *Advances in Cereal Science and Technology*, Vol. III, Pomeranz, Y. (Ed.), American Association of Cereal Chemists, 1980, pp 41–71.
33. Preiss, J., in *Encyclopedia of Plant Physiology*, N.S 13A, Loewus, F. A. and Tanner, W. (Eds), Springer-Verlag, Berlin, 1982, pp 397–417.
34. Palmer, G. H., *J. Inst. Brewing*, **78**, 326–32 (1972).
35. MacDonald, F. D. and Preiss, J., *Plant Physiol.*, **78**, 849–52 (1985).
36. Preiss, J., MacDonald, F. D., Singh, B. K., Robinson, N. and McNamara, K., in *New Approaches to Research on Cereal Carbohydrates*, Hill, R. D. and Munck, L. (Eds), Elsevier, Amsterdam, 1985, pp. 1–17.
37. Lii, C.-Y. and Lineback, D. R., *Cereal Chem.*, **54**, 138–49 (1977).
38. Alexander, D. E. and Creech, R. G., in *Corn and Corn Improvement*, 2nd Edn, Sprague, G. F. (Ed.), Am. Soc. Agron., Madison, Wisconsin, 1977, pp 363–90.
39. Dunn, C. F., *Food Technol. Austr.*, **373**, 110–11 (1985).
40. Nelson, O. E. and Rines, H. W., *Biochem. Biophys. Res. Commun.*, **9**, 297–300 (1962).
41. Nelson, O. E. and Tsai, C.-Y., *Science*, **145**, 1194–5 (1964).
42. Nelson, O. E., Chourey, P. S. and Chang, M. T., *Plant Physiol.*, **62**, 383–6 (1978).
43. Sheere, M., Wessler, S. and Federoff, N., *Cell*, **35**, 225–33 (1983).
44. Eriksson, G., *Hereditas*, **63**, 180–204 (1969).
45. Boyer, C. D. and Preiss, J., *Biochem. Biophys. Res. Commun.*, **80**, 169–75 (1978).
46. Preiss, J., *Carb. Res.*, **61**, 321–34 (1978).

47. Preiss, J. and Boyer, C. D., in *Mechanisms of Saccharide Polymerisation and Depolymerisation*, Marshall, J. J. (Ed.), Academic Press, New York, 1980, pp 161–74.
48. Black, R. C., Loerch, J. D., McArdle, F. J. and Creech, R. G., *Genetics*, **53**, 661 (1966).
49. Lavintman, N., *Arch. Biochem. Biophys.*, **116**, 1–8 (1966).
50. Manners, D. J., Rowe, J. J. M. and Rowe, K. L., *Carb. Res.*, **8**, 72–81 (1968).
51. Carey, E. E., Rhodes, A. M. and Dickinson, D. B., *Hort. Sci.*, **17**, 241–2 (1982).
52. Lavintman, N., Tandecarz, J., Carceller, M. and Cocidini, E. C., *Eur. J. Biochem.*, **50**, 145–55 (1974).
53. Krisman, C. R. and Barengo, R., *Eur. J. Biochem.*, **52**, 117–23 (1975).
54. Budzinski, A., *Stärke*, **34**, 505–9 (1982).
55. Galliard, T., *Ann. Proc. Phytochem. Soc. Eur.*, **26**, 103–15 (1985).
56. Knight, J. W. and Olson, R. M., in *Wheat Starch: Starch Chemistry and Technology*, Whistler, R. L., BeMiller, J. N. and Paschall, E. F. (Eds.), Academic Press, New York, 1984, pp 491–505.
57. Van Vliet, W. F. and Schreimer, W. H., *Eur. Pot. J.*, **3**, 263–71 (1960).
58. Halmer, P. and Bewley, J. D., in *Encyclopedia of Plant Physiology*, N.S. 13A, Loewus, F. A. and Tanner, W. (Eds), Springer-Verlag, Berlin, 1982, pp 748–93.
59. Stitt, M. and Steup, M., in *Encyclopedia of Plant Physiology*, 18, Douce, R., and Day, D. A. (Eds), Springer-Verlag, Berlin, 1985, pp 347–90.
60. Wall, J. S., in *Recent Advances in the Biochemistry of Cereals*, Laidman, D. L. and Wyn-Jones, R. G. (Eds), Academic Press, London, 1979, pp 275–312.
61. Kasarda, D. D., Bernardin, J. E. and Nimmo, C. C., in *Advances in Cereal Science and Technology*, Vol. 1, Pomeranz, Y. (Ed.), American Society of Cereal Chemists, Minnesota, 1976, pp 158–236.
62. Lafiandra, D. and Kasarda, D. D., *Cereal Chem.*, **62**, 314–19 (1985).
63. Shewry, P. R., Parmar, S., Fulrath, N., Kasarda, D. D. and Miller, T. E., *Can. J. Genet. Cytol.*, **28**, 76–83 (1986).
64. Bietz, J. A. and Wall, J. S., *Cereal Chem.*, **50**, 537–47 (1973).
65. Bietz, J. A., Huebner, F. R., Sanderson, J. E. and Wall, J. S., *Cereal Chem.*, **54**, 1070–83 (1977).
66. Ewart, J. A. D., *J. Sci. Food Agric.*, **28**, 843–8 (1977).
67. Forde, J., Malpica, J.-M., Halford, N., Shewry, P. R., Anderson, O., Greene, F. C. and Miflin, B. J., *Nucl. Acids Res.*, **13**, 6817–32 (1985).
68. Jackson, E. A., Holt, L. M. and Payne, P. I., *Genet. Res. Camb.*, **46**, 11–17 (1985).
69. Kasarda, D. D., Autran, J.-C., Lew, E. J.-L., Nimmo, C. C. and Shewry, P. R., *Biochim. Biophys. Acta*, **747**, 138–50 (1983).
70. Kasarda, D. D., Okita, T. W., Bernardin, J. E., Baecker, P. A., Nimmo, C. C., Lew, E. J.-L., Dietler, M. D. and Greene, F. C., *Proc. Nat. Acad. Sci.*, **81**, 4712–16 (1984).
71. Payne, P. I. and Corfield, K. G., *Planta*, **145**, 83–8 (1979).
72. Payne, P. I., Holt, L. M., Jackson, E. A. and Law, C. N., *Phil. Trans. Roy. Soc.*, **B304**, 273–307 (1984).

73. Platt, S. G. and Kasarda, D. D., *Biochim. Biophys. Acta*, **243**, 407–15 (1971).
74. Shewry, P. R., Miflin, B. J., Lew, E. J.-L. and Kasarda, D. D., *J. Exp. Bot.*, **34**, 1403–10 (1983).
75. Shewry, P. R., Field, J. M., Faulks, A. J., Parmar, S., Miflin, B. J., Dietler, M. D., Lew, E. J.-L. and Kasarda, D. D., *Biochim. Biophys. Acta*, **788**, 23–34 (1984).
76. Thompson, R. D., Bartels, D. and Harberd, N. P., *Nucl. Acids Res.*, **13**, 6833–46 (1985).
77. Shewry, P. R. and Miflin, B. J., in *Advances in Cereal Science and Technology*, Vol. 7, Pomeranz, Y. (Ed.), American Association of Cereal Chemists, Minnesota, 1985, pp 1–83.
78. Kreis, M., Shewry, P. R., Forde, B. G., Forde, J. and Miflin, B. J., in *Oxford Surveys of Cell and Molecular Biology*, Vol. 2, Miflin, B. J. (Ed.), Oxford University Press, Oxford, 1985, pp 253–317.
79. Miflin, B. J., Field, J. M. and Shewry, P. R., in *Seed Proteins*, Daussant, J., Mosse, J. and Vaughan, J. (Eds), Academic Press, London, 1983, pp 255–319.
80. Thompson, R. D., Bartels, D., Harberd, N. P. and Flavell, R. B., *Theor. Appl. Genet.*, **67**, 87–96 (1983).
81. Anderson, O. D., Litts, J. C., Gautier, M.-F. and Greene, F. C., *Nucl. Acids Res.*, **12**, 8129–45 (1984).
82. Pomeranz, Y., *J. Sci. Food Agric.*, **16**, 586–93 (1965).
83. Orth, R. A. and Bushuk, W., *Cereal Chem.*, **49**, 268–75 (1972).
84. Mecham, D. K., Cole, E. W. and Ng, H., *Cereal Chem.*, **49**, 62–7 (1972).
85. Huebner, F. R. and Wall, J. S., *Cereal Chem.*, **53**, 258–69 (1976).
86. Field, J. M., Shewry, P. R. and Miflin, B. J., *J. Sci. Food Agric.*, **34**, 370–7 (1983).
87. Payne, P. I., Law, C. N. and Mudd, E. L., *Theor. Appl. Genet.*, **58**, 113–20 (1980).
88. Payne, P. I., Corfield, K. G., Holt, L. M. and Blackman, J. A., *J. Sci. Food Agric.*, **32**, 51–60 (1981).
89. Burnouf, T. and Bouriquet, R., *Theor. Appl. Genet.*, **58**, 107–11 (1980).
90. Orth, R. A. and Bushuk, W., *Cereal Chem.*, **51**, 118–26 (1974).
91. Wrigley, C. W., *Ann. Technol. Agric.*, **29**, 213–27 (1980).
92. Damidaux, R., Autran, J.-C., Grignac, P. and Feillet, P., *Compt. Rend. Acad. Sci. D*, **287**, 701–4 (1978).
93. Bernardin, J., *Bakers Dig.*, **52**, 20–3 (1978).
94. Ewart, J. A. D., *J. Sci. Food Agric.*, **19**, 617–23 (1968).
95. Bloksma, A. H., *Cereal Chem.*, **52**(II), 170–83r (1975).
96. Khan, K. and Bushuk, W., *Bakers Dig.*, **52**, 14–20 (1978).
97. Khan, K. and Bushuk, W., *Cereal Chem.*, **56**, 505–12 (1979).
98. Forde, J., Forde, B. G., Fry, R., Kreis, M., Shewry, P. R. and Miflin, B. J., *FEBS Lett.*, **162**, 360–6 (1983).
99. Field, J. M., Shewry, P. R., Miflin, B. J. and March, J. F., *Theor. Appl. Genet.*, **62**, 329–36 (1982).
100. Shewry, P. R., Tatham, A. S., Forde, J., Miflin, B. J. and Kasarda, D. D., in *Gluten Proteins*, Graveland, A. and Moonen, J. H. E. (Eds), TNO, Wageningen, 1984, pp 51–8

101. Tatham, A. S., Shewry, P. R. and Miflin, B. J., *FEBS Lett.*, **177**, 205–8 (1984).
102. Tatham, A. S., Miflin, B. J. and Shewry, P. R., *Cereal Chem.*, **62**, 405–12 (1985).
103. Urry, D. W., *Methods Enzymol.*, **82**, 673–716 (1982).

3

Effects of Chemical, Genetic and Enzymatic Modifications on Protein Functionality

RAFAEL JIMENEZ-FLORES and THOMAS RICHARDSON

Department of Food Science and Technology, University of California, Davis, California, USA

1. INTRODUCTION

The nutritional quality of a protein as well as its functional characteristics in a food is of paramount importance in the utilization of proteins in food systems. The functionality of a protein results from those combinations of physico-chemical properties which impart desirable behavior in a food. Among required functionalities in food proteins are included good solubility, water binding, fat binding, emulsification and foaming properties depending upon the food application. Because of inherent structural limitations in some food proteins, they lack the requisite physico-chemical characteristics to serve a function in foods. Consequently, there have been numerous efforts to enhance the functionalities of many food proteins using physical, chemical and enzymatic methods. Often the physical properties of a protein can be improved by solvent treatments and/or thermally induced alterations; however, these changes will not be addressed in this review. There is an extensive literature on chemical and enzymatic modifications of proteins to improve their functional behavior in foods. However, because of recent reviews on the chemical and enzymatic modifications of food proteins,[1-6] only contributions to the literature within the last few years will be considered in this article. Moreover, only within the last half-decade has it become possible to manipulate the primary sequence of proteins genetically to alter their structures. This opens the way for the food scientist to work with the molecular geneticists and molecular biologists to modify food proteins structurally to enhance their functionalities. Thus chemical, enzymatic and genetic methods offer the food scientist the tools to understand better the structure–function relationships that govern food protein functionalities. In addition, chemical, enzymatic and genetic studies may result in practical possibilities for incorporating desirable physico-chemical attributes into food protein systems. Consequently, this review will be devoted to current studies employing chemical, enzymatic and genetic methods for structural changes of proteins that might be relevant to the better utilization of food proteins.

2. CHEMICAL MODIFICATIONS

Chemical modifications of food proteins have provided valuable structure–function relationships. However, the practical utilization of chemically modified proteins in foods is limited because often the residues derivatized in the protein may be essential amino acids. This may decrease the availability of the altered amino acids, particularly lysine which is a primary target of electrophilic reagents. In addition, the derivatized protein or its digestion products may be toxic. Moreover, one must consider the toxicity of the reagent used and any residues that may accompany the protein. Nonetheless, studies on the chemical alteration of food proteins continue to derive structure–function information to develop practical methods for the modification of protein functionality and to link limiting essential amino acids covalently into deficient proteins.

In Table 1 are shown[7] the various functional groups in proteins that are principal sites of derivatization. The nucleophilicities of the amino and sulfhydryl groups make them particularly vulnerable to reaction. Thus, the ε-amino group of lysine is a primary point of attack.

Table 1
Amino Acid Side-chains in Proteins Commonly Modified[7]

Side-chain	*Commonly used modifications*
Amino	Alkylation, acylation
Carboxyl	Esterification, amide formation
Disulfide	Reduction, oxidation
Imidazole	Oxidation, alkylation
Indole	Oxidation, alkylation
Phenolic	Acylation, electrophilic substitution
Sulfhydryl	Alkylation, oxidation
Thioether	Alkylation, oxidation
Guanidino	Condensation with dicarbonyls

Discussion of derivatization will be limited to acylations, phosphorylations, glycosylations, mild acid treatments and esterifications. By far the most numerous modifications of food proteins are the acylations, especially those using acetic and succinic anhydrides as the acylating reagents.

2.1. Acylation

Summarized in Table 2 are the effects of acylation on the functional properties of food proteins. In general, the acylating reagents acetic anhydride and succinic anhydride react with the ε-amino groups of lysine residues. In the case of acetic anhydride at the appropriate pH, the positively charged amino group is replaced by a neutral acetyl residue. Thus, the net charge on the protein becomes more negative, which is reflected in a slightly lower isoelectric point and greater electrophoretic mobility toward the cathode. These characteristics are accentuated with succinylation where the amino group is blocked and an additional negatively charged carboxylate is introduced. Succinylated proteins exhibit a definite acidic shift in the isoelectric point. Moreover, the protein solubilities of acylated proteins tend to decrease precipitously near the isoelectric point and the extent of insolubilization is greater than for the control protein. In general, acylation of proteins increases protein solubilities and hydration of modified proteins. This presumably reflects availability of polar residues for association with water. Since acylation blocks positively charged groups and upsets the electrostatic balance of charges in the native protein, the greater net negative surface charge on the protein probably results in unfolding of the protein and dissociation of sub-units due to charge repulsions. This exposes buried hydrophobic residues and alters the balance of available hydrophilic and hydrophobic residues in the protein. The change in hydrophilic–hydrophobic balance can alter the functional properties of proteins, particularly those dealing with surface and interfacial activities. In general, acylation enhances the emulsification characteristics of a protein because exposed hydrophobic residues associate with the oil and the increased negative surface charge on the droplets stabilizes the emulsion via charge repulsion of the particles. Enhanced emulsification properties are also related to increased protein solubilities. Oil absorption seems to be variable and dependent on exposure of hydrophobic residues in the protein. The foaming characteristics of acylated proteins seem to depend on charge, hydrophilic–hydrophobic balance and protein solubility. In general, acylated proteins show relatively good foaming capacity but poor foam stability. Presumably the initial

Table 2
Effect of Acylation on Functional Properties of Treated Proteins Compared to Control Proteins

Acylated protein	*Change in functional properties*	*Ref.*
1. *Vicia faba* protein isolates; acetylation	Little change in emulsion characteristics; thermal gel strengths increase to maximum then decrease with extent of acetylation.	8
2. *Vicia faba* protein isolates; acetylation	Heating (85°C, 20 min) coagulates acetylated proteins; DSC endotherms shifted to lower temperatures with acetylation.	9
3. *Vicia faba* protein isolates; acetylation	Acetylated *Vicia faba* proteins are compatible with gelatin in concentrations up to 50%. Swellability, brittleness of foods and viscosity of aqueous solutions of *Vicia faba* proteins increase with acetylation. Heat denaturation increases viscosity.	10
4. *Vicia faba*; succinylation	Succinylation improves foam formation. Maximum foam capacity at high degree of succinylation in neutral solution, but foam stability is low. Heating at 80°C; high protein concentrations at pH 5·5 favor stable foams from succinylated protein.	11
5. Winged bean flour; acetylation, succinylation	Acylation increased nitrogen solubility in water. Water absorption generally increased. Oil absorption for succinylated protein remained unchanged; for acetylated protein oil absorption increased then decreased. Emulsifying capacity of succinylated protein increased but that for acetylated protein increased slightly then decreased. Foam capacities for acylated proteins generally increased, but foam stabilities generally decreased. Densities generally increased with acylation.	12
6. Rapeseed flour; acetylation, succinylation	63% succinylation or 87% acetylation increased nitrogen solubility and decreased phytic acid extractability.	13
7. Rapeseed flour; acetylation, succinylation	Acylation lowered isoelectric pH of protein, increased the total nitrogen precipitation yield at isoelectric point and reduced phytic acid contamination of protein.	14

8. Rapeseed protein isolate; acetylation, succinylation	Acylated protein isolates from isoelectric precipitation had high nitrogen solubility, emulsifying and fat absorption properties. Acylated isolates had poorer whipping properties than control isolate.	15
9. Oat protein; acetylation, succinylation	Succinylation caused dissociation of oat polypeptides. Acetylation improved solubility, emulsifying properties and fat binding capacity, with succinylation being better. Water holding capacity was decreased. Foamability enhanced slightly but foam stability adversely affected. Bulk densities decreased with acylation.	16
10. Egg white proteins; acetylation, succinylation	Conalbumins iron complexing impaired. Lysozyme inactivated. Succinylated egg white improved heat stability and foamability. No change in angel cake performance. Acetylated egg white decreased heat stability, improved foamability but angel cake performance poor at high levels of acetylation.	17
11. Cottonseed flour; succinylation	Succinylation improved nitrogen and hydrophobic amino acid extraction from flour. Succinylation converts much salt- and alkali-soluble proteins to water-soluble forms. Electrophoretic mobilities of major proteins increased. Emulsion stabilities increased.	18
12. Cottonseed flour; maleylation, succinylation, dimethylglutarylation, acetylation, sodium sulfite	Acylation (except acetylation) increased protein extraction over conventional procedures. Proteins were precipitated isoelectrically at pH 4. Acetylation decreased protein extraction and precipitation. Succinylated, maleylated and dimethylglutarated flours were highly water-soluble and did not coagulate on heating. Acetylation decreased heat coagulation of isolate, but did not affect its water solubility. Sodium sulfite treatment increased heat coagulation of isolate and decreased its solubility. Only dimethylglutaration increased oil absorption.	19

(continued)

Table 2—*contd.*

Acylated protein	*Change in functional properties*	*Ref.*
13. Cottonseed flour protein isolates; succinylation	Succinylation increased water solubility. Heat coagulability was inversely related to degree of succinylation. Above 60% modification, emulsion capacity increased and was positively related to water solubility.	20
14. Cottonseed flour; acetylation, succinylation	Acetylation decreased nitrogen solubility in water or 5% NaCl whereas succinylation increased it. Water absorption capacity did not change due to succinylation, but increased because of acetylation. High levels of acylation decreased oil absorption capacity. Emulsifying capacity increased with acylation then decreased. Foaming capacity increased but foam stability decreased on acylation. Bulk density increased due to modification.	21
15. Sunflower seed protein isolate; acetylation, succinylation	Succinylation decreased bulk density, increased aqueous solubility, decreased isoelectric point and increased oil and water absorption, emulsification and foaming properties. Acetylation improved functional properties of the protein to a lesser extent.	22
16. Sunflower seed protein isolates; succinylation	Water absorption increased with succinylation; oil absorption changed little with acylation. Isoelectric point decreased with succinylation. Generally emulsification properties changed little with succinylation. Generally sunflower globulin had high foaming capacity and stability before and after succinylation.	23
17. Pea protein isolates; acetylation, succinylation	Increased emulsifying capacity, emulsion stability, foaming capacity and stability and water absorption compared to control. Acylation lowered isoelectric point. *In vitro* enzymatic hydrolysis not impaired by acylation.	24

18. Leaf protein concentrates; succinylation	Bulk density decreased then increased with succinylation. Oil and water absorption increased. Solubility and emulsifying activity also increased slowly with succinylation.	25
19. Myofibrillar protein of pale soft exudative porcine muscle; succinylation	Isoelectric point became more acidic. Proteins did not coagulate during heating at 98°C for 10 min. Water absorption ability not improved, but emulsion capacity improved slightly by succinylation.	26
20. Meat protein isolates; succinylation	Succinylated protein isolates from swine liver, lung and bone increased protein solubility, emulsion capacity, oil absorption, water hydration and whippability.	27
21. Soy proteins; acetylation, succinylation	Acetylated and succinylated soy proteins decreased rennet curd strength when added to milk.	28
22. Casein derivatives; succinylation	Acylation decreased relative viscosity, hydration; increased electrophoretic mobility of β-casein.	29
23. Milk proteins; acetylation	Acetylation increased ethanol stability and counteracted destabilization induced by Ca^{2+} addition.	30
24. Casein, sunflower seed and rapeseed protein isolates; acetylation	Acetylation changed solubility profile. Solubility reduced below isoelectric point. Above isoelectric point solubility of sunflower seed protein isolate unchanged, solubility of rapeseed protein isolate reduced. Emulsifying capacity of acetylated proteins unchanged.	31
25. Whey protein concentrates; succinylation	As extender in meat patties, increased cooking yield, retention of fat and moisture, decreased shrinkage. In supplemented wieners, lower emulsion stability and shrinkage increased firmness; less acceptable color and high cooking losses.	32
26. Whey protein concentrate (WPC); succinylation	Substitution of succinylated WPC for non-fat dry milk in ice-cream increased viscosity and resistance to melting and reduced freezing time and overrun. Adding succinylated WPC to instant pudding increased hydration and retarded rate of firming during storage.	33

(continued)

Table 2—*contd.*

	Acylated protein	*Change in functional properties*	*Ref.*
27.	Succinylated whey protein concentrate in bread making	In general succinylation did not improve the application of whey protein concentrate in bread.	34
28.	α_{s1}-Casein; acylation with palmitoyl-*N*-hydroxysuccinimide ester	Covalently incorporated palmitoyl residues into α_{s1}-casein. Palmitoyl proteins showed enhanced tendency to associate.	35
29.	α_{s1}-Casein; acylation with palmitoyl-*N*-hydroxysuccinimide ester	Generally incorporation of palmitoyl residues into α_{s1}-casein improved emulsification and foaming properties.	36
30.	Casein and β-lactoglobulin; acylation with poly-(L-methionine)	Polymethionine residues were covalently linked to casein and β-lactoglobulin using methionine-*N*-carboxyanhydride. Solubility of casein decreased with increased acylation. *In vitro* digestion of modified casein by α-chymotrypsin and trypsin reduced.	37,38
31.	Soybeans; acylation with *N*-hydroxysuccinimide ester of *N*-acetyl-L-methionine (NAM)	NAM in modified soybeans was biologically available, but to a lesser extent than free DL-methionine added to unmodified soybeans at same level.	39
32.	Wheat gluten; succinylation, atraconylation	At neutral pH range, acylated glutens showed greater water solubility, emulsification capacity, water holding capacity and water adsorption. At pH 3 atraconylation was reversible.	40

formation of foams is favored by the exposure of hydrophobic residues and increased solubility of proteins. However, greater charge repulsion in the foam lamellae leads to foam collapse and weakened foam stability. The heat stability of acylated proteins tends to increase, i.e. it requires higher temperatures to coagulate modified proteins. Presumably the more disordered structure in the acylated proteins with accompanying charge repulsions leads to a more thermally stable protein and failure to aggregate at lower temperatures.

Generally the functional properties of acylated proteins depend upon the variations introduced in net surface charge, exposure of hydrophobic residues and the hydrophile–hydrophobe balance of amino acid residues. The availability of polar vs. non-polar residues seems to define the amphipathic nature of the modified proteins which in turn defines its behavior in a food system.

In general, the nutritional quality of acylated proteins *in vivo* is in part dependent on the type of protein, the extent of protein modification and the type of acylating agent used (refs 41, 42 and references therein). In rat feeding trials, the nutritional value of acylated proteins was usually less than that of the untreated protein, particularly at higher levels of acylation. Moreover, the nutritional value of acetylated proteins was more readily available than that of succinylated proteins. In studies by Siu and Thompson[42] the net protein ratio (NPR) in rat feeding trials suggested that whey protein concentrates can be succinylated to a low level (15% E-NH_2 groups succinylated) without affecting the protein quality, probably because whey proteins are good sources of lysine. A medium level of succinylation (37%) can still produce a good quality protein with a higher NPR than casein. High levels of succinylation (74–83%) had deleterious effects on protein quality.

2.2. Phosphorylation

Chemical phosphorylation may prove useful in changing the functional properties of food proteins. The chemical phosphorylations of food proteins have been extensively reviewed in an excellent paper by Matheis and Whitaker[43] in which original citations may be found.

Phosphorus oxychloride ($POCl_3$) is probably the most extensively used reagent for protein phosphorylation. From 0·6% to 3·9% phosphorus has been introduced into various proteins and polypeptides.

Trichloroacetonitrile reacts with phosphoric acid to yield the phosphorylating agent trichloroacetimidoyl phosphate [reaction (1)]. The trichloroacetimidoyl phosphate can react further with phosphoric acid, carboxyl groups and hydroxyl groups [reaction (2)] to yield the corresponding phosphorylated compound. Protein phosphorylation occurs when the ROH represents a serine or threonine residue.

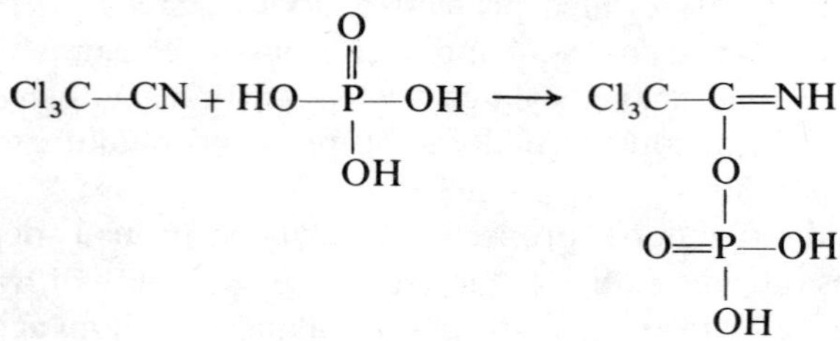

(1)

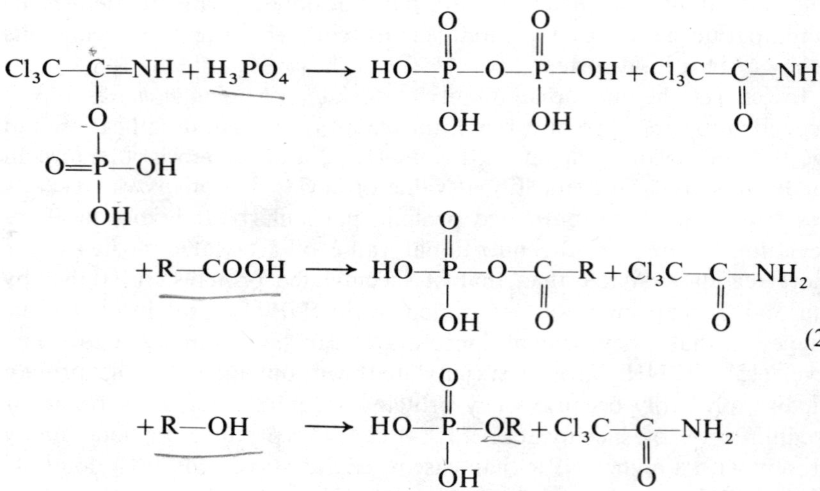

(2)

Proteins kept in a reaction mixture of phosphorus pentoxide dissolved in phosphoric acid undergo phosphorylation; however, this is a rather harsh treatment for food proteins.

Soybean proteins have been phosphorylated with cyclic sodium trimetaphosphate under alkaline conditions.[44] However, it has been questioned whether the protein-phosphate was covalently or non-covalently bound.[43]

A variety of reagents, including monophenyl phosphodichloride, phosphoroamidate and diphosphoimidazole, have been used to phosphorylate a number of proteins.[43]

In general, the type of phosphorylated residues can be characterized by their pH stability, by the loss of protein functional groups, by digestion of protein and analysis for phosphoamino acids and by ^{31}P nuclear magnetic resonance spectroscopy.[43] Phosphorylation generally occurs at serine hydroxyls, lysine amino groups and histidine imidazole nitrogen. In Table 3 are listed some of the changes in functional properties of proteins that occur as a result of phosphorylation.[43]

Because of the hydrophilicity of incorporated phosphate groups, an increase in water solubility would be expected. In general this was observed for soy proteins.[44,45] However, in the work of Hirotsuka *et al.*[45] solubility at the isoelectric point (pH 4·5) increased at high levels of phosphorylation but was comparable to the control at low levels of phosphorylation. However, there was a decreased level of water solubility with some phosphorylated proteins possibly reflecting cross-linking of proteins. The increased viscosity and gel-forming properties may also result from cross-linking reactions. Some proteins exhibit enhanced emulsifying activities whereas others do not. There evidently is no good explanation for these differences.

As one might expect, phosphorylated proteins should have a relatively high affinity for Ca^{2+}. When phosphorylated β-lactoglobulin at pH 5 or 7 was dialyzed against Ca^{2+}, a gel rapidly formed.[50] The viscosity of an emulsion prepared from phosphorylated β-lactoglobulin treated with Ca^{2+} was about twice that prepared with untreated β-lactoglobulin.[50] However, in the case of phosphorylated soy protein isolate[45] the normal sensitivity of soy proteins to Ca^{2+} precipitation decreased directly with phosphorylation.

Phosphate bonds in phosphorylated proteins can be cleaved by phosphatases.[50,54] In general, *in vitro* digestibility studies with proteolytic enzymes indicate that they are readily digestible, with retardation in digestion rates in some cases.[43] The only *in vivo* digestibility study was with phosphorylated casein.[46] In the bioassay used, *Tetrahymena thermophili* grew as well on phosphorylated casein as the control, indicating that phosphorylation did not affect the digestion, absorption and utilization of the amino acids by the organism. However, feeding studies using mammals must be carried out to verify that phosphorylated proteins can be readily utilized and are not toxic.

2.3. Mild Acid Hydrolysis

Treatment of gluten[55] or soy proteins[56] with dilute (0·05N) solutions of HCl heated for varying periods of time deamidates asparagine and glutamine residues. This increases the negative charge on the surface of the protein, resulting in unfolding and exposure of hydrophobic residues. Denaturation of the protein by the acid treatment may also make the

Table 3
Changes in Functional Properties of Chemically Phosphorylated Proteins[43]

Phosphorylated protein	*Change in functionality*	*Ref.*
1. Soybean protein	Increased water solubility Increased solubility at high levels of phosphorylation	44
2. Casein, lysozyme, serum albumin hemoglobin	Decreased water solubility	46, 47
3. Casein, ovalbumin, serum albumin	Increased viscosity	46, 48, 49
4. Casein, lysozyme, soybean protein	Increased water binding	44, 46
5. Ovalbumin, serum albumin	Increased buffering capacity	48, 49
6. β-Lactoglobulin, soybean protein	Increased emulsifying activity	44, 45, 50
7. Casein	Decreased emulsifying activity	46
8. Soybean protein	Increased foam expansion and stability	44
9. Ovalbumin	Loss of heat coagulability	48
10. α_{s1}-, β- and κ-caseins, β-lactoglobulin	Increased calcium binding[50]	51
11. Casein, soy protein isolate, serum globulin, gluten, serum albumin	Gel forming properties[45,52,53]	46, 47
12. β-Lactoglobulin	Gel forming properties in presence of Ca^{2+}	50
13. Soybean protein isolate	Decreased sensitivity to Ca^{2+}	45

protein more hydrophobic. The increased amphiphilicity of the proteins generally results in enhanced functional properties.

In the case of deamidated gluten[55] the helix content decreased with increased degree of deamidation. The surface tension decreased in proportion to the degree of deamidation, reflecting an increased amphiphilicity. On the other hand, the surface hydrophobicity of gluten increased in proportion to the extent of deamidation. The emulsifying properties of gluten were improved and correlated with the surface hydrophobicity.

The surface hydrophobicity of acid-modified soy protein[56] increased rapidly at an early stage of the mild acid treatment, then gradually increased up to 30 min of heating in the presence of 0·05N HCl. The surface tension of the acid-modified soy protein decreased in a manner that reflected the increase in surface hydrophobicity, suggesting that deamidation unfolded the soy proteins, exposing hydrophobic groups. In general, emulsifying activity and emulsion stability, along with foaming power and foam stability, increased up to a heating time of 30 min in 0·05N HCl at 95°C. The results suggest that the improvement in the functional properties of acid-modified soy protein may be mainly due to the increased surface hydrophobicity resulting from acid treatment.

Mild acid hydrolysis of ovalbumin[57] in 0·03N HCl heated *in vacuo* at 115°C yields large deamidated polypeptide fragments. This treatment hydrolyzes peptide bonds on either side of aspartic acid residues. Upon standing in air at room temperature for 16 h the peptides associated, presumably via hydrophobic bonding, and polymerized by formation of disulfide bonds. These polymerized products exhibited good functional properties such as solubility, emulsifying and foaming properties. Presumably enhanced functionality resulted from deamidation with exposure of negative charges, better amphiphilicity by appropriate arrangement of hydrophilic and hydrophobic residues, and high molecular weight polymers formed through hydrophobic interactions and disulfide bonds.

2.4. Glycosylation

Glycoproteins function in food systems to provide excellent emulsion and foam stability.[58,59] Because carbohydrates impart desirable functional properties to proteins, several carbohydrates have been covalently attached to β-lactoglobulin to modify its functionality.[60,61]

Maltose and β-cyclodextrin were coupled to β-lactoglobulin using the cyclic carbonate procedure which activated the sugar.[60] The amino groups of β-lactoglobulin were the primary site of glycosylation, accounting for 70–77% of the groups modified. Sulfhydryl and hydroxyl groups were also partially glycosylated. An average of 2·2 maltose residues per group modified was observed at the highest level of modification. Evidently the maltose cyclic carbonate reacted with maltose residues previously coupled to the protein. The coupling efficiency of activated β-cyclodextrin was higher than that of the activated maltose at the same molar ratio of carbohydrate to β-lactoglobulin, i.e. 3·8 vs. 1·3 sites modified, respectively. Since the molecular weight of the β-cyclodextrin is larger than the molecular weight of maltose, the mass of carbohydrates linked to β-lactoglobulin was considerably larger for the β-cyclodextrin derivative.

In addition to coupling maltose and β-cyclodextrin to β-lactoglobulin, glucosamine and glucosamineoctaose were also used to modify β-lactoglobulin after activation of carboxyl groups in the protein with a carbodiimide.[60] The carboxyl groups of β-lactoglobulin were the primary sites of modification, accounting for 75–89% of the groups modified. Again, the coupling efficiency of the amino oligosaccharide to β-lactoglobulin was higher than for glucosamine at the same molar ratio of reactants. The mass of carbohydrates linked to β-lactoglobulin was greater for glucosamineoctaose compared to glucosamine because of the increased molecular weight of the oligosaccharide. In general, the coupling efficiency of the active compound was higher for the carbodiimide process than for the cyclic carbonate reaction.

Gel electrophoretic analyses indicated that a heterogeneous population of protein molecules was formed upon glycosylation.[60] The electrophoretic mobility of maltosyl-β-lactoglobulin derivatives was essentially unchanged while the mobility of glucosaminyl-β-lactoglobulin derivatives decreased as the extent of modification increased.

The maltosyl and glucosaminyl derivatives of β-lactoglobulin exhibited altered physico-chemical properties compared to the native protein.[62] The chemical and structural properties of proteins affect their functional properties in food systems; however, insufficient information is currently available to correlate the basic properties of proteins to their functional properties.

In general, the viscosity of glycosylated β-lactoglobulin derivatives increased with increased extent of derivatization. It is likely that the viscosities of these proteins increased because the mass of the protein molecule increased with some unfolding of the protein coupled with extensive hydrophilic interactions between the glycosylated proteins and water. The ultraviolet difference spectra and the intrinsic fluorescence of the modified proteins revealed that the microenvironments of aromatic amino acids of β-lactoglobulin were increasingly exposed to the surface of the protein as the extent of modification increased. The aromatic amino acid residues for the maltosyl derivatives were more exposed to ultraviolet radiation than those of β-lactoglobulin or glucosaminyl derivatives. Intrinsic fluorescence measurements indicated that for the maltosyl-β-lactoglobulin the environments of the tryptophyl residues were more polar. On the other hand, intrinsic fluorescence of some glucosaminyl derivatives revealed that the environments of the tryptophyl residues were less polar than in the native protein. Apparently the glucosamine residues permitted the tryptophyl residues to be slightly more exposed to the surface of the protein but the tryptophyl residues remained in relatively non-polar environments. Thus it appears that the microenvironments of tryptophyl residues of glycosylated derivatives of β-lactoglobulin were changed more after modification with maltose than after modification with glucosamine.

The surface hydrophobicities of the maltosyl derivatives decreased as the extent of modification increased, whereas the hydrophobicities of the glucosaminyl derivatives were relatively unchanged. Evidently the surface hydrophilic carbohydrate residues interfered with the binding of the hydrophobic probe, leading to lowered hydrophobicities, particularly for the maltosyl derivative.

From circular dichroism measurements, the maltosyl derivatives of β-lactoglobulin had a decreased amount of α-helical structure and a correspondingly increased amount of unordered structure compared to similarly modified glucosaminyl derivatives. Thus, ordered secondary structures of extensively derivatized β-lactoglobulin were partially unfolded.

Enzymatic hydrolysis studies on maltosyl and glucosaminyl derivatives of β-lactoglobulin revealed differences in *in vitro* digestibilities by trypsin and α-chymotrypsin.[63] The rate and extent of proteolysis of β-lactoglobulin were affected by the extent of modification, by the type of carbohydrate residue coupled to β-lactoglobulin and by the type of proteolytic enzyme. As the number of residues modified increased, the rate of hydrolysis increased. However, the rates of hydrolysis for the glucosaminyl derivatives were lower than those for the maltosyl derivatives. Presumably this was *related* to a more expanded and open structure for the maltosyl derivatives. α-Chymotrypsin was more effective than trypsin in hydrolyzing the maltosyl and glucosaminyl derivatives since sites for chymotrypsin hydrolysis were more exposed because of unfolding.

In a more recent study by Kitabatake *et al.*[61] on glycosylation of β-lactoglobulin, gluconic and melibionic acids were covalently attached to amino groups of β-lactoglobulin by using a water-soluble carbodiimide for activation of the carboxyl groups of the carbohydrates. Up to 8 moles of glycosyl residues could be attached per monomer of β-lactoglobulin. The glycosylated proteins exhibited excellent solubility even at low ionic strength and at the isoelectric pH of β-lactoglobulin. The stability of the glycosylated derivatives to heat coagulation was increased. The solubility and heat stability increased as the degree of glycosylation increased. In general, on a molar basis, melibionic acid was more effective than gluconic acid in enhancing solubility and resistance to heat coagulation.

Glycosylation of proteins leads to substantial changes in their physico-chemical characteristics. These alterations are probably reflected in changes in the functional properties of the modified proteins.

2.5. Esterification

As we have seen, the most widely studied derivatization of food proteins involves the acylation (or alkylation) of the ε-amino groups of lysine residues. However, changes in lysine residues can be detrimental to the nutritional quality of a protein. As an alternative to derivatizing the side-chain amino groups of the essential amino acid lysine, there have been a number of recent studies exploring modifications of the side-chain carboxyl groups of the non-essential amino acids aspartic acid and glutamic acid.[64–68]

Since these acidic acids are non-essential for humans, the nutritional quality of the protein in question should not be adversely affected by their derivatization. Esterification of the side-chain carboxyl groups of acidic amino acids in a protein will make the protein more basic and thus yield a protein with a greater net positive charge at a given pH. Esterified proteins with a positive charge, particularly where the isoionic points may be in the region of 9–10, possess a number of potentially useful physico-chemical properties. For example, at pH 7 most proteins possess a net negative charge.[68a] It becomes possible then that such negative proteins can interact electrostatically with a protein containing a net positive charge, leading to complex formation. If the formation of such complexes could be properly controlled, they might prove useful in formulating new food products.

The following discussion will emphasize studies of polycationic ester derivatives of β-lactoglobulin, a major protein in milk.[65–67] We will see how the physico-chemical properties of a protein can be markedly altered by changes in the surface charge of a protein.

In the presence of an appropriate alcohol containing an acid catalyst such as HCl, proteins can be esterified simply by suspending them in the medium. Esters of carboxyl groups in proteins treated with methyl and ethyl alcohol readily form derivatives with an enhanced net positive charge [reaction (3)]. However, with longer chain alcohols it becomes

$$\mathrm{P}\begin{matrix}\diagup \mathrm{NH_3^+} \\ \diagdown \mathrm{COO^-}\end{matrix} \xrightarrow[\mathrm{ROH}]{\mathrm{H^+}} \mathrm{P}\begin{matrix}\diagup \mathrm{NH_3^+} \\ \diagdown \mathrm{COOR}\end{matrix} + \mathrm{H_2O} \qquad (3)$$

increasingly difficult to prepare esters directly because of the hydrophobicity of the alcohols and their relative unreactivity. It may be possible to prepare long-chain alcohol esters of proteins by first preparing the methyl esters and then transesterifying in the presence of an acid catalyst with the long-chain ester alcohol. Esterification of β-lactoglobulin with methanol or ethanol occurs over a period of several days at 4°C with the protein actually dissolving in the alcohol as the esterification proceeds. The alcohols and acid catalysts can be readily removed by dialysis of the reaction mixture and the proteins in the retentate can be recovered after freeze-drying. Esterification and other chemical modifications can result in alterations in the physico-chemical properties of proteins.

2.5.1. Isoionic Points

A simple mixed-bed ion exchange resin technique allows the easy estimation of isoionic points of the purified proteins.[69,70] Counter-ions associated with the proteins are bound by the mixed-bed resin added to a few milliliters of a 1% weight/volume protein solution. A pH meter is used to monitor the supernatant protein solution until the pH becomes stable at a value equivalent to the isoionic point of the protein. With this simple technique it is possible to determine isoionic points that are within 0·03 pH unit of isoelectric points of the same protein determined with more sophisticated methods. In our studies with methyl or ethyl esters of β-lactoglobulin, isoionic points varied between 7 and 10, depending upon the extent of esterification. This is compared to 5·2 for native β-lactoglobulin.

2.5.2. Protein Interactions

Soluble or insoluble complexes can be formed when polycationic proteins resulting from esterification of carboxylic groups in the proteins can interact electrostatically with more negatively charged proteins. The structure and properties of the interacting proteins define the physical characteristics of the complex. However, protein matrices of value in the development of new or modified food products may be derived from controlled formation of insoluble protein complexes. Esterified polycationic proteins, for example, have been proposed as electrostatic coagulants for the proteins in milk to form cheese-like structures.[64]

When solutions of methyl or ethyl esters of β-lactoglobulin with pI values in the range 8·7–9·5 are mixed with solutions of caseins from milk at pH 7, insoluble complexes are formed. Evidently the polyanionic caseins interact with the polycationic esters of β-lactoglobulin to form insoluble protein aggregates that can be easily sedimented by centrifugation of the mixture at 2000 *g* for 20 min. Electrostatic interactions between the proteins can be quantified by essentially titrating a constant amount of the caseins in solution with increasing amounts of β-lactoglobulin esters. Resultant insoluble complexes are removed by centrifugation and, by using the mixed-bed resin technique discussed in the foregoing section, the isoionic point of the excess protein remaining in the supernatant solution can be determined. From these data a type of titration curve can be obtained from which an equivalence point can be estimated as shown in Fig. 1.[65]

It is evident then that protein structures potentially useful in product development can be created by exploiting such protein interactions. The forms that the protein complexes will take, as mentioned previously, are dictated by certain structural features of the proteins in solution. Caseins have a unique amphiphilic structure which likely plays a crucial role in the formation of insoluble protein aggregates as just described. Thus it is possible to visualize a system wherein the repulsive anionic groups in the caseins are neutralized upon addition of the polycationic esters of β-lactoglobulin. This creates electrostatic interactions between the casein and β-lactoglobulin to form complexes that interact through residual hydrophobic bonding on their surfaces to favor the formation of weak but numerous hydrophobic bonds to complete the aggregation process. In the development of food products, the proper manipulation of such non-covalent interactions between proteins is important. Often the manipulation of non-covalent interactions is done on a trial-and-error basis such as heating a product at various temperatures for different periods of time until a satisfactory result is obtained. It has been only in recent years that attempts have been made to understand the underlying molecular mechanisms governing protein behavior in foods.[71,72] A better understanding of the relationships between food protein structure and their functions will allow us more intelligently to utilize conventional as well as novel proteins in foods.

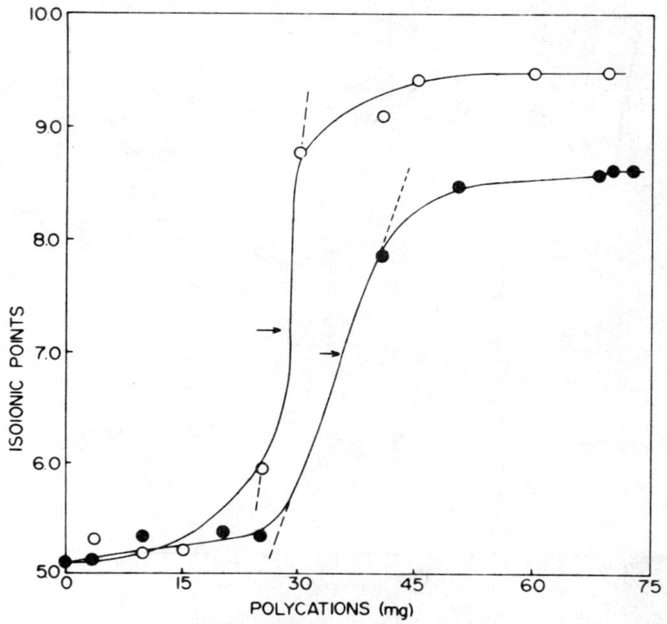

Fig. 1. Equivalence points for precipitation of a given amount of soluble casein upon the addition of increasing quantities of positively charged proteins. The isoionic points are for proteins remaining in the supernatant fluids. Ethyl-esterified β-lactoglobulin (●), methyl-esterified β-lactoglobulin (○). Equivalence points are indicated by arrows; $n = 2$ reported as mean values.[65]

2.5.3. Circular Dichroism (CD)

Upon passage of circularly polarized light through a solution of an optically active substance, the absorptivity depends on the handedness of the light. For example, some absorption bands may absorb more strongly the right and others may absorb more strongly the left circularly polarized beam. Circular dichroism may be defined at a given wavelength as $\Delta\varepsilon = \varepsilon_L - \varepsilon_R$, the difference in extinction coefficients.[73] A plot of molar ellipticity, θ (a measure of circular dichroism), of a protein solution versus wavelengths between 185 and 250 nm yields a CD spectrum from which the contributions of various conformational states in a protein may be calculated. Certain characteristic conformations in a protein (e.g. the α-helix) give rise to characteristic CD bands allowing an estimate of the contribution of a particular conformation to protein structure (Fig. 2).

CD bands in the near-ultraviolet regions between 260 and 320 nm when

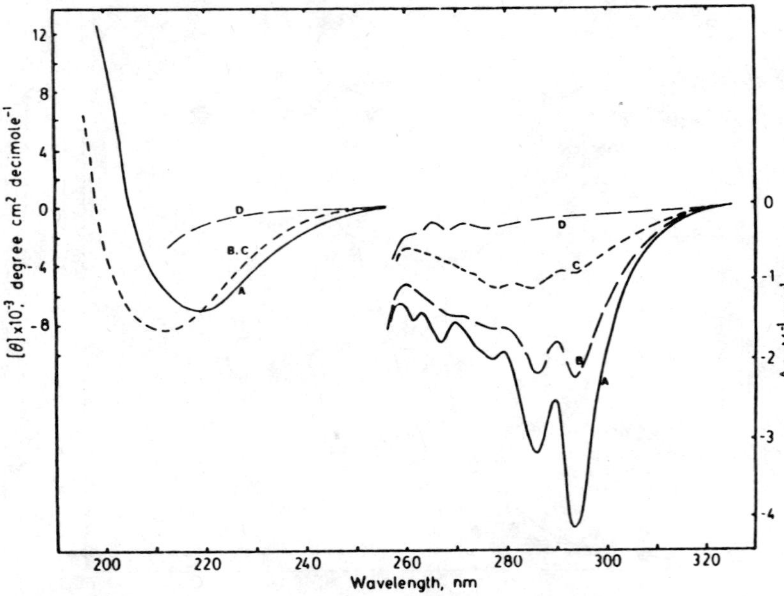

Fig. 2. Circular dichroism spectra of β-lactoglobulin (A), ethyl-esterified β-lactoglobulin (B), methyl-esterified β-lactoglobulin (C), and β-lactoglobulin in 5M guanidine hydrochloride (D).[67]

$\Delta\varepsilon$ is plotted against wavelength (Fig. 2) result from the non-random orientation of aromatic amino acid chromophores (e.g. tryptophan) within the protein. From the analysis of CD spectra as shown in Fig. 2, β-lactoglobulin in the 185–250 nm region is likely comprised of 10% α-helix, 40% antiparallel chain β and 50% unordered structure.[74] There is a general displacement of the CD spectra to lower wavelengths and an increased intensity of the negative absorption bands upon denaturation of β-lactoglobulin at high pH values.[74]

Thus modifications in the conformation of proteins resulting from various chemical and physical treatments that may be important in food processing can be detected from changes in the CD spectra of the proteins.

The protein denaturant guanidine completely disrupts the native structure of β-lactoglobulin and, as shown in Fig. 2,[67] there results a disappearance of the CD bands in the aromatic chromophore region (260–320 nm). On the other hand, the spectrum for native β-lactoglobulin in this region yields very intense CD bands. Esterification of the protein with ethanol or methanol results in decreased intensities, suggesting a more random conformation for the esterified proteins. The shift of the CD bands in the far-UV to shorter wavelengths in addition to increased intensity[67] also indicates a more random structure for β-lactoglobulin upon esterification. Phosphorylated[75] and amidated[67] β-lactoglobulins exhibit similar spectral changes. These data are consistent with the concept that chemical derivatizations that alter surface charge characteristics result in disruption of the three-dimensional structure of the native protein. With unfolding of the protein molecule through charge repulsion effects there is probably an exposure of hydrophobic residues, giving the protein more amphipathic character.

2.5.4. *Protein Surface Hydrophobicity*

Partial unfolding of polypeptide chains resulting from chemical or physical modification of the proteins probably exposes hydrophobic amino acid residues normally shielded from water within the interior of the globular protein. Hydrophobic interactions are extremely important in protein folding and in protein associations.[76] Thus chemical or physical perturbations that alter the surface hydrophobicity of proteins can have profound effects on protein functionalities.

cis-Parinaric acid and 1-anilinonaphthalene-8-sulfonic acid (1,8-ANS) have been used as hydrophobic probes to study the surface hydrophobicity of esterified proteins.[66,67] There is an enhancement in the fluorescence of these hydrophobic probes as they bind to hydrophobic patches on the protein.[77,78] Plots of the initial slopes (S_0) of fluorescence enhancement of *cis*-parinaric acid versus increased protein concentration lead to a good correlation with the effective hydrophobicity of the protein as determined by hydrophobic partition experiments using poly(ethylene glycol)palmitate.[78]

Esterification of β-lactoglobulin with methanol, ethanol or butanol results in greater apparent relative hydrophobicities (increased S_0) of the proteins compared to the native β-lactoglobulin when 1,8-ANS was used as a hydrophobic probe (Fig. 3). Relative to the native protein, the largest increases in fluorescence occurred with the methyl and ethyl esters of β-lactoglobulin. The butyl ester of β-lactoglobulin possessed an enhanced fluorescence, but not of the magnitude noted for the methyl and ethyl esters of the proteins (Fig. 3). Combination of the surface hydrophobicity information obtained using hydrophobic probes with CD spectra as discussed in the foregoing section suggests that esterification of β-lactoglobulin increases the net positive surface charge on the protein. Partial unfolding of the β-lactoglobulin results from repulsive forces

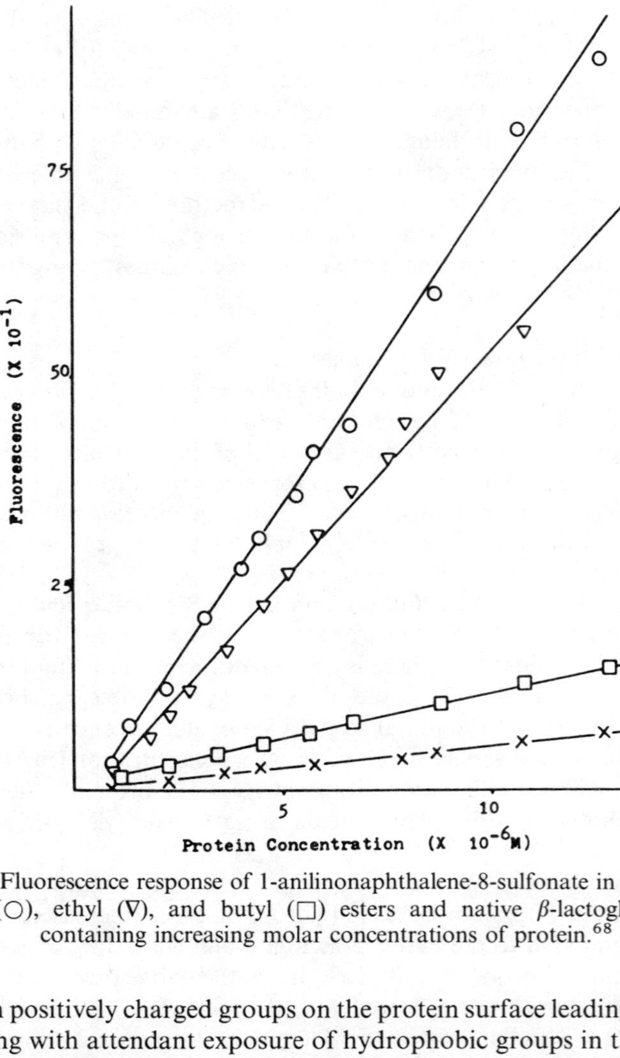

Fig. 3. Fluorescence response of 1-anilinonaphthalene-8-sulfonate in solutions of methyl (○), ethyl (∇), and butyl (□) esters and native *β*-lactoglobulin (×) containing increasing molar concentrations of protein.[68]

between positively charged groups on the protein surface leading to partial unfolding with attendant exposure of hydrophobic groups in the protein. Analogous alterations in conformations of food proteins resulting in changes in their functionality can likely occur as a result of thermal perturbations used in processing of foods.

The partitioning of non-ionic, non-polar molecules such as heptane into solutions of proteins has also been used as an indicator of the relative hydrophobicities of proteins.[79] An enhanced hydrophobicity of the proteins as a result of esterification is indicated by the greater partitioning of heptane into solutions of esterified β-lactoglobulin compared with those containing native β-lactoglobulin.[66,67] From studies with a large number of proteins,[78] there was a significant negative correlation between interfacial tension and protein hydrophobicity. At the same time, there was a significant positive correlation between emulsifying activity index and hydrophobicity.[78] Thus it follows that the enhanced hydrophobicity of a protein which decreases the interfacial tension between oil and water yields a better emulsifying agent. Although difficult to quantify, the enhanced surface hydrophobicity probably leads to a more effective hydrophobic–hydrophilic balance (amphiphilicity) in the modified protein. If this is true, then the esterified proteins should reduce interfacial and surface tension as measured at a water–oil interface and water–air interface, respectively.

2.5.5. *Interfacial and Surface Tensions*

The effects of proteins on interfacial tensions and vegetable oil–water interfaces and surface tensions at the air–water interface can be readily measured using a DuNuoy Ring Tensiometer.[80] The surface tension at the air–water interface was reduced by native β-lactoglobulin as well as its methyl, ethyl and butyl esters. At protein concentrations of 0·2% w/v, mean surface tension values over duplicate trials ($n = 3$) for native β-lactoglobulin and methyl, ethyl and butyl esters of β-lactoglobulin were $64{\cdot}6 \pm 0{\cdot}4$, $50{\cdot}3 \pm 0{\cdot}4$, $54{\cdot}3 \pm 0{\cdot}4$ and $56{\cdot}4 \pm 0{\cdot}3$ dynes cm^{-1}, respectively.[68] The surface tension of distilled water under comparable conditions was $73{\cdot}6 \pm 0{\cdot}7$ dynes cm^{-1}. Thus native β-lactoglobulin reduced the surface tension by 12%, whereas the methyl ester of β-lactoglobulin reduced the surface tension by 32%.

Interfacial tensions of 0·2% w/v protein solutions at an oil–water interface were also measured to ascertain the effect of chemical modifications of β-lactoglobulin on changes in surface functionality. A mean interfacial tension of $20{\cdot}0 \pm 0{\cdot}8$ dynes cm^{-1} over duplicate trials ($n = 3$) compared with a mean value of $33{\cdot}2 \pm 0{\cdot}9$ dynes cm^{-1} at a water–corn oil interface was obtained with solutions of native β-lactoglobulin. On the other hand, the esterified proteins were more effective in lowering the interfacial tension. Respective mean values for methyl, ethyl and butyl esters of β-lactoglobulin over duplicate trials ($n = 3$) were $14{\cdot}5 \pm 0{\cdot}5$, $16{\cdot}7 \pm 0{\cdot}2$, and $17{\cdot}8 \pm 0{\cdot}3$ dynes cm^{-1}.[68] The methyl ester reduced the tension at the oil–water interface 56% relative to the system with no protein. It may be that some portion of the surface activities of the

Table 4
Summary of Some Consequences of Chemical Modifications of Proteins

Introduced relative side-chains, e.g. thiols, phosphate
Altered surface charge
Changed pI; prepare more negatively or positively charged proteins
Altered equilibrium between protein subunits
Induced more randomness in structure of protein from charge repulsions at surface
Stabilized against thermal denaturation?
Enhanced surface hydrophobicity of protein by exposing buried hydrophobic residues
Increased amphiphilic behavior of proteins; more detergent-like?
Modified proteins stabilized by intramolecular thiol–disulfide interactions

modified proteins may be directly due to the introduction of alcohol residues into the protein. However, it is likely that major contributions to the surface activities also result from conformational alterations of the protein as a result of exposure to the alcohol and to changes in the net surface charge on the proteins. The ability of a protein to reduce interfacial tension and to retain some of its native structure in forming cohesive films are important to the formation and stability of foams.[58] It is widely understood that reduction of interfacial tension at the oil–water interface by various emulsifying agents is important to the formation of stable emulsions. Among the important functional characteristics of food proteins are the emulsifying properties of the proteins. As seen from the preceding examples, chemical derivatization of proteins can be used to derive valuable structure–function relationships in food proteins. As shown in Table 4, a variety of changes at the molecular level that can affect protein functionality can result from the introduction of various functional groups into proteins. Functional groups introduced by chemical modification of proteins can participate in subsequent reactions. For example, yogurt-like gels can be obtained when phosphorylated β-lactoglobulin is exposed to Ca^{2+}.[50] In addition, the secondary effects resulting from chemical modifications may play a major role in altering the functional properties of proteins. The reaction conditions alone in protein modifications (e.g. pH, solvent, ions, etc.) may actually be dominant factors in controlling the changes in protein conformations that alter functional characteristics. However, as shown in Table 4, there could result a series of interrelated changes in functional properties stemming from something as simple as a change in the net surface charge on the protein.

Table 5
Structural Properties and Interactions of Polypeptides which Influence their Functionality[72]

Amino acid composition (major groups)
Amino acid sequence (segments/polypeptides)
Secondary/tertiary conformation (compact/coil)
Surface charge, hydrophobicity, polarity
Size, shape (topography)
Quaternary structures
Secondary interactions (intra- and inter-peptide)
Hydrogen bonding, ionic, van der Waals, hydrophobic and electrostatic interactions
Disulfide/sulfhydryl content
Environmental conditions (pH, O/R status, salts, temperature)

Although we have discussed only a few of the factors that affect the nutritional properties of proteins, there are numerous structural features of proteins that influence their functionality (Table 5). Most of these structural features depend upon the amino acid composition and sequence. Consequently, a method for systematically changing the primary structure of food proteins to study functionality would complement studies involving chemical derivatization. This brings us to the systematic genetic manipulation of the primary structures of food proteins. Indeed, molecular biologists and synthetic oligonucleotide chemists have provided us with the tools to study structure–function relationships in food proteins systematically.

3. GENETIC MODIFICATIONS

Chemical modifications of proteins have provided much useful structure–function information on the functional behavior of food proteins. Although providing useful structure–function information on proteins, chemical derivatization often results in polydisperse protein products because of more-or-less random reactions with protein functional groups of varying reactivity. Thus, specificity of modification is lost, which can hamper interpretation of structure–function information.

Recent developments in recombinant DNA technology can be used systematically to alter single amino acids in the primary sequences of proteins.[81–84] The functional characteristics of proteins can be markedly altered by single changes in amino acids. For example, the conformation and physical properties of valine (sickle-cell) hemoglobin are totally altered compared to the normal glutamate hemoglobin.[85] It thus becomes possible to consider food protein functionality in the broader context of relatively large changes in protein structure from chemical derivatization or in the limited context of highly selective alterations in the primary sequence of the proteins using genetic engineering techniques.

On the other hand, in the genetic engineering of enzymes, the precision of modification inherent in recombinant DNA techniques is required to alter the K_m, V_{max}, specificity etc. of the enzyme. It is now feasible genetically to engineer enzymes to change their thermal inactivation, kinetic characteristics, and specificity to make them more suitable for use in food processing and analyses. By the same token, manipulation of the functionality of food proteins for the most part will probably not require such precision, and it becomes possible to alter their properties with more generalized changes. This report will discuss the use of oligodeoxynucleotide directed mutagenesis to modify bovine caseins specifically to improve their functional behavior in cheese-making.

3.1. Oligodeoxynucleotide Site-directed Mutagenesis

Oligodeoxynucleotide site-directed mutagenesis has been used to make specific changes in the primary sequences of proteins in general and of enzymes in particular.[81–84,86–89] Seven general conditions are required for oligodeoxynucleotide site-directed mutagenesis.[81,82] A protein can be restructured by:

(1) cloning the relevant gene into an appropriate vector that is capable of autonomous replication (e.g. plasmids and viral genomes);
(2) obtaining expression of the cloned gene;
(3) determining the primary sequence of the DNA insert;
(4) making the site to be mutagenized available in a single-stranded form;
(5) using an appropriate, specifically-priming synthetic oligodeoxynucleotide containing base mismatches for altering the cloned sequence for specifically annealing to the single-stranded site;
(6) performing site-directed mutagenesis wherein the sequences that are to remain wild-type must be regenerated with fidelity;
(7) identifying mutant colonies with provisions for isolating and characterizing the mutant DNA.

To achieve oligodeoxynucleotide site-directed mutagenesis, several strategies are available.[81–84] The strategy depicted[82] in Fig. 4 modifies

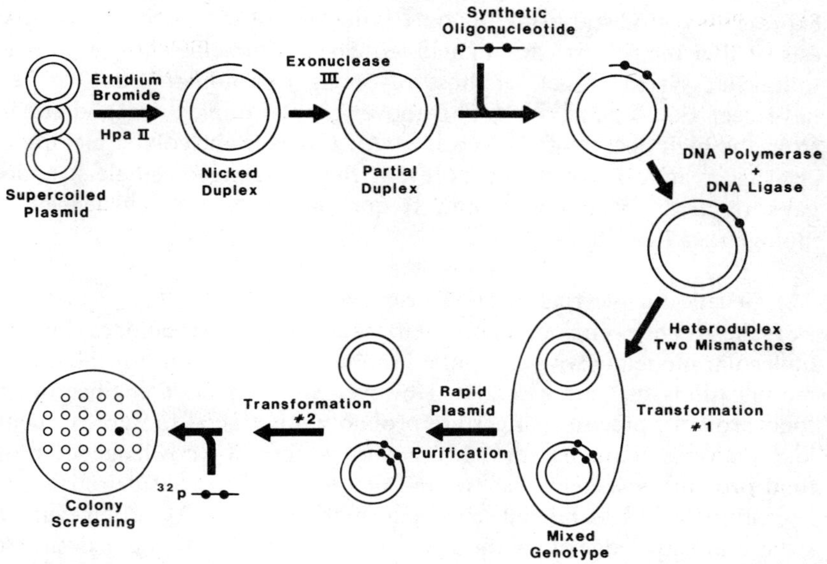

Fig. 4. A strategy for oligodeoxynucleotide site-directed mutagenesis involving a plasmid.[82]

a plasmid to yield a single-stranded segment for site-directed mutagenesis. The initial supercoiled plasmid circles are nicked in one strand by a restriction endonuclease in the presence of ethidium bromide. A partial single-stranded segment is obtained by digestion of the nicked strand from the 3′ end with exonuclease III. The gapped circles are annealed or hybridized with a synthetic, homologous oligodeoxynucleotide (16 bases) carrying, by design, two mismatches. The synthetic oligodeoxynucleotide heteroduplex plasmid circles help prime *in vitro* DNA synthesis. Molecular cloning and *in vivo* DNA replication generate homoduplexes, some of which have the sequence of the synthetic primer oligodeoxynucleotide. Purified plasmids can be used to transform *Escherichia coli* a second time. Colony screening, with the same synthetic oligodeoxynucleotide, labeled with ^{32}P as a hybridization probe, allows identification of the desired mutant colony regardless of its phenotype.

Since detection of mutants is at the level of DNA and involves only colony hybridization, the foregoing method can be used as a general strategy. Consequently, the procedure can be applied to any DNA sequence and does not depend on the phenotype of the mutant. Thus, a protein can be modified by inserting its cDNA into an appropriate expression vector and using oligodeoxynucleotide site-directed mutagenesis to alter the primary amino acid sequence. More efficient site-directed mutagenic systems, such as those involving recombinant M-13 phage, have been developed.[81,83,84] The above 'gapped duplex' method suffers from the high degree of site specificity of the endonucleolytic nicking or cleavage reaction (which may prevent the targeting of single-stranded gaps to areas of interest), and frequently yields low efficiencies of mutagenesis (less than 10%).[81]

3.2. Genetic Engineering of Food Proteins

For the genetic engineering of enzymes, X-ray crystallography and molecular modeling with computer graphics are important for the precise modifications that are essential. However, in the genetic engineering of food proteins, precise changes are probably not crucial to improvements in functional characteristics. Moreover, structural knowledge of most food proteins is sketchy at best, and one is forced to consider structural alterations at a less refined level than with enzymes. Although detailed structural knowledge of the food protein to be modified is highly desirable, it may still be possible to manipulate the functionality of food proteins with only limited information.

One of the best characterized groups of food proteins is the bovine milk proteins.[90,91] The caseins, in particular, are very functional milk proteins and serve as the basis for a major segment of the dairy industry, particularly in the manufacture of cheese. Because of large domains of disordered structure, it is not possible to crystallize the caseins. Consequently, no X-ray crystallographic data on their three-dimensional structure exist. On the other hand, primary sequences of the bovine casein are known,[90,91] and much has been inferred about their three-dimensional structure from various physico-chemical studies. As a result, alterations in the primary sequence of the caseins which could prove useful in enhancing their behavior in dairy processing and storage may be postulated. It may eventually be possible to engineer the dairy cow or its mammary gland[92] either to overproduce selected normal caseins to modify the behavior of milk or to superimpose the production of modified, novel caseins on the biosynthesis of normal caseins to change the characteristics of dairy products, possibly in producing new foods. However, in this latter case, novel caseins may induce an immunogenic response in the cow.

Bovine caseins are a heterogeneous group of phosphoproteins precipitated at pH 4·6 and 20°C. The caseins are comprised of four major polypeptide families: α_{s1}-, α_{s2}-, β- and κ-casein.[90] As phosphoproteins, they are phosphorylated, containing 8, 10–13, 5 and 1 seryl phosphate per monomer, respectively. In addition, some of the κ-casein molecules are glycosylated, while others are not, resulting in electrophoretic heterogeneity. The caseins tend to associate via hydrophobic interactions to form small spherical sub-micellar complexes.

Macromolecular protein aggregates termed 'casein micelles' are formed when seryl phosphate residues in α_{s1}-, α_{s2}- and β-casein at the surface of the sub-micelles interact with inorganic amorphous calcium phosphate complexes. Casein micelles are roughly spherical and stabilized in colloidal suspension by κ-casein concentrated on their surfaces.[93] The casein micelles maximize the amount of calcium phosphate and protein in a limited volume for effective nutrition of the young.

Characteristics of many dairy products are determined by the physical stability and behavior of colloidal casein micelles. For example, physical association of micelles induced by thermal treatments[94] can lead to unwanted gelation problems in heated milk products. On the other hand, enzymatic action on the κ-casein stabilizing the micelles can result in desired gelation in the preparation of cheese curd.[95] The structural properties of the milk proteins determine the physical interactions of the milk proteins in dairy processing. The primary amino acid sequences of the caseins largely dictate their secondary structures, their surface hydrophobicity and their sites of phosphorylation and glycosylation.[90] Consequently, the properties of the caseins and their complexes are defined to a great extent by the primary sequence of amino acids. Since it is possible to alter the primary sequence of a protein using genetic engineering techniques,[81–84] it follows that changes in the primary sequence of the caseins can alter their behavior toward enzymes and physical treatments such as heating. It is now within the realm of possibility to change the primary sequences of food proteins systematically to modify the physical properties or functionalities of those proteins. Recombinant DNA technology may eventually prove useful in designing not only novel enzymes but new food proteins as well.

It is now feasible to produce engineered enzymes and proteins in microorganisms.[81–84,86–89,96–103] Detailed basic studies of the structure–function relationships of isolated proteins are possible through systematic alterations in the primary sequences of proteins. In the case of caseins, this basic information will furnish background needed eventually to engineer the dairy cow or its mammary gland[92] to produce milk proteins with more desirable properties. It may become possible to induce the mammary gland to overproduce certain of the caseins to modify the behavior of the milk protein system. In addition, biosynthesis of novel caseins with altered primary sequences might be superimposed on normal casein biosynthesis to engineer desirable properties into the milk protein complex. Although these possibilities are long-term (perhaps 15–20 years), research on the genetic engineering of caseins from various species is progressing. The remainder of this section focuses on the genetic engineering of the caseins, gives some hypothetical examples of casein modifications and tries to anticipate how the food scientist can interact with the molecular biologist and geneticist to design more desirable food proteins.

3.3. Cloning Casein cDNAs

Casein cDNAs (DNAs complementary to the mRNA) derived from various species have been cloned into *Escherichia coli*. However, no one has yet inserted the casein cDNAs into expression vectors to achieve production of the caseins by microorganisms.

(1) *Rat and other species*. cDNAs coding for caseins from rat,[104–107] from mouse,[108,109] from rabbit,[110] and from guinea pig[111] have been cloned.

DNA sequences coding for the rat caseins have been studied in greatest detail of any species. A major thrust of this research is to define DNA segments responsible for the hormonal regulation of casein biosynthesis.

Restriction enzyme mapping and DNA sequencing[104–107] have been used to characterize full-length rat casein cDNA clones for α, β and γ rat caseins. From derived amino acid sequence data, the rat β-casein is about 38% homologous with bovine β-casein whereas the α-casein is approximately 31% homologous with bovine α_{s1}-casein. However, the rat γ-casein appears to have a primary sequence unrelated to the bovine caseins.

More recently, the actual genomic structural genes for rat γ-casein[112] and β-casein[113] were isolated and characterized in detail from rat genomic libraries. They are both very large complex genes containing long intervening sequences between the exons.

(2) *Bovine casein cDNAs*. Bovine casein cDNAs coding for α_{s1}-casein[114,115] and κ-casein[114] have been cloned. Resultant plasmids were used to transform *E. coli*. In addition to the foregoing full-length cDNAs, cDNA fragments coding for portions of the caseins have been cloned.[116–119] Although none of the bovine cDNA sequences has been specifically inserted into expression vectors to produce proteins for further

study, this opens the way for systematic structure–function studies of the caseins.

(3) *Potential for engineering caseins*. Once the appropriate cDNA in an expression vector has been incorporated into a host cell such as *E. coli* it would be possible for the primary sequences of the caseins to be produced by microorganisms. One can then obtain sufficient protein to examine the effects of structural changes elicited using the aforementioned oligodeoxynucleotide site-directed mutagenesis.

In the longer term, reliance must be placed on the possibility of stable incorporation of casein structural genes into the bovine genome followed by appropriate development and expression of the gene. For example, it may be possible to inject the structural gene along with its controlling elements into the bovine embryo and have the gene stably integrated into the bovine genome for expression in the adult under appropriate circumstances.[120–123] Once the gene is stably integrated into the genomic DNA, it is possible that it will be transmitted to the progeny via the germ cells.[120,122,123] However, much research remains to be done before this potential becomes a reality. It is also unlikely that caseins will be produced by microorganisms for food use because of unfavorable economics and relatively low yields of proteins. It would probably be more favorable economically to prepare casein fractions from acid casein rather than to rely on microbial production.

There are numerous modifications possible in altering the caseins to be more desirable from a functional point of view. Based on current knowledge of the caseinate system in milk, it is possible to propose changes in casein structure which may eventually prove useful.

For example, the ripening of cheese involves the proteolytic breakdown of the protein matrix to yield a more desirable texture. Conversion of α_{s1}-casein to α_{s1}-I-casein and an amino terminal peptide by cleavage of a Phe23–Phe24 or Phe24–Val25 bond by residual clotting enzyme (chymosin or other acid protease) is thought to be one of the primary changes in the caseins during maturation of cheddar cheese.[124] Additional bonds with enhanced chymosin sensitivity to accelerate the rate of textural development in cheese may be inserted into caseins. Conversion of Ile71 to Phe71 (by changing one nucleotide base) in α_{s1}-casein would generate an additional Phe71–Val72 bond in a region of α_{s1}-casein disordered by a series of seryl phosphates, other charged amino acid residues and an adjacent prolyl residue.[90] The Val72 can also be converted to a Phe (with one base change) or to Tyr or Trp with two base changes. This bond should be readily available for cleavage by residual acid proteases with a propensity for cleaving bonds adjacent to aromatic amino acids. Cleavage of a bond more toward the middle of the α_{s1}-casein (as at position 71) should maximize the rheological effects of proteolysis in promoting faster ripening of the cheese. More rapid textural development in cheese would be highly desirable from an economic storage point of view.

The thermal stability of the caseinate system in milk is important in thermal processing of dairy products. Factors that minimize the aggregation of the casein–colloidal calcium phosphate complexes (micelles) prevent unwanted thermally induced gelation of milk. It is known that κ-casein plays an important role in inhibiting thermally induced gelation of the caseins.[94] Thus genetic manipulation of the bovine genome to insert κ-casein genes to overproduce κ-casein might enhance the thermal stability of casein aggregates in milk.

We have seen how oligodeoxynucleotide site-directed mutagenesis can be used to engineer the primary sequences of proteins. In part, this has been a somewhat speculative review with an attempt to introduce the reader to the potential that genetic engineering holds forth upon successful application. Much additional research is necessary before successful genetic manipulation of plants and animals to tailor-make more functional proteins becomes a reality. As more is learned about the molecular basis for protein functionality in foods, it becomes increasingly possible to design proteins for particular uses. Consequently, the future holds the possibility of the food scientist working with the molecular biologist and geneticist to design food proteins with desired functionalities.

4. ENZYMATIC MODIFICATIONS

Protein functionality can be modified to suit food processing and final product characteristics by the action of hydrolytic enzymes. Protein substrates undergo partial hydrolysis of the peptide backbone to alter protein functionality. There are also two other very important modes of action of hydrolytic enzymes. One is the production of inter- or intramolecular cross-links. Cross-linking can be used to induce coagulation, gelation, texture alteration, and to increase viscosity in protein solutions. The other is to attach specific chemical groups to proteins, such as essential amino acids to increase their nutritive quality or to incorporate hydrophobic or hydrophilic amino acids to enhance the amphiphilicity of the protein.

The advantages of using enzymes to perform functional modifications of proteins have been recognized for some time.[125] Among the most important advantages are the specificity, mild conditions and low enzyme concentration required to perform the reaction. As one might expect, there is a minimal occurrence of toxic byproducts.

The following review on functional modifications of food proteins by the action of enzymes has been divided into two major areas. The first part deals with the degradative action or hydrolysis performed by proteases on milk proteins, meat (animal) proteins and plant (vegetable) proteins. In the second portion the biosynthetic aspects of enzymes are considered. This includes plastein and plastein-like reactions, transglutaminase-catalyzed reactions and cross-linking reactions of polyphenol oxidases and peroxidases.

4.1. Proteases

The proteolytic action of proteases has been by far the most widely used feature in the enzymatic modification of protein functionality. The literature in this area is voluminous. Because of limitations in space, this review will be limited to recent applications of proteases in the modification of specific functional properties of proteins.

4.1.1. Milk Proteins

The action of proteolytic enzymes on milk proteins has been reviewed by Visser[126] and on whey proteins by Kester and Richardson.[1] Basic studies on enzyme kinetics of the chymosin-catalyzed proteolysis of κ-casein, of major importance in the cheese industry, have been recently carried out.[127]

Some of the recent advances on enzymatic alteration of milk protein functionality are summarized in Table 6. The information in Table 6 attempts to relate as closely as possible the action of the enzyme(s) on a protein to function in a food system. In the first example, enhanced emulsifying properties of the *N*-terminal peptide obtained from α_{s1}-casein by hydrolysis with several proteinases (pepsin, chymosin, plasmin) were observed. Proteolysis thus yields a readily isolated fragment identified as the *N*-terminal group of amino acids 1 to 23 of α_{s1}-casein. This fragment displays very similar emulsification activity (EAI) to whole α_{s1}-casein. This fragment has the advantage of having a higher EAI at acid pH (3–5) whereas α_{s1}-casein does not.

The *N*-terminal peptide is adsorbed remarkably well onto an oil surface. This suggests to the authors that the mechanism is somewhat different

Table 6
Proteolytic Enzymes and their Action on Functionality of Milk Proteins

Substrate	*Enzyme*	*Action*	*Functionality*
α_{s1}-Casein[128]	Pepsin, chymosin or plasmin	Limited hydrolysis of the α_{s1}-casein to isolate *N*-terminal peptide constituted of residues 1–23	The *N*-terminal peptide displayed remarkable fat absorption and showed emulsifying activity even at acid pH
β-Casein[129,130]	Metalloprotease from *Streptococcus dyacetylactis*	Specific endopeptidase that cleaves Pro7–Phe8, Pro186–Ile187, Ala189–Phe190	By the liberation of smaller peptide fragments than those of chymosin, the texture and flavor of cheese during ripening is developed by these microbial exopeptidases
	Metalloproteases from *Penicillium caseicolum* and *Penicillium roqueforti*	Specific endopeptidase that cleaves Lys28–Lys29, Pro70–Glu91, Glu100–Ala101	
β-Casein (bovine and human)[131]	Several proteases	Specific hydrolysis yielding peptides with specific neuroactive sequence Tyr-Pro-Phe-Pro-Gly-Pro-Ile	Opioid effect on rat brain opiate receptors; several fragments show high affinity
κ-Casein and β-lactoglobulin[132]	Chymosin	Acceleration of the thermally induced gelation of the complex formed between κ-casein and β-lactoglobulin only with the addition of the enzyme to the thermalized mixture; produce high breaking stress gel	At protein concentration of 3%, gelation of a mixture of κ-casein and β-lactoglobulin was obtained
κ-Casein and α-lactalbumin[133]	Trypsin and chymotrypsin pronase	Hydrolysis of thermalized complex of κ-casein with α-lactalbumin	The thermally induced gel of κ-casein and α-lactalbumin was readily hydrolyzed by the enzymes

from that with α_{s1}-casein. The surface-active nature of this peptide might account for the increase in the EAI of α_{s1}-casein during digestion with pepsin.[128]

During cheese ripening, the metalloproteases have relevance to aging and to the development of flavor and texture. Two studies have shown specific action of metalloproteases on β-casein and have identified the sites of proteolysis.[129,130] This suggests that during cheese ripening metalloproteases, shown to be endopeptidases, complete the proteolysis initiated by the residual chymosin by attacking the peptides released by the latter protease. The short fragments thus produced would, in turn, be hydrolyzed by other microbial exopeptidases.

Gel formation has always been important in the production of cheese, yogurt and other formulated foods. Chymosin action enhanced the thermo-gelling capabilities of κ-casein and β-lactoglobulin, lowering the protein concentration required for gelation (from 8% to 3% protein).[132] From these studies it is concluded that self-association of κ-casein is due to a hydrophobic interaction. This hydrophobic association, in addition to a thiol–disulfide interchange, participates in the formation of a κ-casein–β-lactoglobulin complex which gelled in the presence of chymosin. Gelation of κ-casein with α-lactalbumin by heat treatment facilitated the enzymatic hydrolysis of the protein complex.[133]

Pepsin hydrolysis of modified whey protein concentrate resulted in products with improved foaming capacity that could be used in mixtures or by themselves to replace egg white in food processing.[134]

4.1.2. Meat Proteins

Protein functionality in meat proteins depends primarily on the amount of soluble protein present in the system to be processed. Therefore, the common test performed on modified meat proteins involves changes in solubility. In addition, other functionalities that are important in meat systems include water absorption, viscosity and gelation, elasticity, emulsification, fat absorption, and foaming.

In Table 7 are listed recent studies on the effects of proteases on meat proteins. The action of endogenous enzymes (such as calcium activated proteases and catepsins) on meat tenderization has not been considered. In general, the use of enzymes in meat products is directed to the hydrolysis and solubilization of the proteins. A main component of meat protein is myosin; its functionality has been recently reviewed.[139] Artificial tenderization is a main use of enzymes in meat systems. However, there are often areas where the protein functionality can be modified extensively and in a

Table 7
Hydrolysis of Meat Proteins and its Effect on Functionality

Substrate	*Enzyme*	*Functional property*
Bovine blood globin[135]	Pepsin	Pepsin hydrolyzates of bovine globin did not have enhanced emulsifying activity, but the addition of carboxymethylcellulose to the pepsin-digested globin increased the emulsifying activity and emulsion stability
Fowl myofibrillar proteins[136]	Commercial enzyme mixture Milenzyme AFP 2000	Partial hydrolysis of mechanically deboned fowl (MDF) improved the protein solubility, emulsifying capacity, and gelation when compared to myofibrillar protein controls
Bovine heart myofibrils[137]	Ficin	Hydrolyzates of different treatment times on a model system using heart myofibrils and ficin yielded increased solubility and emulsification; the usefulness of this procedure was tested in normal and low-salt frankfurters (inhibitor of this enzyme is toxic)
Bovine tendon and porcine skin collagen[138]	Pepsin and pancreatin	Hydrolysis at 37°F for up to 4 h at pH 1·5 of unscalded pigskin and on heated tendon were highly solubilized (75–94%) when incubated with pepsin and hydrochloric acid; this reaction is very sensitive to change in pH

more varied way. Some examples are the proteins from blood, fish and collagen.

4.1.3. Plant Proteins

Plant protein modification is to a great extent directed towards its nutritional improvement and as a substitute for more expensive proteins of animal origin. Protein modifications can result in profound changes in a variety of functional properties of the protein. A review[140] has considered in detail the changes in plant proteins by immobilized proteases.

Rather than giving an exhaustive review, some examples of the effects of proteases on plant proteins of enzymatic modifications of plant proteins are given in Table 8.

Soybean proteins have been extensively reviewed.[148] These proteins are of great importance given their low cost and high nutritive value (once inhibitor and limiting amino acids problems have been solved). Also advantageous is the fact that approximately 90% of the soybean proteins, mainly globulins, exist as dehydrated storage proteins.[148] Therefore it is not surprising that soybean proteins are a preferred group of proteins to be altered, especially by the action of proteases, in an attempt to incorporate them into high-protein food systems.

The examples shown follow the same principles as previous hydrolytic processes. Cross-linking of soy proteins is a method that will be treated in the plastein reaction section. An excellent example of enzyme action and the mechanism by which it changes the function of 11s soy globulin has been given.[143] A mechanism for the thermally and enzymatically induced gelation of 11s globulin was deduced from the results. These workers proposed that the acidic subunits surround the basic subunits in 11s globulin, and the heat treatment dissociates the 11s molecules which are maintained in their globular conformation. Then they associate and form a high molecular weight aggregate (8×10^6). When this complex is treated with bromelain the characteristics change to form a gel. The association is believed to be due mainly to disulfide bond formation because the aggregate dissociates into smaller fragments (15 000 MW) in the presence of sodium dodecyl sulfate or mercaptoethanol.

4.2. Plastein Reaction

The term plastein has been applied to high molecular weight polypeptides produced after partial hydrolysis of proteins (typically soy bean protein) by papain or chymotrypsin. This reaction has been extensively reviewed.[149,150] The mechanism of the plastein reaction is shown in Fig.

Table 8
Hydrolysis of Plant Protein and its Effect on Functionality

Substrate	*Enzyme*	*Functional property*
Soy protein[141]	Trypsin	Tryptic solubilization increased the emulsion viscosity and stability, as well as foam volume and stability, but when used in frankfurters decreased the firmness and cook stability
Soybean glycinin[142]	Trypsin	The digest of native glycinin was as good an emulsifier as heat-denatured glycinin; the novel way that emulsification was measured was flocculation strength, which is the rigidity of the flocculated structure as defined by the minimum detergent concentration for the dissociation of flocculation
Soybean protein and 11s globulin[143]	Bromelain	Thermally induced gelation on reaction with bromelain for 15 min; some polymerization via disulfide bonds was observed by sodium dodecyl sulfate/polyacrylamide gel electrophoresis
Protein concentrate from oats[144]	Trypsin	Modified proteins by trypsin hydrolysis had increased emulsifying capacity, activity and stability, hydration capacity and foaming; the fat binding capacity was decreased
Rapeseed proteins[145,146]	Pepsin, trypsin, papain	Hydrolysis of rapeseed flour followed by ultrafiltration yielded nitrogenous material having 'net protein ratio' and 'apparent digestibility coefficient' similar to those of casein
Vicia faba proteins[147]	Thermitase protease from *Thermoactinomyces vulgaris*	Depending on the degree of hydrolysis, the solubility, heat coagulability, foaming power and water absorption are increased; emulsifying capacity and foam stability are not affected

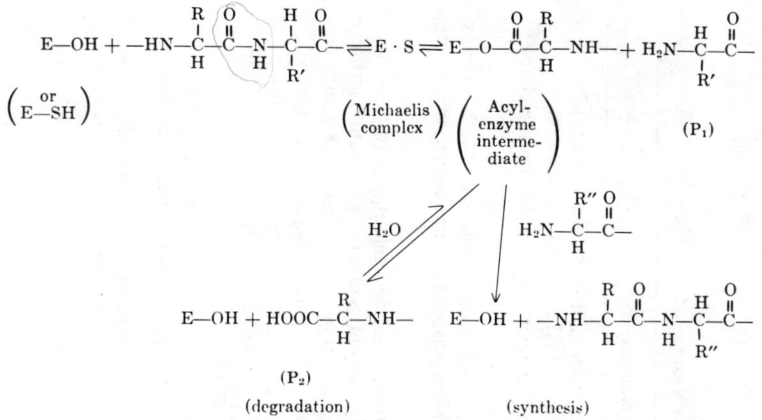

Fig. 5. The mechanism of the plastein reaction.[3] (Reprinted from Whitaker and Puigserver (1982) with permission.)

5. In the normal hydrolysis of proteins, low concentration and optimum pH for enzyme action yield low molecular weight peptides. This is shown in Fig. 5 in the path going from P1 to P2. A change in conditions (usually an increase in peptide concentration and a change in pH from 4 to 7) yields plasteins which possess different functional properties from the original proteins. The nature of these changes is unclear and may be due either to transpeptidation (formation of one new peptide bond for every broken one), as interpreted by the absence of change in the average molecular weight of the peptides,[154,159,161,165] or to hydrophobic interaction and aggregation.[160,161]

Based on rheological studies, the plastein reaction of hydrolyzed soy protein in the presence of α-chymotrypsin and/or papain does not cause any change in the average molecular weight. Structural changes reflected in the apparent viscosity of the mixture were best explained by the probable existence of two steps. The first is connected to transpeptidation which is catalyzed by the enzymes for several minutes. The second would be connected to the interaction of the polypeptide chains (with increased hydrophobicity), in which covalent bonds are not formed. These latter interactions can last for several days.

Regardless of the nature of the changes in the protein due to the plastein reaction, whether it is modified (one-step) or adjusted to the nature of protein and enzyme used, it appears to have considerable potential in food processing. In Table 9 is a summary of recent plastein reactions involving

Table 9
Effects on Functionality of Plastein and Plastein-like Reactions

Substrate	*Enzyme*	*Functional protein*
Gelatin + C_4–C_{12} alkyl esters of L-leucine[151–158]	Papain	Application of the plastein-like reaction called the 'one-step process' results in the incorporation of the n-alkyl esters of L-leucine; the optimization of different functional properties depended on the chain length of the alkyl group C_4–C_6 Increased whippability, surfactants in processed foods such as ice cream; foaming stability, viscosity, and elasticity were better than egg white C_{10}–C_{12} The emulsifying capacity of modified gelatin behaved as that of non-ionic, low molecular weight synthetic surfactants; although the mechanism for emulsification was different, the results were comparable to Tween-80. Temperature dependence studies showed the existence of a liquid crystalline structure of a lamellar type at 30–80°C in the phase behavior of emulsified systems with C_{12} derivatives Regulation of ice nucleation in oil/water emulsions of C_{12} derivatives was observed and called 'anti-freeze protein'
Soy protein + L-methionine ethyl ester[159,160]	Papain	Nutritional enhancement was the main purpose of this modification; the forces involved in the methionine attachment are not clear, and could be covalent[159] or hydrophobic[160]

Ovalbumin hydrolyzate[161,162,170]	α-Chymotrypsin	Aggregation of particles in plastein action as studied by scanning electron microscopy, and molecular weight variation in the peptides along the reaction; the increase in viscosity is due to aggregation of peptides helped by some transpeptidation
Horse bean protein concentrate + DL-methionine ethyl esters[163,164]	Thermitase	The incorporation of methionine into the proteins was directly proportional to the degree of hydrolysis, and was stereospecific (only mechanistic studies were made)
Casein + L-methionine methyl ester[165,166]	α-Chymotrypsin	The incorporation of methionine and digestibility of the modified casein were studied; the methionine-enriched plastein is digestible by physiological proteases and peptidases just as well as casein (nutrition studies to be made)
Fish frame protein hydrolyzates[167,168]	Pepsin or α-chymotrypsin	Degree of hydrolysis, choice of enzyme, time of reaction and pH were studied during the plastein reaction using fish protein hydrolyzate (physico-chemical and functional properties to be published); the gel forming ability and insolubility were found[168] to be due to hydrophobic interactions
Hydrolyzates[169] of (1) zein, (2) egg albumin	Pepsin	The plastein reaction was studied on each protein, to determine optimal conditions and enzyme; holding water capacity increased 3-fold for albumin and 10-fold for zein; oil binding increased 50% in albumin and 70% in zein; gel strength improved for both These results showed that electron microscopy can be useful in observing ultrastructures in enzymatically modified proteins

various proteins, enzymes and conditions of reaction. Watanabe *et al.*[151] have worked extensively on the addition of n-alkyl esters of L-leucine to gelatin to create a proteinaceous surfactant. The range or variation in the C chain of n-alkyl was 4 to 12; different functionality is observed in the C chain length of the n-alkyl attached to the L-leucine. Its usefulness in food systems has been recently studied as an emulsifying agent[153] or foaming agent.[154,155] Its possible use in food items such as mayonnaise, bread, snow-jelly and ice-cream was tested.[152] Furthermore, physico-chemical properties such as interfacial tension, surface properties compared to non-ionic detergents, dynamic viscoelasticity, differential thermal analysis, and birefringence (in aqueous systems) with a polarizing microscope were examined.[155–157] These modified proteins have also been shown to regulate ice nucleation in emulsion systems.[158] The chemical properties of enzymatically modified soy proteins with different levels of methionine have recently been reported.[159,160] There are divergent views as to the forces involved in the incorporation of L-methionine ethyl ester into the soy protein. Covalent and/or hydrophobic forces may be responsible for incorporation of L-methionine.

4.3. Transglutaminase

Transglutaminase catalyzes a calcium-dependent acyl transfer reaction, in which one substrate is the γ-carboxamide group(s) of glutaminyl residue(s) of proteins (acyl donors). The second substrate (acyl acceptors) is a nucleophile, which could be free amino groups as in lysyl side-chain.[171]

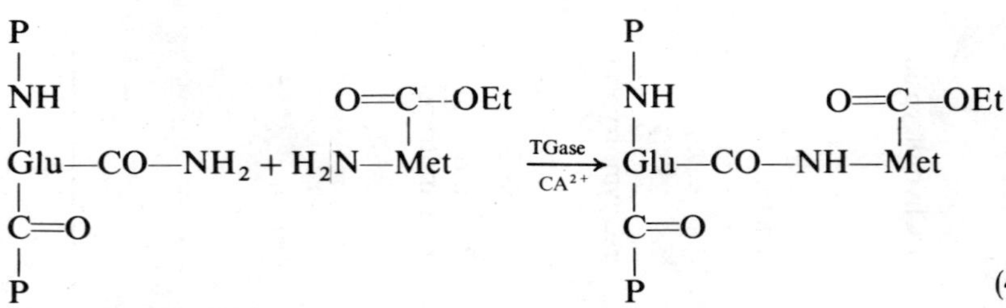

(4)

The ethyl alcohol group on the methionyl ester can be readily removed by holding at pH 8–9 for a few hours.[171] The two main sources of transglutaminase that have been used in the study of modification of food proteins are from guinea pig liver and bovine plasma.

Transglutaminase has been used to incorporate methionine ethyl esters into food proteins [reaction (4)]. Ikura *et al.*[172] proved the feasibility of increasing the methionine content of casein, soybean and gluten by its incorporation using guinea pig liver transglutaminase. According to Ikura *et al.*,[172] methionine and lysine incorporated into food proteins by the action of transglutaminase appear to be biologically available. Another way to use transglutaminase in the modification of protein functionality is by promoting polymerization involving covalent cross-links.

Ikura *et al.*[173] and Motoki and Noriki[174] studied the formation of homologous and heterologous polymers of acetylated α_{s1}-casein with itself and other food proteins. The action of transglutaminase was also studied for meat systems by Kurth and Rogers.[175] They determined that cross-linking of myosin with casein, gluten or soy protein modified the texture of meat systems. Measurement of the functional properties of covalently polymerized proteins by transglutaminase showed a decrease in solubility and emulsifying capacity. However, the most significant result was that the hydration abilities of the polymerized proteins were greater than those of the native proteins. Therefore, such modified proteins may be useful for the production of intermediate moisture protein foods.

Transglutaminase also catalyzes the formation of gels at sufficiently high protein concentrations and at pH 7·5.[176,176a] As mentioned earlier, the sources of the transglutaminase for the experiments mentioned have been guinea pig liver and bovine plasma. Evidently the potential commercial usefulness of these reactions would be enhanced if transglutaminase was produced from a more economical source, possibly using recombinant DNA technology.

4.4. Polyphenol Oxidase and Peroxidase

Matheis and Whitaker[177] have recently discussed the reactions of polyphenol oxidase (PPO) and peroxidase (POD) in some detail. PPO oxidizes mono- and di-hydroxy-phenols to benzoquinones which then can react non-enzymatically with amino, sulfhydryl, thioether and *N*-terminal propyl groups in proteins. The initial products may be colorless protein-bound *o*-diphenols, which after oxidation cause cross-linking by the addition of two side-chain groups to a single benzoquinone molecule. Figure 6 illustrates some of the possible reactions of PPO. Peroxidase (POD) can act upon the tyrosyl residue of proteins in the absence of low molecular weight hydrogen donors, and in presence of H_2O_2. In such cases the presence of di- and ter-tyrosine polymers indicates that the action of the enzyme is directed toward the tyrosyl residues of the cross-linked protein. However, definitive evidence is still lacking and some controversy exists as to the basic mechanism with POD.[177]

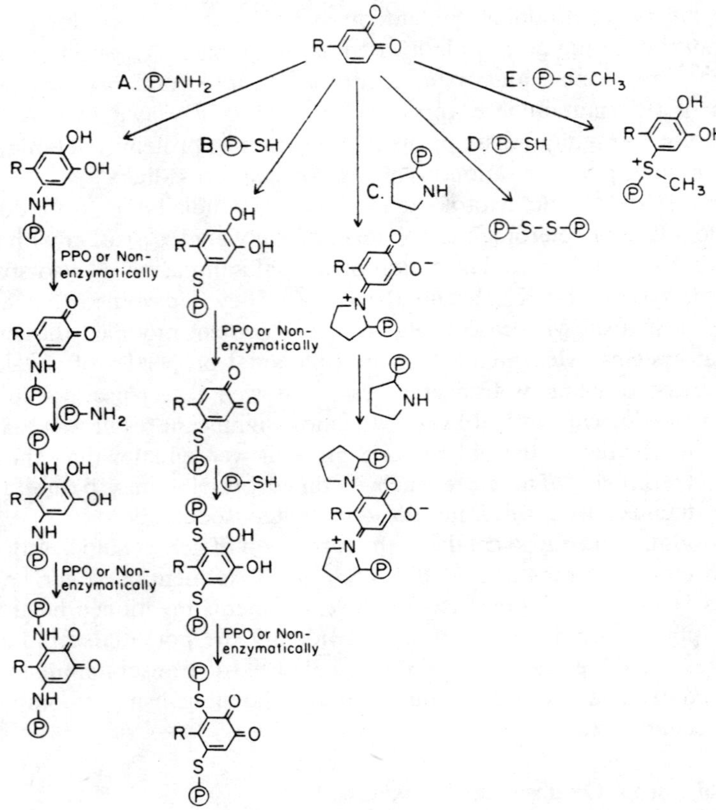

Fig. 6. Possible reactions of proteins with enzyme-generated *o*-quinones.[177]

Cross-linking of proteins with each other or with other food constituents (pentosanes or essential amino acids) can be desirable. One example is the dough-improving effect[178] or the gelation of some proteins.[179] However, some problems may also result from the action of PPO and POD, such as the impairment of the nutritional properties or uncontrolled reactions during protein isolation.[177,180]

A recent example of cross-linking by the action of POD is the cross-linking of casein with soluble soy protein.[177] This experiment yielded a polymerized casein–casein or casein–soy protein. Further studies are needed to determine the exact mechanism of the polymerization and its effect on the functionality of the protein or proteins thus polymerized.

ACKNOWLEDGEMENT

This review was made possible by support from the College of Agriculture and Environmental Sciences, University of California, Davis.

REFERENCES

1. Kester, J. J. and Richardson, T., *J. Dairy Sci.*, **67**, 2757 (1984).
2. Feeney, R. E., Yamasaki, R. B. and Geohagan, K. F., in *Modification of Proteins: Food, Nutritional and Pharmacological Aspects*, Feeney, R. E. and Whitaker, J. R. (Eds), Am. Chem. Soc., Washington, DC, 1982.
3. Whitaker, J. R. and Puigserver, A. J., in *Modification of Proteins: Food, Nutritional and Pharmacological Aspects*, Feeney, R. E. and Whitaker, J. R. (Eds), Am. Chem. Soc., Washington, DC, 1982.
4. Shukla, T. P., in *Food Protein Deterioration: Mechanisms and Functionality*, Cherry, J. P. (Ed.), Am. Chem. Soc., Washington, DC, 1982.
5. Feeney, R. E. and Whitaker, J. R., *New Protein Foods*, Vol. 5, Academic Press, 1985.
6. Fox, P. F., Morrissey, P. A. and Mulvihill, D. M., in *Developments in Food Proteins*, Vol. 1, Hudson, B. J. F. (Ed.), Applied Science Publishers, London, 1982, pp. 1–60.
7. Feeney, R. E., in *Food Proteins: Improvement through Chemical and Enzymatic Modifications*, Feeney, R. E. and Whitaker, J. R. (Eds), Am. Chem. Soc., Washington, DC, 1977, p. 3.
8. Schmandke, H., Maune, R., Schuhmann, S. and Schultz, M., *Nährung*, **25**, 99 (1981).
9. Schmandke, H., Danilenko, A. N., Bekbov, T. M., Maune, R., Grinberg, V. J. and Tolstoguzov, V. B., *Nährung*, **25**, K27 (1981).
10. Schmandke, H., Bottcher, H., Plaschnick, D. and Kuhrt, A., *Nährung*, **26**, 385 (1982).
11. Schwenke, K. D., Rauschal, E. J. and Robowsky, K. D., *Nährung*, **27**, 335 (1983).
12. Narayana, K. and Rao, M. S. N., *J. Food Sci.*, **49**, 547 (1984).
13. Thompson, L. U. and Cho, Y. S., *J. Food Sci.*, **49**, 771 (1984).
14. Cho, Y. S. and Thompson, L. U., *J. Food Sci.*, **49**, 765 (1984).
15. Thompson, L. U. and Cho, Y. S., *J. Food Sci.*, **49**, 1584 (1984).
16. Ma, C. Y., *J. Food Sci.*, **49**, 1128 (1984).
17. Ball, H. R. Jr and Wenn, S. E., *Poultry Sci.*, **61**, 1041 (1982).
18. Choi, Y. R., Lusas, E. W. and Rhee, K. C., *J. Am. Oil Chem. Soc.*, **58**, 1044 (1981).
19. Choi, Y. R., Lusas, E. W. and Rhee, K. C., *J. Food Sci.*, **47**, 1713 (1982).
20. Choi, Y. R., Lusas, E. W. and Rhee, K. C., *J. Food Sci.*, **48**, 1275 (1983).
21. Rahma, E. H. and Rao, M. S. N., *J. Agric. Food Chem.*, **31**, 352 (1983).
22. Kafirullah, M. and Wills, R. B. H., *J. Food Technol.*, **17**, 235 (1982).
23. Schwenke, K. D. and Rauschall, E. J., *Nährung*, **10**, 1015 (1983).
24. Johnson, E. A. and Brekke, C. J., *J. Food Sci.*, **48**, 722 (1983).

25. Cho, Y. S. and Kim, J. K., *J. Korean Soc. Food Nutr.*, **12**, 251 (1983). Cited in *Food Sci. Technol. Abstr.*, **17**, 6G21 (1985).
26. Sung, S. K., *Korean J. Food Sci. Technol.*, **16**, 353 (1984). Cited in *Food Sci. Technol. Abstr.*, **17**, 5S201 (1985).
27. Song, I. S., Yoo, I. J., Kang, T. S. and Song, K. W., *Korean J. An. Sci.*, **26**, 288 (1984). Cited in *Food Sci. Technol. Abstr.*, **17**, 5S138 (1985).
28. Lee, Y. H. and Marshall, R. T., *J. Dairy Sci.*, **67**, 263 (1984).
29. Korolczuk, J., *NZ J. Dairy Sci. Technol.*, **19**, 107 (1984).
30. Horne, D. S. and Parker, T. G., *J. Dairy Res.*, **49**, 449 (1982).
31. Schmandke, H., Hartmann, B. and Schultz, M., *Nährung*, **25**, 479 (1981).
32. Thompson, L. U., Reniers, D. J. and Baker, L. M., *J. Dairy Sci.*, **65**, 1715 (1982).
33. Thompson, L. U., Reniers, D. J., Baker, L. M. and Siu, M., *J. Dairy Sci.*, **66**, 1630 (1983).
34. Thompson, L. U. and Baker, L. M., *Cereal Chem.*, **60**, 71 (1983).
35. Haque, Z. and Kito, M., *J. Agric. Food Chem.*, **31**, 1225 (1983).
36. Haque, Z. and Kito, M., *J. Agric. Food Chem.*, **31**, 1231 (1983).
37. Gaertner, H., Lupi, N. and Puigserver, A., *Lait*, **62**, 578 (1982).
38. Gaertner, H. F. and Puigserver, A. J., *J. Agric. Food Chem.*, **32**, 1371 (1984).
39. Matheis, G., Sen, L. C., Clifford, A. J. and Whitaker, J. R., *J. Agric. Food Chem.*, **33**, 39 (1985).
40. Baber, K. J. and Warthesen, J., *J. Agric. Food Chem.*, **30**, 930 (1982).
41. Eisle, T. A., Brekke, C. J. and McCurdy, S. M., *J. Food Sci.*, **47**, 43 (1981).
42. Siu, M. and Thompson, L. U., *J. Agric. Food Chem.*, **30**, 1179 (1982).
43. Matheis, G. and Whitaker, J. R., *J. Agric. Food Chem.*, **32**, 699 (1984).
44. Sung, H.-Y., Chen, H.-J., Lui, T. Y. and Su, J. C., *J. Food Chem.*, **48**, 716 (1983).
45. Hirotsuka, M., Taniguchi, H., Narita, H. and Kito, M., *Agric. Biol. Chem.*, **48**, 93 (1984).
46. Matheis, G., Penner, M. H., Feeney, R. E. and Whitaker, J. R., *J. Agric. Food Chem.*, **31**, 379 (1983).
47. Salak, J., Vodraza, Z. and Cejka, J., *Coll. Czech. Chem. Commun.*, **30**, 1036 (1965).
48. Heidelberger, M., Davis, B. and Treffers, H. P., *J. Am. Chem. Soc.*, **63**, 498 (1941).
49. Mayer, M. and Heidelberger, M., *J. Am. Chem. Soc.*, **68**, 18 (1946).
50. Woo, S. L. and Richardson, T., *J. Dairy Sci.*, **66**, 984 (1983).
51. Yoshikawa, M., Sasaki, R. and Chiba, H., *Agric. Biol. Chem.*, **45**, 909 (1981).
52. Boursnell, J. C., Dewey, H. M. and Wermall, A., *Biochem. J.*, **43**, 84 (1948).
53. Ferrel, R. E., Olcott, H. S. and Fraenkel-Conrat, H., *J. Am. Chem. Soc.*, **70**, 2101 (1948).
54. Willmitzer, L. and Wagner, K. G., *Eur. J. Biochem.*, **59**, 43 (1975).
55. Matsudomi, N., Kato, A. and Kobayashi, K., *Agric. Biol. Chem.*, **46**, 1583 (1982).
56. Matsudomi, N., Sasaki, T., Kato, A. and Kobayashi, K., *Agric. Biol. Chem.*, **49**, 1251 (1985).
57. Matsudomi, N., Sasaki, T., Tanaka, A., Kobayashi, K. and Kato, A., *J. Agric. Food Chem.*, **33**, 738 (1985).

58. Kinsella, J. F., *Food Chem.*, **7**, 273 (1981).
59. Basch, J. J., Farrell, H. M. Jr and Greenberg, R., *Biochim. Biophys. Acta*, **488**, 589 (1976).
60. Waniska, R. D. and Kinsella, J. E., *Int. J. Peptide Protein Res.*, **23**, 573 (1984).
61. Kitabatake, N., Cuq, J. L. and Cheftel, J. C., *J. Agric. Food Chem.*, **33**, 125 (1985).
62. Waniska, R. D. and Kinsella, J. E., *Int. J. Peptide Protein Res.*, **23**, 467 (1984).
63. Waniska, R. D. and Kinsella, J. E., *J. Agric. Food Chem.*, **32**, 1042 (1984).
64. DiGregorio, F. and Sisto, R., *J. Dairy Res.*, **48**, 267 (1981).
65. Mattarella, N. L. and Richardson, T., *J. Dairy Sci.*, **65**, 2253 (1982).
66. Mattarella, N. L. and Richardson, T., *J. Agric. Food Chem.*, **31**, 972 (1983).
67. Mattarella, N. L., Creamer, L. K. and Richardson, T., *J. Agric. Food Chem.*, **31**, 968 (1983).
68. Halpin, M. I. and Richardson, T., *J. Dairy Sci.*, **68**, 3189 (1985).

68a. Malamud, D. and Drysdale, J. W., *Anal. Biochem.*, **86**, 620 (1978).

69. Janus, J. W., Kenchington, A. W. and Ward, A. G., *Research*, **4**, 247 (1951).
70. Ho, C. and Waugh, D. F., *J. Amer. Chem. Soc.*, **87**, 110 (1965).
71. Kinsella, J. E., *Crit. Rev. Food Sci. Nutr.*, **8**, 219 (1976).
72. Kinsella, J. E., in *Food Protein Deterioration: Mechanism and Functionality*, Cherry, J. P. (Ed.), Am. Chem. Soc., Washington, DC, 1982, pp 301–26.
73. van Holde, K. E., *Physical Biochemistry*, Prentice-Hall, Englewood Cliffs, New Jersey, 1981, p. 205.
74. Timasheff, S. N., Susi, H., Townend, R., Gorbunoff, M. J. and Kumosinski, T. F., in *Conformations of Biopolymers*, Vol. I, Ramachandran, G. N. (Ed.), Academic Press, London, 1967, pp 185–90.
75. Woo, S. L., Creamer, L. K. and Richardson, T., *J. Agric. Food Chem.*, **30**, 65 (1982).
76. Janin, J. and Chothia, C., in *Protein Structure, Function and Industrial Applications*, Hofman, E., Pfeil, W. and Curich, H. (Eds), Pergamon Press, New York, 1978, p. 227.
77. Brand, L. and Gohlke, J. R., *Ann. Rev. Biochem.*, **41**, 843 (1972).
78. Kato, A. and Nakai, S., *Biochim. Biophys. Acta*, **624**, 13 (1980).
79. Mohammadzadeh, K., Smith, L. M. and Feeney, R. E., *Biochim. Biophys. Acta*, **194**, 256 (1969).
80. Zuidema, H. H. and Waters, G. W., *Ind. Eng. Chem. Anal. Edn*, **13**, 312 (1941).
81. Craik, C. S., *BioTechniques*, **3**, 12 (1985).
82. Dalbadie-McFarland, G., Cohen, L. W., Riggs, A. D., Morin, C., Itakura, K. and Richards, J. H., *Proc. Nat. Acad. Sci.*, **79**, 6409 (1982).
83. Zoller, M. J. and Smith, M., *Methods Enzymol.*, **100**, 468 (1983).
84. Botstein, D. and Shortle, D., *Science*, **229**, 1193 (1985).
85. Stryer, L., *Biochemistry*, 2nd Edn, Freeman, New York, 1981, pp 92, 93.
86. Maugh, T. H. II, *Science*, **223**, 269 (1984).
87. Rastetter, W. H., *Trends Biotechnol.*, **1**, 80 (1983).
88. Ulmer, K. M., *Science*, **219**, 666 (1983).
89. Winter, G. and Fersht, A. R., *Trends Biotechnol.*, **2**, 115 (1984).

90. Swaisgood, H. E., in *Developments in Dairy Chemistry*, Vol. 1, Fox, P. F. (Ed.), Applied Science Publishers, New York, 1982, p. 1.
91. Eigel, W. N., Butler, J. E., Ernstrom, C. A., Farrell, H. M. Jr, Harwalker, V. R., Jenness, R. and Whitney, R. McL., *J. Dairy Sci.*, **67**, 1599 (1984).
92. Patton, S., Welsch, U. and Singh, S., *J. Dairy Sci.*, **67**, 1523 (1984).
93. Schmidt, D. G., in *Developments in Dairy Chemistry*, Vol. 1, Fox, P. F. (Ed.), Applied Science Publishers, New York, 1982, p. 61.
94. Fox, P. F. (Ed.), in *Developments in Dairy Chemistry*, Vol. 1, Applied Science Publishers, New York, 1982, p. 189.
95. Dalgleish, D. G., in *Developments in Dairy Chemistry*, Vol. 1, Fox, P. F. (Ed.), Applied Science Publishers, New York, 1982, p. 157.
96. Fersht, A. R., Shi, J.-P., Knill-Jones, J., Lowe, D. M., Wilkinson, A. J., Blow, D. M., Brick, P., Carter, P., Waye, M. M. Y. and Winter, G., *Nature*, **314**, 235 (1985).
97. Perry, L. J. and Wetzel, R., *Science*, **226**, 555 (1984).
98. Fersht, A. R., *Enzyme Structure and Mechanism*, 2nd Edn, Freeman, New York, 1985, pp 105, 324, 386.
99. Winter, G., Fersht, A. R., Wilkinson, A. J., Zoller, M. and Smith, M., *Nature*, **299**, 756 (1982).
100. Wilkinson, A. J., Fersht, A. R., Blow, D. M. and Winter, G., *Biochemistry*, **22**, 3581 (1983).
101. Wilkinson, A. J., Fersht, A. R., Blow, D. M., Carter, P. and Winter, G., *Nature*, **307**, 187 (1984).
102. Fersht, A. R., Shi, J.-P., Wilkinson, A. J., Blow, D. M., Carter, P., Waye, M. M. and Winter, G. P., *Angew. Chem. Int. Ed. Engl.*, **23**, 467 (1984).
103. Craik, C. S., Largman, C., Fletcher, T., Roczniak, S., Bart, P. J., Fletterick, R. and Rutter, W. J., *Science*, **228**, 291 (1985).
104. Blackburn, D. E., Hobbs, A. A. and Rosen, J. M., *Nucl. Acids. Res.*, **10**, 2295 (1982).
105. Hobbs, A. A. and Rosen, J. M., *Nucl. Acids Res.*, **10**, 8079 (1982).
106. Richards, D. A., Rodgers, J. R., Supowit, S. C. and Rosen, J. M., *J. Biol. Chem.*, **256**, 526 (1981).
107. Richards, D. A., Blackburn, D. E. and Rosen, J. M., *J. Biol. Chem.* **256**, 533 (1981).
108. Mehta, N. M., Raafat-El Guelly, M., Josh, J., Heling, R. B. and Bannerjee, M. R., *Gene*, **15**, 285 (1981).
109. Hennighausen, L. G. and Sippel, A. E., *Eur. J. Biochem.*, **125**, 131 (1982).
110. Suard, Y. M. L., Tosi, M. and Kraehnenbuhl, J.-P., *Biochem. J.*, **201**, 81 (1982).
111. Craig, R. K., Hall, L., Parker, D. and Campbell, P. N., *Biochem. J.*, **194**, 989 (1981).
112. Yu-Lee, L.-Y. and Rosen, J. M., *J. Biol. Chem.*, **258**, 10794 (1983).
113. Jones, W. K., Yu-Lee, L.-Y., Clift, S. M., Brown, T. L. and Rosen, J. M., *J. Biol. Chem.*, **260**, 7042 (1985).
114. Stewart, A. F., Willis, I. M. and Mackinlay, A. G., *Nucl. Acids Res.*, **12**, 3895 (1984).
115. Nagao, M., Maki, M., Sasaki, R. and Chiba, H., *Agric. Biol. Chem.*, **48**, 1663 (1984).

116. Maki, M., Nagao, M., Hirose, M. and Chiba, H., *Agric. Biol. Chem.*, **47**, 441 (1983).
117. Willis, I. M., Stewart, A. F., Caputo, A., Thompson, A. R. and Mackinlay, A. G., *DNA*, **1**, 375 (1982).
118. Kershulite, D. R., Ivanov, V. N., Kapelinskaya, T. V., Kaledin, A. S. and Gorodetsky, I., *Dokl. Akad. Nauk SSSR*, **271**, 1502 (1983).
119. Ivanov, V. N., Kershulite, D. R., Bayer, A. A., Aklundova, A. A., Submova, G. E., Judmkova, E. S. and Gorodetsky, S. I., *Gene*, **32**, 381 (1984).
120. Watson, J. D., Tooze, J. and Kurz, D. T., *Recombinant DNA: A Short Course*, Freeman, New York, 1983, pp 176, 200.
121. Kucherlapati, R. and Skoultchi, A. I., *CRC Crit. Rev. Biochem.*, **16**, 349 (1984).
122. Palmiter, R. D. and Brinster, R. L., *Cell*, **41**, 343 (1985).
123. Hammer, R. E., Pursel, V. G., Rexroad, C. E. Jr, Wall, R. J., Bolt, D. J., Ebert, K. M., Palmiter, R. D. and Brinster, R. L., *Nature*, **315**, 680 (1985).
124. Creamer, L. K. and Olson, N. F., *J. Food Sci.*, **57**, 631 (1982).
125. Whitaker, J. R., Enzymatic modifications of proteins applicable to foods, in *Food Proteins: Improvement through Chemical and Enzymatic Modification*, Feeney, R. E. and Whitaker, J. R. (Eds), American Chemical Society, Washington, DC, 1977.
126. Visser, S., *Neth. Milk Dairy J.*, **35**, 65 (1981).
127. van Hooydonk, A. C. M., Olieman, C. and Hagedoorn, H. G., *Neth. Milk Dairy J.*, **37**, 207 (1984).
128. Shimizu, M., Lee, W. S., Kaminogawa, S. and Yamauchi, K., *J. Food Sci.*, **49**, 1117 (1984).
129. Zevaco, C. and Desmazeaud, M., *J. Dairy Sci.*, **63**, 15 (1980).
130. Trieu-Cuot, P., Archieri-Haze, M.-J. and Gripon, J. C., *Lait*, **62**, 234 (1982).
131. Yoshikawa, M., Yoshimura, T. and Chiba, H., *Agric. Biol. Chem.*, **48**, 3185 (1984).
132. Doi, H., Ideno, S., Kuo, F. H., Ibuki, F. and Kanamori, M., *J. Nutr. Sci. Vitaminol.*, **29**, 679 (1983).
133. Doi, H., Hiramatsu, M., Ibuki, F. and Kanamori, M., *J. Nutr. Sci. Vitaminol.*, **31**, 77 (1985).
134. To, B., Helbig, N. B., Nakai, S. and Ma, C.-Y., *Can. Inst. Food Sci. Technol. J.*, **18**, 150 (1985).
135. Nakamura, R., Hayakawa, S., Yasuda, K. and Sato, Y., *J. Food Sci.*, **49**, 102 (1984).
136. Smith, D. M. and Brekke, C. J., *J. Agric. Food Chem.*, **33**, 631 (1985).
137. Smith, D. M. and Brekke, C., *J. Food Sci.*, **49**, 1525 (1984).
138. Retesward, A. L., *J. Food Technol.*, **20**, 129 (1985).
139. Asghar, A., Samejima, K. and Yasui, T., *CRC Crit. Rev. Food Sci. Nutr.*, **22**, 27 (1985).
140. Lee, J. W. and Lopez, A., *CRC Crit. Rev.*, **21**, 289 (1984).
141. Paulson, A. T., Tung, M. A., Garland, M. R. and Nakai, S., *Can. Inst. Food Sci. Technol. J.*, **17**, 202 (1984).
142. Kamata, Y., Ochiai, K. and Yamauchi, F., *Agric. Biol. Chem.*, **48**, 1147 (1984).
143. Mohri, M. and Matsushita, S., *J. Agric. Food Chem.*, **32**, 486 (1984).

144. Ma, C.-Y., *Can. Inst. Food Sci. Technol. J.*, **18**, 79 (1985).
145. Lacroix, M., Amiot, J. and Brisson, G. J., *J. Food Sci.*, **48**, 1644 (1983).
146. Lacroix, M., Amiot, J., Bigue, N. and Brisson, G., *Can. Inst. Food Sci. Technol.*, **16**, 242 (1983).
147. Behnke, U., Schultz, M., Ruttloff, H. and Schmandke, H., *Nährung*, **26**, 313 (1982).
148. Wolf, W. J., *J. Am. Oil Chem. Soc.*, **54**, 112A (1977).
149. Yamashita, M., Arai, S. and Fujimaki, M., *J. Agric. Chem.*, **24**, 100 (1976).
150. Fujimaki, M., Arai, S. and Yamashita, M., Enzymatic protein degradation and resynthesis for improvement, in *Food Proteins: Improvement through Chemical and Enzymatic Modification*, Feeney, R. E. and Whitaker, J. R. (Eds), American Chemical Society, Washington, DC, 1977.
151. Watanabe, M., Toyokawa, H., Shimada, A. and Arai, S., *J. Food Sci.*, **46**, 1467 (1981).
152. Watanabe, M., Shimada, A., Yazawa, E., Kato, T. and Arai, S., *J. Food Sci.*, **46**, 1738 (1981).
153. Watanabe, M., Fujii, N. and Arai, S., *Agric. Biol. Chem.*, **46**, 1587 (1982).
154. Yazawa, E., Shimada, A., Kowazima, M. and Arai, S., *Nippon Shokuhin Kogoyo Gakkaishi*, **31**, 145 (1984).
155. Arai, S., Watanabe, M. and Fujii, N., *Agric. Biol. Chem.*, **48**, 1861 (1984).
156. Shimada, A., Yamamoto, I., Sase, H., Yamazaki, Y., Watanabe, M. and Arai, S., *Agric. Biol. Chem.*, **48**, 2681 (1984).
157. Yamamoto, I., Kushuhara, K., Matsumoto, M., Shimada, A., Watanabe, M. and Arai, S., *Agric. Biol. Chem.*, **48**, 2689 (1984).
158. Arai, S., Watanabe, M. and Tsusi, R. F., *Agric. Biol. Chem.*, **48**, 2173 (1984).
159. Aso, H., Kimura, H., Watanabe, M. and Arai, S., **49**, 1649 (1985).
160. Noar, S. R. and Shipe, W. F., *J. Food Sci.*, **49**, 1316 (1984).
161. Gololobov, M. Yu. and Belikov, V. M., *Nährung*, **27**, K7-K8 (1983).
162. Gololobov, M. Yu., Belikov, V. M., Vitt, S. V., Paskonova, E. A. and Titova, E. F., *Nährung*, **25**, 961 (1981).
163. Ludwig, E. and Winkler, H., *Nährung*, **27**, 319 (1983).
164. Winkler, H., Notzold, H. and Ludwig, E., *Nährung*, **28**, 1029 (1984).
165. Delincee, H. and Hajos, Gy., *Acta Alimentaria*, **13**, 309 (1984).
166. Noack, J. and Hajos, Gy., *Acta Alimentaria*, **13**, 205 (1984).
167. Montecalvo, J. Jr, Constantinides, S. M. and Yang, C. S. T., *J. Food Sci.*, **49**, 1305 (1984).
168. Heck, N. D., *Diss. Abs. Int.*, **B44**, 2403 (1984).
169. Schmidl, M. K., Shipe, W. F., Chabot, J. F. and Hood, L. F., *J. Food Process. Preserv.*, **7**, 131 (1983).
170. Gololobov, M. Yu., Antonova, T. V., Belikov, V. M. and Pasakonova, E. A., *Nährung*, **26**, 427 (1982).
171. Whitaker, J. R., personal communication (1985).
172. Ikura, K., Yoshikawa, M., Sasaki, R. and Chiba, H., *Agric. Biol. Chem.*, **45**, 2587 (1981).
173. Ikura, K., Goto, M., Yoshikawa, M., Sasaki, R. and Chiba, H., *Agric. Biol. Chem.*, **48**, 2347 (1984).
174. Motoki, M. and Noriki, N., *J. Food Sci.*, **48**, 561 (1983).
175. Kurth, L. and Rogers, P. J., *J. Food Sci.*, **49**, 573 (1984).

176. Nio, N., Motoki, M. and Takinami, K., *Agric. Biol. Chem.*, **49**, 2283 (1985).
176a. Motoki, M., Nio, N. and Takinami, K., *Agric. Biol. Chem.*, **48**, 1257 (1984).
177. Matheis, G. and Whitaker, J. R., *J. Food Biochem.*, **8**, 137 (1984).
178. Kunimori, T., Nishiyama, J. and Matsumoto, H., *Cereals Food World*, **23**, 500 (1978).
179. Hoseney, R. C. and Faubion, J. M., *Cereal Chem.*, **58**, 421 (1981).
180. Matheis, G. and Whitaker, J. R., *J. Protein Chem.*, **3**, 35 (1984).

4

Large Scale Separation and Isolation of Proteins

M. P. BROCKLEBANK

John Brown Engineers and Constructors Ltd, Portsmouth, UK

INTRODUCTION

Proteins are a family of complex biological macromolecules produced by and found only in plants and microbial and animal cell tissues. They perform a variety of tasks in living systems, depending on their amino acid content, structure and association with atoms and other molecules of biological origin. Over the centuries, proteins have been recovered, usually in very crude forms, from animal and plant tissues and used for mainly domestic and food processing applications. In the last 20 years or so much attention has been focussed on their production and usage as essential dietary supplements, their potential for the conversion of non-oil feedstocks such as carbohydrates, their potential for modifying existing food products and, in very pure form, their health care applications. Consequently the following usage categories may be identified:

(1) The utilisation of their amino acid content to provide the essential and supplementary dietary components of human and animal foods.
(2) The application of their useful physical properties, such as elasticity, water retention and adhesion, in technologies such as baking, cooking and adhesives manufacture.
(3) The use of their specific catalytic properties to interact with other biological and synthetic organic chemicals and bring about cleavage, additive or spatial molecular changes.
(4) The use of the specific biologically active properties of some

Table 1
Sources, World Outputs and Market Values of Typical Protein Products

Category	*Feedstock*	*Protein*	*Production process*	*Annual output (t)*	*Approximate price (£/t)*
1. Nutritional	Soya bean	Soya bean meal	Extraction	$>20 \times 10^6$	130
2. Functional	Soya bean	Soya isolate	Extraction	$>1 \times 10^5$	1 500
	Whey	Whey protein	Ion exchange recovery	500	3 000–6 000
3. Reactive	Starch	Amylases	Fermentation	600	15 000–30 000
	Animal tissue	Rennin	Extraction	60	10^8
4. Health care	Blood	Albumin	Precipitation	200	2×10^6
	Sugars	Human growth hormone	Fermentation	20 kg(?)	1×10^{10}
	Calf serum	Monoclonal antibodies	Tissue culture	10 kg(?)	3×10^9

discrete protein molecules in aqueous solution as diagnostic and therapeutic agents in human and animal health applications.

It is the demand for the high purity levels required by the category (4) biologically active proteins such as insulin, hormones and monoclonal antibodies which has focussed recent attention on the often complex problems which may arise in protein recovery and purification processes.

Production levels of protein products vary enormously (Table 1). The wide amino acid spectrum of protein derived from microorganisms, termed single-cell protein (SCP), provides a non-agricultural source of nutritional category (1) products. Enzymes are extracted from animal sources, e.g. trypsin, and isolated from microorganisms. In fact, it is no understatement to say that genetic and protein engineering,[1] that is the altering of DNA to yield useful proteins, has become possible only because of rapid developments in analytical scale separation and purification techniques. This has allowed the isolation to high purity of the many specific enzymes required to cleave DNA 'strands'. Health care proteins are very high value, low output products. Some such as albumin and hormones are extracted from human and animal sources, e.g. blood or glands. For many others the use of genetically engineered microorganisms for production is a reality, and the challenge is now in the scale-up of processes to recover these from the fermenter broth at minimum cost and with maximum therapeutic purity and yield.

Protein market values reflect feedstock costs, production and recovery process costs, R&D expenses and scale. Typical values are given in Table 1 where it can be seen that significant increases occur from category (1) through to category (4). In general categories (1) and (2) protein prices are directly related to feedstock costs, whereas recovery of R&D expenses, low outputs and the often complex separation and purification procedures lead to high category (3) and (4) prices.

2. THE NATURE OF PROTEINS

2.1. Chemistry

Proteins are polymers of amino acids (molecular weight range 10^4–10^9 daltons) linked by amide (peptide) bonds. The cumulative interactions of the amino acid side-chains, together with some specific interactions, e.g. disulphide bridges, determine the protein structure (conformation) and the specific chemical/biological functions of a protein relate to the interaction of this tertiary structure and particularly its exposed surface groups with adjacent molecules or structures. Protein conformation results from and is maintained by weak non-covalent bond (hydrophobic, hydrogen and ionic) interactions between internal amino acid residues and between exposed surface groups and its surrounding medium. Hydrophobic interactions between non-polar groups are considered to represent the most important contribution to specific protein conformations, and it is postulated that, if these cause folding towards the centre of the protein structure and away from its aqueous surroundings, the folded state is the most stable.[2]

Since non-covalent bonds are very weak, protein structures are easily shifted away from their minimum free energy conformation. Thus moderate (60°C) temperatures are sufficient to break the non-covalent bonds and denature the protein. Also, if the internal hydrophobic groups can be attracted to the surface of the structure by similar groups in the surrounding medium, the system may change to a new minimum free energy denatured state. Thus the addition of non-polar solvents as well as hydrophobic reagents such as urea may denature some proteins.

2.2. Behaviour in Solution

The solubility and behaviour of proteins in aqueous solution are largely determined by the net properties of their exposed surface amino acid residues. Hence proteins of amino acids with polar and neutral side-chains containing electronegative atoms (N, S, O) are hydrophilic and are readily soluble in water and dilute salt solutions because of the formation of hydrogen bonds with adjacent water molecules (the 'albumins'). Those proteins with a preponderance of exposed non-polar side-chains consisting of C and H atoms have a hydrophobic nature and are

Table 2
Isoelectric Points of Some Proteins

Protein	*pI*	*Protein*	*pI*
Pepsin	<1·0	Haemoglobulin	6·8
Egg albumin	4·6	Myoglobulin	7·0
Serum albumin	4·9	Ribonuclease	7·8
Urease	5·0	Chymotrypsinogen	9·5
β-Lactoglobulin	5·2	Cytochrome C	10·7
Fibrinogen	5·8	Lysozyme	11·0
γ_1-Globulin	6·6		

sparingly soluble in water (the 'globulins') although soluble in dilute aqueous salt solutions.

As with amino acids, the exposed ionisable surface groups give proteins a titratable acid–base behaviour and an isoelectric point (pI) (see Table 2). The interaction between protein surface groups and the surrounding solution is fundamental to virtually all physico-chemical separation and purification techniques. Thus the solubility of a protein in solution depends upon the properties of its exposed surface groups, solvent pH, solvent temperature, the presence of ionic species in solution, solution polarity level (dielectric constant), the presence of oppositely charged surfaces, and the presence of chemicals in solution which have a biological affinity for the protein.

Hence protein solubility is primarily created by associations between water molecules and its hydrophilic surface groups as well as solution conditions which maximise surface charge and so prevent aggregation. Conversely, protein aggregation, precipitation and other interactions which remove proteins from solution are encouraged by the following:

(1) Lowering solution temperatures to below saturation level.
(2) Adjusting the solution pH to the protein pI point in order to neutralise electrical repelling charge and thus encourage aggregation.
(3) Addition of quantities of non-polar solvents to reduce the attraction of surface polar groups with water and so encourage protein–protein hydrogen bonding between surface polar groups as well as some denaturation leading to hydrophobic group interaction.
(4) Addition of large quantities of very polar solvents which cause unfolding by bonding with internal groups; this results in protein–protein hydrophobic interactions and H-bonding between surface polar groups.
(5) Increasing the levels of salt ions in the surrounding solution which bond more readily with water molecules and so allow hydrogen bonding of exposed protein–protein polar groups, this leading to aggregation and precipitation. This is termed 'salting out'. (The opposite effect is 'salting in' whereby the solubility of proteins is increased in low concentrations of neutral salts since these are thought to interact with surface charged groups, thus stabilising them and preventing aggregation.)
(6) Raising solution temperature in order to denature proteins, thus exposing hydrophobic groups which then interact forming agglomerates.

(7) The presence of charged and other species in heterogeneous systems (e.g. chromatography gels) which attract protein molecules and remove them from their aqueous solution.

3. RECOVERY PROCESS CONCEPTS

The objective of a protein recovery process is to produce, at acceptable cost, the maximum yield of protein of sufficient purity and possessing the desired compositional and functional properties. Therefore the following constraints and observations apply to the physical and chemical operations selected to recover a protein.

(1) Their severity must be consistent with the desired product performance; e.g. recovery by heat denaturation and then flocculation is acceptable for dietary proteins but completely unacceptable for biologically active proteins.
(2) Product usage determines the level of purification required; e.g. very little purification is needed to recover a gluten protein extract from wheat, but separation techniques offering high levels of fractionation must often be used to produce analytical enzymes from animal or microbial sources.
(3) The recovery process scheme depends very much on the feedstock; typical contaminants to be removed from the desired protein

Table 3
Protein Concentrations in Recovery Process Feedstocks

Feedstock	*Protein*	*Concentration in feedstock*
Fermentation broth	Single cell protein	30 g/litre
Fermentation broth	Starch enzymes	0·1–3 g/litre
Fermentation broth	Glucose oxidase	0·2 g/litre
Fermentation broth	β-Glucosidase	0·001 g/litre
Fermentation broth	Insulin	0·08 g/litre
Fermentation broth	Protein A	0·5 g/litre
Fermentation broth	Interferon	1–10 mg/litre
Tissue culture	Monoclonal antibodies	1–5mg/litre
Whey	Whey protein	0·9% w/v
Leaf matter	Leaf protein extract	5% w/w
Soya bean	Soya meal	70% w/w
Meat processing wastes	Various	0·1% w/v
Vegetable processing wastes	Various	1–3% w/w

Table 4
Common Recovery Process Techniques

Process step function	*Unit operations employed*	*Denaturation effect on protein*	*Examples*
Insoluble contaminant removal from protein solution	Centrifugation	Low	Removal of cell matter
	Dead-end filtration		
	Cross-flow filtration		
	Gravity sedimentation		
	Two-phase aqueous extraction		
Extraction of soluble proteins from solid feedstocks	Maceration/dissolution	Low	Leaf protein extraction
	Cell disruption		Intracellular protein recovery
Water removal	Evaporation	Medium/high	Whey powder production
	Ultrafiltration	Low	Bulk enzyme production
	Protein precipitation	Low/high	$(NH_4)_2SO_4$ precipitation of enzymes
	Thermal aggregation	High	Leaf protein concentration
	pI aggregation	Medium/high	Soya protein isolate precipitation
Separation of dissimilar soluble contaminants from protein solutions	Ultrafiltration	Low	Bulk enzyme recovery
	Precipitation/aggregation	As above	Nucleic acid removal
	Gel filtration	Low	Salt removal
	Two-phase aqueous extraction	Low	Intracellular enzyme recovery
	Electrodialysis	Low	Salt removal
Protein–protein fractionation	Ion exchange gel chromatography	Low	Health care protein purification
	Fractional precipitation	Low/medium	Plasma fractionation
	Electrophoresis	Low	Tissue enzyme purification
	Affinity chromatography	Low	Monoclonal antibody purification
	Hydrophobic interaction chromatography	Low	Health care protein purification
	High performance liquid chromatography	Medium/low	Limited use to date

include insoluble solids, inorganic and organic soluble (small) molecules, soluble macromolecules and water.

(4) The major bulk separation problem, particularly with fermentation and 'wastes' feedstocks, concerns the removal of relatively small concentrations of the desired protein and often its contaminants from large volumes of water.

(5) The major purification problem concerns the isolation at high purity of the desired protein from similar proteins and other large macromolecules.

(6) Similar physico-chemical techniques used to remove impurities and isolate and purify proteins from a wide variety of feedstock sources occur in a large variety of recovery processes.

(7) Since recovery processes are often batch and multi-stage, yields are low.

Constraints (3), (4), (5) and (7) are particularly relevant to biotechnology, i.e. the production of proteins using microbial fermentation and cell culture, since protein concentrations in fermenter broths are very low (Table 3). As a comparison, Table 3 also provides protein concentrations in other feedstock sources. The range of common recovery process techniques is given in Table 4. Their process order is primarily dependent on the number, type, concentration and relative locations within the feedstock of the contaminants and the desired protein. Also the limitations of these unit operations and the equipment employed are often determined by the scale of operation. (It must be noted that the recent usage of the term 'large scale' relates to the scale-up from laboratory bench to commercial plant of processes recovering fragile biologically active proteins from genetically engineered microbial and animal cell feedstocks. Implicit in this are the scale-up constraints on those separation techniques, such as gel chromatography, developed specifically for labile, low charge proteins.)

4. SEPARATION TECHNIQUES

4.1. Precipitation

The precipitation and consequent recovery of soluble proteins by the addition of reagents to their solutions is widely practised in many industries. It is a low cost, high yield procedure usually used to separate a crude protein isolate from solution although, with exacting control of system parameters such as pH, temperature and reagent concentration, a degree of purification by fractional precipitation is achievable.[3] The biochemical aspects of precipitation have been studied in depth by Dunnill and co-workers.[4-6] Protein precipitation is achieved either by altering the aqueous characteristics of the native solution by the addition of high concentrations of neutral chemicals or by adding reagents to alter the chemical or physical characteristics of protein molecules. The usual procedures adopted to achieve these include:

(1) Addition of neutral salts (e.g. ammonium sulphate), termed salting out.
(2) Alteration of solution dielectric constant (polarity) by addition of non-polar solvents (e.g. ethanol).
(3) Addition of acids or alkalis to adjust the solution to the protein pI value.
(4) Addition of non-ionic polymers, e.g. polyethylene glycol.[7]
(5) Addition of charged polyelectrolytes, e.g. polyacrylic acid.[8]
(6) Addition of metal ions.
(7) Addition of reactive reagents creating insoluble complexes, e.g. Rivanol.[9]

The first three methods are most widely used, particularly the addition of salts which encourage strong hydrophobic interactions between protein molecules. Ammonium sulphate is the salt of choice because it can be used at high concentration to improve yields and it has very little denaturing effect on protein structures. Its disadvantages relate to the isolation of health care proteins, particularly on the large scale since it has some toxicity to humans and unlike ethanol is not toxic to bacteria. The degree of protein precipitation is affected by salt or solvent concentration and total protein concentration. Typical levels of reagents promoting precipitation are given in Table 5. Protein solubility in solution depends upon temperature, pH and reagent ionic strength, and these may be empirically related according to:

$$\log_e S = \beta - KR$$

where S is protein solubility, β is a constant (dependent on temperature and pH), K is an independent constant, and R is ionic strength.

Consequently, to achieve any degree of fractionation by precipitation, values of pH, temperature and R must be selected which minimise S for each of the components of the protein mixture.[3] Protein precipitation by pH adjustment is effective provided that the proteins have a low hydration point, high surface hydrophobicity, and the acid anions chosen do not

Table 5
Protein Precipitation by Reagent Addition

Protein (concn)	*Solution pH*	*Reagent (strength)*	*Total yield*	*Ref.*
Soya protein isolate (25 g/litre)	4·8	H_2SO_4	84–87%	35
β-Galactosidase (9 g/litre)	n.a.	$(NH_4)_2SO_4$ (40% saturation)	93%	40
Plasma protein (51 g/litre)	7·2	Ethanol (8% w/v)	5–10% initial (fraction I)	18
Final plasma fraction (25 g/litre)	5·2	Ethanol (40% w/v)	n.a.	18
Glucose oxidase (n.a.)	7·6	Polyethyleneimine (0·25–0·5% w/v)	n.a.	37

damage the proteins irreversibly. Large changes of solubility within one pH unit of the isoelectric point may occur, so accurate process control (±0·03 pH unit) may be essential.[10]

Batch, CSTR and tubular reactors may be used for protein precipitation, and important factors to consider for the design of these and the subsequent solid–liquid separation step include:

(1) The use of precipitation agents, temperatures and pH conditions which do not denature or otherwise inhibit the desired product properties.
(2) The identification of reagent conditions which yield the desired precipitate of adequate purity and at high yield.
(3) The economic usage with recycling if possible of precipitation chemicals.
(4) The avoidance of denaturing shear regions in reactors *and* transport equipment, e.g. pumps handling precipitates.
(5) The absence of air entrapment in agitation systems since liquid/air interfaces denature proteins.
(6) Effective heat transfer techniques to create uniform, closely controlled temperatures.
(7) The need to avoid possible denaturation effects on products where precipitation times are long.
(8) The creation of a high population of large, dense, shear-stable agglomerates.
(9) Achievement of maximum density differences between the supernatant and the precipitate. (Centrifugal recovery using ethanol

is therefore significantly easier than from ammonium sulphate solutions.)

(10) The balance between short residence times (resulting from high growth rates caused by high degrees of mixing) and small average particle size leading to difficult recovery procedures.

(11) The addition method for the precipitating agent.

Test data from a stirred tank will not necessarily apply to a tubular reactor. The evaluation of parameters such as mixing intensity (4) also assists precipitation reactor design and suggests that an ageing parameter (called the Camp number) should be investigated which minimises the break-up of the protein agglomerates under the shear conditions expected in the precipitate recovery equipment. Acoustic conditioning is an alternative, fast response, ageing technique for increasing agglomerate strength and size. This involves the passage of the precipitate slurry between two vibrating paddles, of different amplitude.[5,11] Effective conditioning techniques result in dryer centrifuge and filter cakes.

4.2. Cell Disruption

Cell disruption is an important unit operation used to break open microbial and animal cell wall structures in order to release intracellular enzymes and proteins into free solution and so facilitate their recovery. Other uses for disruption in biotechnology include the rupture of bacterial SCP to facilitate protein agglomeration for centrifugal recovery processes, the mild rupture of cell walls for rDNA experiments on genetic fragments, and the reduction of particle size of filamentous microorganisms to assist separation. Cell disruption is commonly achieved using:

(1) The high shear fields developed in bead-mills and homogenisers.

(2) Pressure waves and cavitation effects resulting from high-frequency sonification.

(3) Enzymic degradation of cell wall structures.

(4) Thermal degradation of cell wall structures.

Intracellular autolytic enzymes can also break down the cell wall and cytoplasmic membrane.[12]

Bead mills are commonly used at laboratory and pilot plant scale where experiments have indicated that the disruption of yeast to release the soluble cytoplasmic enzyme, alcohol dehydrogenase, can be correlated as follows:[13]

$$\log_e \frac{R_m}{R_m - R} = kt$$

where k is a constant for fixed mill type and operating conditions, R_m is the maximum weight of protein released per unit weight of yeast, R is the weight of protein released under experimental conditions per unit weight of yeast, and t is the residence time.

Tests also show that the efficiency of milling, i.e. the rate of protein release per unit energy input, decreases as the amount of protein released increases, and a residence time[14] of 800 s appeared optimum to achieve, in a single pass, 95% of the theoretical release of protein from a thawed sample of brewer's yeast in a 0·6 litre mill (usable volume 0·28 litre). On this basis the throughput for large bead mills (200 litre chamber) may be limited to approximately 400 litres/h, since frictional heat removal is difficult. Because of this most workers favour the pressure homogeniser in which cell suspensions under high pressure (usually up to 550 bar) are forced through a very narrow annulus to 'impact' onto an annular surface and rupture. At cell concentrations up to 20% dry solids and constant flow rate, the protein release from baker's yeast is correlated to be:[15]

$$\log \frac{R_m}{R_m - R} = KNP^{2\cdot 9}$$

where K is a constant (function of temperature), N is the number of passes through the unit, and P is the upstream pressure.

Yeast cells, being spherical in shape, are more difficult to rupture than rod-shaped bacteria or filamentous particulates. Scale-up is relatively easy, with release yields increasing in larger machines because of higher average shear pressures. Since single homogenisers are available with throughput rates up to 6 m^3/h, they provide the favoured technique for the high rate rupture, at the industrial scale, of recombinant bacteria.

4.3. Solid–Liquid Separations

4.3.1. Usage

Solid–liquid separation is a key requirement of protein recovery processes; the separations to be made are threefold, namely:

1. Separation of insoluble matter including cells and cell debris from protein solutions.
2. Separation of precipitated impurities, e.g. nucleic acids.
3. Separation of precipitated proteins from solutions.

Care must be taken to select separation equipment consistent with the properties of the protein and its recovery process. In particular, equipment that avoids the shear denaturation effects of foaming and air entrapment at solid–liquid interfaces, which limit the effects of small temperature variations on precipitation processes, and minimise protein precipitate break-up, may be required when separating protein precipitates. Other major factors influencing the selection of solid–liquid separation equipment include:

(1) Particle size and density difference with mother-liquor.
(2) Throughput rate.
(3) Feed solids content.
(4) Process requirement, e.g. sterile, sterilisable, containment.
(5) Precipitate or cake handling characteristics.
(6) Recovered solids or liquid feedstock fraction.

The equipment types used for the above separations include centrifuges, pressure and vacuum 'dead-end' filters and cross-flow filtration systems.

4.3.2. Centrifugation

Centrifuges are widely used to separate a variety of solids, ranging from food proteins such as casein, gluten and yeast, bacteria and cell debris from fermentation broths containing functional enzymes and precipitates of monoclonal antibodies, and therapeutic proteins.[16,17] In pilot and commercial scale processes, five different types are used, their key characteristics being summarised in Table 6.

Tubular centrifuges are used in low-throughput processes particularly where small particles (below 0·5 μm) such as cell debris or chemical precipitates are present in small quantities. For 16 000g machines nominal throughput rates of 0·01–0·1 m^3/h and maximum bowl solids capacities of up to 6 litres are available.

Disc stack centrifuges are commonly used to remove microbial cells from fermentation broths. As a general guide, 2–3% dw/v baker's yeast (12 μm dia.) suspensions can be separated in nozzle bowl machines at 70% of their nominal hydraulic throughput rate. Self-cleaning disc units operating at 8000g remove 1–3% dw/v bacteria (1–3 μm) at 8% of their nominal hydraulic throughput whilst at 14 500g self-cleaning disc stack machines separate cell debris (down to 0·5 μm) at 10–20% of their rated throughput.[18] Chamber centrifuges, which operate at around 3000g, develop insufficient centrifugal force to remove cells but are used for the batchwise recovery of protein precipitates. One demanding application is the recovery without denaturation of precipitates of biological proteins

Table 6
Centrifuge Characteristics

Centrifuge type	*Usage examples*	*Advantages*	*Disadvantages*
Tubular	Pilot and laboratory scale separations of cells, cell debris, nucleic acid and protein precipitates	High 16 000–60 000*g* forces Low denaturation effect Can be located in constant temperature 'chamber'	Low throughput Batch Low solids capacity; manually removed
Chamber	Polishing broths after filtration Recovery of blood proteins	Low denaturation effect Constant temperature operation (bowl cooling)	Low *g* force (to 3500*g*) Batch Manual solids removal
Decanter	Removal of filamentous fungi from broths Recovery of gluten	High throughputs/machine Continuous solids discharge High solids loading in feedstream	Low *g* force (to 4000*g*) Wet solids Foaming of solids at discharge
Nozzle bowl	Yeast separation	High liquid throughputs High solids capacity (to 5% v/v in feed) Moderate *g* force (8000*g*) Enclosed centrate discharge	Feed must have high solids content (above 2%) Poor dewatering Precipitate denaturation may occur
Self-cleaning bowl	Bacteria separation Above 0·5 μm cell debris separation Soya protein isolate	High liquid throughput Steam-sterilisable High *g* force (to 15 000*g*) Good dewatering Enclosed centrate discharge Designs available for minimum entry shear effects on precipitates	Low solids throughput Precipitate denaturation may occur at exit Expensive

such as albumin.[19] In 25 litre bowl capacity units, up to 20 kg of 20–50% dw solids cake can be achieved. Decanter centrifuges developing 3000–5000*g* are used for continuous recovery of large proteinaceous solids such as gluten from aqueous streams and to separate filamentous fungi (50–100 μm) from fermentation broths with simultaneous solvent extraction of antibiotics.[20]

Analytical expressions for different centrifuge configurations relate liquid and solid throughputs with maximum particle diameter, particle and liquid physical properties and machine geometrical parameters.[21,22] The use of such expressions provides a guide for machine performance,[17] although particle–particle interactions and non-ideal particle shapes and flow conditions may result in actual throughputs being 75% below calculated values. Throughput rate (Q) and particle/fluid properties are related for all centrifuge types according to:

$$Q = Kd^2 \Delta\rho\eta^{-1}$$

where K is a machine constant, d is the particle diameter, $\Delta\rho$ is the solid/liquid density difference, η is the dynamic fluid viscosity.

The exact machine geometry determines the form and complexity of the machine constant K term. Transfer of separation performance data from a centrifuge of one type to another is best carried out by application of the underlying analytical expressions. Sigma analysis is best used for scale-up of a particular machine type.[21]

Adaptations of basic centrifuge designs to meet the needs of protein recovery processes include feed inlet shear minimisation and the avoidance of excessive aeration of sludge discharge,[4] temperature control[19] and steam sterilisation. A typical pilot-scale steam-sterilisable disc stack centrifuge used to separate recombinant or virulent microorganisms from fermenter broths is shown in Fig. 1.

4.3.3. *Dead-end Filtration*

In dead-end filtration, feed slurry impacts directly onto the filtration media with all of the mother-liquor removed through the deposited solid cake and the filtration media. Effective filtration rates can only be achieved with either incompressible feed solids or the periodic removal of deposited compressible solids cakes from the filter media. The latter technique is utilised in rotary vacuum filters.[24] These are mainly used to remove bacterial and filamentous microorganisms (e.g. bacillus species and *Aspergillus niger*) from fermentation broths in large-scale enzyme plants (up to 80–250 m^3 of whole broth per batch). In the usual arrangement,

Fig. 1. Steam-sterilisable disc stack self-cleaning centrifuge unit.

the filtration medium is a thick layer (up to 100 mm) of siliceous earth (precoat) deposited on a rotating drum, and a thin layer of this plus cell deposit is removed by a knife every revolution. Typical feed rates for fermenter broths are 100–200 litres/m^2-h. Filter sizes range from pilot models at 1 m^2 up to $>100\,m^2$ of filtration area for large production scale units. A typical precoat usage is 0·5 kg/m^2-h. An alternate reverse cycle pressure filter has been developed by one manufacturer.[23] Food-compatible precoats are available (e.g. potato starch) to recover for e.g. baker's yeast. Biologically active protein precipitates are not commonly separated by filtration due to the denaturing effects of precipitate/air interfaces and hygienic considerations.

4.3.4. Cross-flow Filtration

Cross-flow filtration systems are increasingly being used to separate whole cells and cell debris from fermentation broths and homogenates. A batch cross-flow separation process (Fig. 2) consists of a porous membrane across which the feed slurry passes at high velocity (2–6 m/s) and through which a fraction of the broth passes. The extent to which insoluble particles and large molecules pass through the membrane is limited by

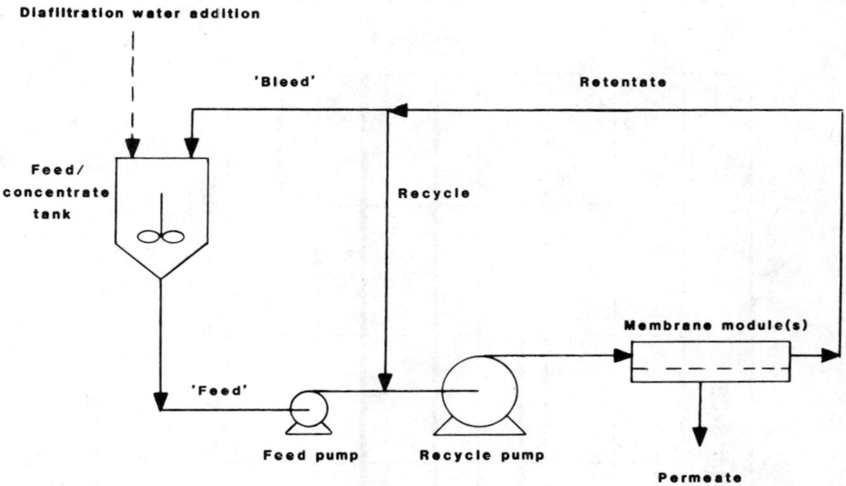

Fig. 2. Batch, closed loop cross-flow membrane filtration system.

its actual pore size (cut-off). The application of cross-flow filtration to protein separation processes is well covered in the literature.[24–26] The three basic membrane processes are microfiltration (MF), ultrafiltration (UF) and reverse osmosis (RO). A comparison of cross-flow and other separation techniques based on particle size is given in Fig. 3.

Membrane materials used in cross-flow processes include: for MF, sintered metal, porous thermoplastics (e.g. PTFE, nylon), and inorganic oxides (e.g. aluminium oxide); for UF, polysulphones and polyacrylamides; for RO, cellulosic and other synthetic polymers. Cross-flow membrane configurations include tubular, spiral sheet, flat sheet and hollow fibre, the last two being commonly used for laboratory and pilot scale (say $10\,m^2$ area) cell–broth separations. Since proteins and other biological source organics foul membranes, it is critically important that membrane system size is based on extensive experimental tests including repeated CIP routines to determine permeate flux rates.

The removal of bacterial cells and cell debris from fermenter broths at laboratory (10–25 litre batches) and pilot scale (50–2000 litre batches) has been the major application area for MF processes to date, particularly since they are better suited for containment and sterile[27] separations than disc stack centrifuges and may also be cheaper. A few instances of cell–broth separations on a large scale ($>50\,m^3$ of broth) are reported by membrane system manufacturers.[28] Cross-flow filtration is claimed to

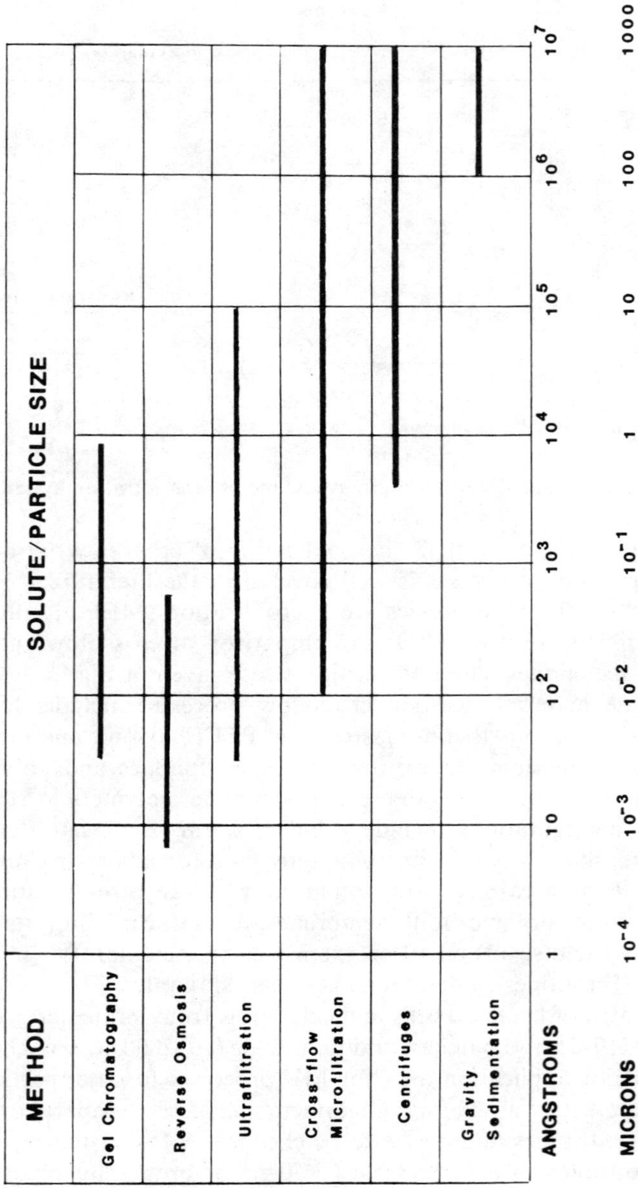

Fig. 3. Size-dependent separation techniques.

offer lower operating costs than rotary vacuum filtration primarily because the need for expensive filter aids is eliminated.[24] Typical flux rates for UF and MF cross-flow cell–broth separations are in the range 15–50 litres/m^2-h. The use of UF membranes for cell debris–soluble enzyme separation also allows some simultaneous molecular scale purification although molecular cut-off levels may be substantially lower than theoretical values because of concentration polarisation 'filtration' effects at the membrane surface.[29,30] In addition, cross-flow systems are preferred to centrifuges where protein adherence to cell debris occurs.[31]

Protein precipitate concentration using cross-flow filtration is not commonly practised. In one laboratory evaluation, using hollow fibre membranes with >130 g/litre retained protein concentration, flux rates of 20–23 litres/m^2-h were measured.[5] Single-pass multi-stage separations are recommended for fragile protein precipitates. The use of MF and UF membrane separations for continuous (cell recycle) fermentation processes enables higher fermenter productivities to be achieved.[32] The future prospects for this application are high, particularly with the advent of steam-sterilisable inorganic membranes.

4.4. Two-phase Aqueous Liquid–Liquid Extraction

Analogous techniques to the solvent extraction of antibiotics[33] have been investigated[34] at laboratory and pilot scale for protein recovery, utilising phase components which, unlike non-polar solvents, exhibit low interfacial surface tensions and dielectric constants and consequently do not denature biomolecules. Macromolecules such as proteins and particulates such as cells and cell debris can be concentrated in one of the phases by careful selection of the type of water-soluble polymer used (such as polyethylene glycol), concentration and molecular weight, solution pH, addition of inorganic ions (e.g. chlorides, phosphate) and temperature. The effective recovery of a macromolecule is favoured by selecting conditions which maximise the concentration and/or volume of the product-rich phase. The degree of separation is defined as:

$$G = K(V_U/V_L)$$

where K is the protein concentration ratio between the upper and lower phases, V_U is the volume of the upper phase, V_L is the volume of the lower phase. The two phases can be separated by centrifugation or gravity.

Potential applications for two-phase systems in fermentation processes include extractive fermentation,[34] microbial extraction and purification,[35–37] including affinity purification.[38] These techniques are still in various stages of development and, whilst some usage in commercial processes will undoubtedly occur, their universal adoption as a recovery technique analogous to solvent extraction of some metabolites remains to be seen.

4.5. Ultrafiltration and Reverse Osmosis

Ultrafiltration is widely used to remove water, inorganic and small organic molecules from protein and enzyme solutions.[39–41] However, whilst UF membranes are in principle capable of fractionating molecules according to their sizes, this application is often restricted because of the random sieving effects of both the concentration polarisation gel layer (accumulated solute molecules) at the membrane surface and molecules adsorbed within the membrane pores. The two major application areas for UF separations requiring the largest plant sizes (500–2000 m^2 membrane area) are the concentration and fractionation of whey proteins and the concentration of extracellular enzymes in fermentation broths. The advantages of UF as a protein concentration and primary purification process include:

(1) Minimum usage of chemical agents
(2) Low operating temperatures
(3) No phase transition
(4) Batch or continuous operation
(5) Easy scale-up
(6) Potential for sterile separations.

At laboratory and pilot plant scales (up to 2–10 m^2 membrane area) hollow fibre and flat sheet plate stack membrane geometries are commonly used. On the industrial scale, and particularly for whey and vegetable juice concentration, tubular membranes (10 mm to 15 m diameter) as well as plate stack systems are used.

Analytical treatment of cross-flow processes relates solvent and solute flux rates during concentration and diafiltration with system parameters such as cross-flow velocity, solution kinematic viscosity, transmembrane pressure and membrane pore diameter using combinations of the pore model (for low solute concentrations) and the gel polarisation model for high solute concentrations.[39] Since such equations utilise and rely on the experimental determination of a mass transfer coefficient, in most practical cases experimental correlations of flux rates with concentration and cross-flow velocities are usually made. Typically, flux rate increases with transmembrane pressure drop and cross-flow velocity but falls with increasing solute concentration and fouling. Hence a minimum area system will require high cross-flow velocities and, within the limits of gel polarisation layer resistance, a high transmembrane pressure drop. However, reduced capital costs will be offset by increasing pumping costs.

The batch feed and bleed configuration (Fig. 2) may be modified into a continuous multi-stage cascade configuration. Feed and bleed operation is advantageous since cross-flow velocities are independent of feed rate. Batch concentration is preferred to continuous operation since the lower average solute concentration results in a higher flux and lower surface area system. However, where continuous outputs or low residence times are required, continuous operating modes may be used. Membranes with nominal cut-offs of 10 000 MW and above are used for enzyme concentration. Solutions are chilled to 5–10°C, low-shear centrifugal or positive displacement pumps are preferred and transmembrane pressure drops are typically 2–5 bar. Flux rates of 20–50 litres/m^2-h are reached at the commencement of concentration but often fall to 5–10 litres/m^2-h at solute concentrations of 15–30% w/w. Concentration ratios depend on the initial solute levels but values of 10–50 are common. Energy usages, which vary with concentration and membrane module geometry, are typically 6–16 kW/m^3 permeate-h, the higher figures occurring in tubular systems. Recovery, after concentration, of over 95% of the initial enzyme activity can often be achieved. It may be possible to increase the flux rates by judicious choice of pH, salt concentration, the presence of uncharged small molecules and more importantly different membrane materials, since these may have varying degrees of protein adsorption.[39]

Since retained enzyme or protein purity increases during concentration as small molecules are removed, membranes with the highest MW cut-off should be selected. Once the limiting retentate concentration has been reached, further impurity reductions can only be achieved by adding water to replace that lost with the small molecule impurities in the permeate. The alternative use for this operating mode, termed diafiltration, occurs when high recoveries of membrane-permeable species are required, for example intracellular enzyme recovery from cell homogenates. The optimum strategy selected for diafiltration relates to a requirement for minimum membrane area or minimum water usage.[41] Other uses for UF in protein recovery processes from fermentation broths include salt removal, e.g. after ion exchange chromatography, and concentration, e.g. after gel filtration or before lyophilisation.

Reverse osmosis is usually used to concentrate, without purification, protein solutions particularly when maximum dry solids levels are required in the retentate, e.g. whey powder. It is more economical than evaporation at solute concentrations up to 20% w/w. Transmembrane pressure drops of 50–70 bar are required and typical flux rates are 20–50 litres/m^2-h. Examples include the pre-concentration of egg white and raw whey.

4.6. Chromatography

4.6.1. Introduction

Although synthetic polymer (e.g. polystyrene, polymethacrylic acid) ion exchange resins have protein adsorption capabilities, their highly charged and hydrophobic nature restricts their protein fractionation capability and tends to denature labile functional proteins. Therefore purification by tedious and relatively ineffective precipitation techniques was the only large-scale method available until, in the late 1950s, gels based on the natural polymers cellulose and dextran, and later agarose, were produced. These hydrophilic matrices have low denaturing effects on proteins, are highly porous (pore access to >1 000 000 MW macromolecules), can support suitable ion exchange and other functional sorbent groups and by cross-linking can be made to have sufficient mechanical strength to be used in bead form in columns up to 100 cm in depth. A wide range of gel chromatographic techniques are available for protein recovery and purification,[42–46] and these, their method of operation, and usage are summarised in Table 7. These techniques can be relatively expensive and so are primarily used to obtain high value, high purity proteins from fermentation, cell culture and blood plasma feedstocks. However, one process for the recovery of whey protein, for use as an egg-white replacer, is based on the use of cellulosic ion exchange media, as is the process to recover protein from meat processing wastes (see Section 5).

Column chromatographic procedures are essential for gel filtration, HPLC and chromatofocussing, and usually used for ion exchange purification. To maximise effectiveness, feedstocks should be particulate-free, have relatively low viscosity in order to minimise frictional drag on the 'soft' natural polymer resins, be free of components that may prevent protein sorption or decrease the useful life of the resins, and be low in organic solvent and salt content. A wide range of column components, for use in both single and multiple stacks systems, are marketed for laboratory to production scale (1000 litre resin volumes), together with feed pumps, vessels and a variety of automated sample collectors and analytical devices.

Table 7
Chromatographic Techniques for Protein Recovery

Technique	*Separation basis*	*Sorption media*	*Usage examples*
Adsorption	van der Waals forces, H bonds, polarities, dipole moments	Hydroxyapatite Titania Calcium phosphate	Sorption from crude feedstocks Fractionation
Ion exchange	Molecular charge	Cellulose, dextran, agarose matrices Inorganic sorbent matrices with reactive coatings	Concentration by sorption from crude feedstocks Fractionation
Gel filtration	Molecular size	Dextran polymers	Desalting Removal of solvents Fractionation
Hydrophobic interactions	Protein surface hydrophobicity	Agarose gels with hydrophobic groups	Sorption from crude feedstocks Fractionation
Chromatofocussing	Isoelectric point	Agarose plus buffers	Fractionation
Affinity chromatography	Specific surface structure	Agarose matrices with affinity ligands	Fractionation Sorption from crude feedstocks
Reverse phase liquid chromatography (RPLC)	Hydrophilic and hydrophobic interactions	Silica with hydrocarbon coatings	Fractionation

4.6.2. *Ion Exchange Chromatography*

Proteins are reversibly bound to ion exchange resins in solutions of pH usually at least one unit away from their pI point. The choice of solution pH depends on protein stability, competing impurities and binding capacities. At pH levels above the pI point, negatively charged proteins bind to anion exchange resins, and below the pI point positively charged proteins bind to cation resins. Test-tube experiments under varying pH conditions are used to determine the conditions for acceptable binding levels of either the desired protein or all of those present. To facilitate effective elution, the loading pH chosen should be as near to the binding point as possible. Since weak ion exchangers are charged over only a limited pH range (6–9) and labile proteins are affected by very acid conditions, weak ion exchangers, based on diethylaminoethyl (DEAE) and carboxymethyl (CM) functional groups, are often used. Strong ion exchangers are used for total protein binding and/or where extremes of pH are needed to ionise groups on a protein to give it a net charge. The selection of the matrix will depend upon the protein molecular weight, the flow rate required, ionic strength and resolution. In general, ionic strengths should be no higher than that required to bind the desired protein since this maximises resolution (purity). Elution is achieved by combinations of pH changes (towards the protein isoelectric point) and increasing ionic strength (to compete for exchanger binding sites). Continuous or step change gradients of pH and ionic strength are commonly used.

Ion exchange procedures may be used for bulk protein removal (concentration) or fractionation. For the latter, purification factors may be typically 5–30 × but this will often be accompanied by dilution during elution. Ion exchange gels have binding capacities of typically 10–50 g of protein per litre of resin. Bed heights of 15–30 cm are often sufficient for effective recoveries.[48] Scale-up is achieved by increasing column cross-sectional areas. Ion exchangers based on cellulose and dextran polymer matrices have poor physical strength and consequently are often used in batch mode, i.e. dispersed in a solution of the desired product. Following binding, gels are removed by filtration or centrifugation[47] and desorbed in batch or continuous mode. Agarose-based resins, with highly cross-linked matrices and increased gel strength, can tolerate flow-rates of up to 12 bed volumes/h in deep columns.[44]

4.6.3. *Gel Filtration*

Separation is achieved by the partial retention of small molecules which diffuse into the matrix pores whereas large molecules pass more quickly through the bed with the carrier solvent. Since gel filtration leads to separation with dilution, ultrafiltration is commonly used afterwards to concentrate the product.

Gel filtration competes with ultrafiltration and dialysis techniques for salts removal together with separations of molecules much larger or smaller than the desired product. It is the only technique readily available for separating similar protein molecules of equal charge but differing by relatively small size variations. Sample volumes vary between 2% and 30% of the bed volume and bed heights vary between 30 cm and 100 cm.[48] Gel filtration is often used after ion exchange processes as an additional purification stage or on the small scale as a preliminary separation stage between macromolecules and salts.

4.6.4. Adsorption Chromatography

A number of inorganic materials exhibit natural protein adsorption and fractionation capabilities, including kieselguhr, bentonite, titania, porous glass, alumina, calcium phosphates and their derivatives, barium sulphate and magnesium pyrophosphate.[49] The recovery of whey protein[50] and bromelaine enzyme from pineapple juice are potential applications, as is the use of these robust steam-sterilisable media, which can be modified into ion exchange and affinity chromatography forms,[51] to recover proteins from clarified and unclarified fermenter broths.

Crystalline hydroxylated calcium phosphate complex (hydroxyapatite) is an inorganic material which adsorbs protein either on its own or when combined with other media. Sorption capacities of up to 50 g/litre have been measured with bovine plasma, albumin and intracellular *E. coli* enzymes.[52,53] However, when made in a crystalline form, it has limited robustness and suffers from loss of flow-rate.[54] Hydroxyapatite in high-pressure liquid chromatographic columns can be used to recover monoclonal antibodies.[55] The use of silica-based adsorbents in fluidised bed systems has been proposed for the recovery of intracellular enzymes from unclarified cell homogenates.[56]

4.6.5. Hydrophobic Interaction Chromatography

In this sorption technique, proteins with similar or negligible charge may be separated by virtue of the different binding strengths of their surface hydrophobic groups to a hydrophobic gel.[57,58] This is in essence a salting-out procedure, so relatively high ionic strength salt solutions are used to encourage solute–sorbent attachment. Ammonium phosphate and sulphate salts are effective in promoting hydrophobic interactions. Elution conditions leading to fractionation include changing the salt to one having a reduced salting-out effect, lowering the ionic strength, lowering the eluate polarity by adding a mildly hydrophobic solute, and raising the solution pH. A large number of proteins exhibit hydrophobic interaction properties with modified agarose and acrylamide gels. The combination of batch hydrophobic interaction chromatography in two-phase polymer systems for protein extraction from unclarified broth homogenates has been demonstrated.[59]

4.6.6. High-performance Liquid Chromatography (HPLC)

HPLC is a relatively new column flow procedure which utilises very small rigid beads (10–100 μm dia.) to provide much higher levels of fractionation of multicomponent feed streams than conventional ion exchange or gel filtration media. Whilst originally an analytical technique, HPLC equipment is now available with resin beds of up to 100 litres volume which are capable of handling kg/h solute flow rates.[60] HPLC is seen as a powerful technique which is capable of separating a wide variety of fermentation products.[61,62] Applications involve difficult separations of molecules of similar size and composition, e.g. purifying a synthetic peptide to chemical homogeneity.

HPLC separations are achieved using two main modes, namely normal and reverse phase, with the latter normally used for protein separations. In this mode the retention of a protein molecule in the mobile phase is dependent on its size and hydrophobic and hydrophilic group interactions with the stationary phase and the carrier solvent. In general, elution varies with protein hydrophobicity, with those proteins containing the most hydrophobic residues showing the greatest attraction for the column

Table 8
Examples of Large-scale Applications of Affinity Chromatography

Purified product	*Ligand/Matrix*
Lectins	*N*-Acetyl-D-glucosamine/Agarose gel
Albumin	Triazine dye/Agarose gel
Antithrombin III	Heparin/Agarose gel
Glycerokinase enzyme	Triazine dye/Agarose gel
Interferons	Monoclonal antibodies/Agarose gel
Monoclonal antibodies	Monoclonal antibodies/Agarose gel
Immunoglobulin	Protein A/Agarose gel

packing.[63] Fractionation may be achieved under isocratic conditions (constant mobile phase composition), but gradient elution involving gradually increasing organic solvent concentration (e.g. acetonitrile) in an acidic aqueous phase may achieve sharper chromatographic peaks. Typical column packings for proteins are octadecylsilyl and butylsilyl functions bonded to silica supports. Some biologically active proteins, e.g. enzymes, can be separated by reverse-phase liquid chromatography but the use of organic solvents for elution may be detrimental to the most labile proteins. In this instance hydrophobic interaction chromatography should be tried.

4.6.7. Affinity Chromatography

Affinity separation techniques are based on the 'natural' binding forces which occur between biological macromolecules such as proteins and specific complementary molecules or structures, called ligands.[64–66] These binding forces, which include van der Waals, electrostatic, hydrogen bonding and hydrophobic forces, must be strong enough for binding and yet sufficiently reversible for product recovery. Affinity chromatography is a widely employed column or batch technique used at the research level and commercially to recover at high purity levels a range of proteins (see Table 8). It is the only separation technique available for labile proteins which can achieve the very high purity levels (approaching homogeneity) required for health care applications. Purification factors of 500–1000 × are typical for many protein–ligand systems.[55] However, binding capacities are low, examples indicating 2–7·5 g/litre of gel for interferon purification.[31,65]

There are four major problems limiting the wider use of affinity chromatography, these being ligand cost, adequate ligand attachment methods to the matrix, the effort required to determine the most appropriate combination of matrix, ligand and its mode of attachment for a specific separation, and the need to prepare relatively pure protein solutions prior to affinity techniques being used. Affinity media costs are staggering at present: £10 000/g for protein A,[67] production costs of $1200–1500/g for monoclonal antibodies, and costs of more than £230 000/litre of monoclonal antibody ligand gel.[55]

In order to overcome the above problems for large-scale applications, interest has focussed on non-specific affinity chromatographic media based on immobilised triazine dyes.[68,69] In the use of such media for the extraction of proteins from clarified cell homogenates, purification factors are often an order of magnitude lower than those for protein ligand affinity separations, with media binding capacities of around 2–3 g/litre.[54] Batch adsorption and elution techniques, often carried out in the same vessel, are common with short wide columns providing rapid throughputs and low process times.[65]

A wide range of other experimental affinity separation techniques are possible including: affinity partitioning; affinity precipitation; ultrafiltration affinity purification; high-performance liquid affinity chromatography (HPLAC).

Affinity partitioning utilises two-phase aqueous liquid–liquid extraction systems to separate protein product and cell debris, ionic species, nucleic acids, etc. By complexing the protein to one of the polymers, it enhances the degree of separation and protects the protein from bulk contaminants in the other phase. Thus phase partition coefficients can be raised significantly, e.g. from 1 to above 20 for glucose-6-phosphate dehydrogenase in a two-phase system utilising PEG bound Cibacron blue dye ligand.[64] Scale-up is relatively straightforward. Processing times are fast, and high binding capacities per unit volume are possible. Industrial scale recovery and purification of phosphofructokinase, an intracellular enzyme from baker's yeast, utilises this technique. Affinity precipitation leads to the precipitation of the desired protein by allowing it to form insoluble complexes with so-called bis-ligands or with affinity ligands attached to polymers.[66] In ultrafiltration affinity chromatography, the product solution and the solution containing a water-soluble affinity ligand–polymer complex are separated in cross-flow mode by an ultrafiltration membrane. This provides a possible means to protect the affinity complex from particulates and large molecule contaminants. HPLAC displays high resolution at analytical levels but low binding capacities, e.g. 0·33 mg/g on silica Procion blue affinity media.

4.7. Protein Separations in Electrical Fields

Two electrical field effects are used to purify soluble proteins. Electrodialysis is used for both salting out (precipitating) proteins by the addition of neutral salts, and desalting (with salt recovery) protein solutions. Electrophoresis and isoelectric focussing separate protein molecules according to their net charge.

4.7.1. Electrodialysis

In electrodialysis, an electric current transfers ionic components into or out of protein solutions flowing between parallel alternate stacks of

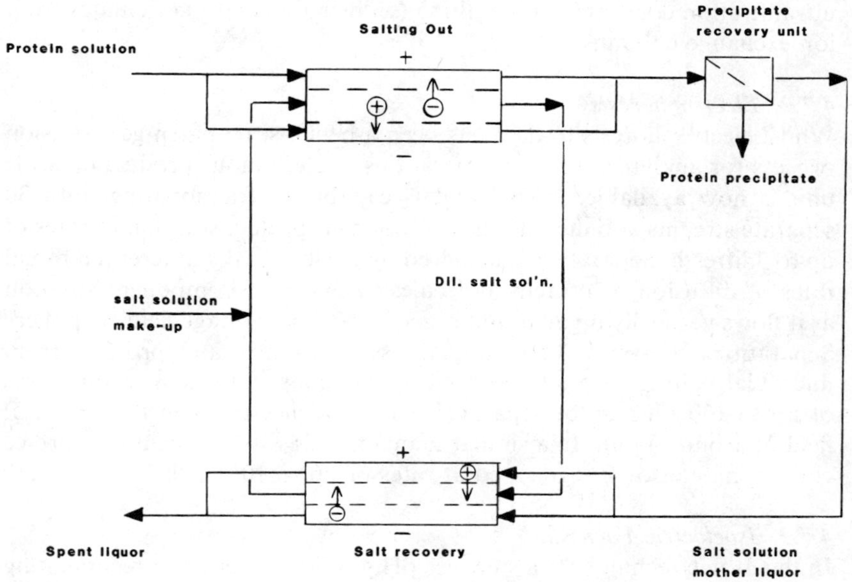

Fig. 4. Scheme for protein salting-out using electrodialysis.

anionic and cationic membranes (see Fig. 4). Since these have molecular weight exclusion limits of approximately 400 daltons, no protein transfer occurs, membrane fouling is minimal and flux rates remain relatively constant during processing. Electrodialysis is an alternative to batch and continuous salt addition for salting-out proteins, and to dialysis, ultrafiltration and gel permeation chromatography for desalting, although the last three must be used for the recovery of solvents and other non-ionic precipitating agents, e.g. polyethylene glycol. The many advantages claimed for electrodialysis in protein separations include fast and controlled removal of salts, no product dilution, low membrane area requirements, negligible product adsorption, easy salt recovery in the same unit, independence of diffusion and protein concentration, and the use of lower concentrations of salting-out agents than are required by conventional direct addition processes.[70]

On a large scale, electrodialysis is used for demineralising whey. Other applications in the recovery and purification of high-value therapeutic proteins from microbial fermentation processes include: salting out (with salt recovery) of impure products from clarified homogenates; desalting ultrafiltration concentrates; desalting (with salt recovery) of eluates from ion exchange columns.

4.7.2. Electrophoresis

Whilst electrophoresis to date has been a bench-scale batch gel diffusion process for analytical protein separations, a continuous production scale unit is now available.[71] This unit is capable of fractionating, into 30 separate streams, a buffered multi-component protein solution at rates of up to 3 litres/h. Separation is achieved by exploiting the differential radial rates of diffusion of protein molecules in the multi-component solution as it flows vertically up an annulus across which a voltage field is applied. Separations of factor VIII, animal tissue enzymes and proteins from microbial cultures are all potential applications.[72] Protein throughputs of up to 100 g/h may be separated with residence times in the electrical field of around 1 min. In a similar manner, cells with different net surface charges may also be segregated at rates of up to 10^{13} cells/h.

4.7.3. Isoelectric Focussing

In this bench technique[73] a stepwise pH gradient is set up in recirculating ampholytes flowing in parallel channels between microporous membranes in an electric field. Thus different protein molecules added to one of the channels will move according to their net charge towards the anode or cathode until they remain in a channel in which the pH approximates to their pI values. Electrodialysis can be used to recover the expensive ampholytes.

4.8. Novel and Experimental Protein Separation Techniques

The procedures outlined above are all used, to a greater or lesser extent, in commercial scale recovery processes and analytical scale separations. In various combinations they can be used to separate and purify essentially any protein. In addition to these, a number of other separation techniques, both speculative and already demonstrated at laboratory scale are possible. These include: foam flotation;[74] separations using magnetic materials;[75] combined field separations such as electrically enhanced membrane processes, and chromatography in electrical fields.[76]

There is probably great potential for combined field techniques for high-value protein recovery processes, the justification being that reductions in the number of stages will invariably reduce capital costs and increase yields. Which techniques will become economically feasible and be used on a large scale remains to be seen.

5. PROTEIN RECOVERY PROCESSES: NON-FERMENTATION FEEDSTOCKS

The purpose of the remainder of this article is to illustrate how the different separation methods described previously are effectively combined together in typical protein recovery processes. This section considers protein recovery from non-fermentation feedstocks.

5.1. Recovery of Waste Proteins from the Meat Industry

The meat processing industries generate large tonnages of recoverable low-value waste products[77] including proteins and fats found in waste blood and water wash streams at concentrations of approximately 0·1%. In the Alvatech process,[77] purified sodium lignosulphonate is used to precipitate, at pH 3, soluble proteins which, together with insoluble fats, are removed by dissolved air flotation. The solids overflow contains 5–15% dry matter and a 70–90% reduction in effluent BOD is achieved. In other processes[78–80] insoluble fats are removed by air flotation, and soluble proteins are precipitated by adding acid to pH 3–5, with flocculants added to aid sedimentation. The remaining 0·1% level of soluble proteins are removed in cellulose gel ion exchange columns (binding capacity 50 g protein/kg gel), the eluate being neutralised, coagulated by heat shock and dried. In a more recent three-stage process, chemical dosing, flocculation and tilted plate separators incorporating dissolved air flotation are used for continuous recovery of the proteins and fats in a 240 m^3/h chicken processing factory effluent stream.[81] BOD, suspended solids and fats are reduced by 80%, 95% and 95% respectively. The flotation layer consisting of 15% w/w solids is sold as an animal feed and provides income for a 2 year capital pay-back period. It is apparent that economically viable processes for the recovery of food proteins from waste streams containing only low concentrations of proteins must utilise relatively simple techniques such as chemical precipitation.

5.2. Proteins from Milk

Three protein products are recovered from milk and milk waste streams: casein, whey powder, and whey protein isolate.

5.2.1. Casein

Casein, a mixture of related phosphoproteins rich in essential amino acids, comprises around 3% of bovine milk. It is widely used in the food industry as a functional additive. It is isolated from skimmed milk in a

Fig. 5. Typical large scale ultrafiltration installation. Danmarks Protein, Denmark: ultrafiltration of whey; capacity 1300 tonnes/day. (Courtesy DDS RO-division).

three-stage process consisting of precipitation at acid pH, precipitate dissolution and spray drying. However, the casein products are found to be denatured and so do not fully exhibit the properties of the native state found in milk. In a recent patent[82] native casein may be produced by concentrating skimmed milk up to six times by ultrafiltration (55 000 daltons MW cut-off), freezing and storage at $-8°C$, followed by slow thawing and 5000*g* centrifugation to recover the cryocasein precipitate.

5.2.2. Whey Powder

Whey, the aqueous liquid waste stream from cheese production, consists of around 5% lactose, 0·5% ash and 0·8–1% protein. Whey powder used as an animal or human food consists of whey protein, lactose and some ash, and is produced by spray drying whey concentrate. Reverse osmosis (RO) is usually used for the initial water removal stages, followed by either direct spray drying or further concentration using multiple-effect evaporators. The optimum RO concentration range is from 6% to 12% dry solids basis; around 4 kW/m^3-h of water permeate is required and 1979 operating costs are given at £1·43/m^3 of permeate.[83] Ultrafiltration is now commonly used to produce whey concentrate solids enriched with up to 75% protein and significantly reduced lactose sugar levels (Fig. 5).

5.2.3. Whey Protein Isolate

Whey protein isolate is a functional protein increasingly used by the food industry as a substitute for egg white. Since very low lactose levels and relatively mild process conditions are required, primary recovery from whey using chromatographic techniques is required. Stauffer in 1972 developed[84] a fairly complicated process based on dextran molecular sieves. More recently, Bio-Isolates have developed a simpler, more robust process (see Fig. 6) to recover proteins directly from whey using cellulosic ion-exchange media in batch adsorption mode. Whey throughputs of approximately 5000 m^3/day yield 1 tonne/day of protein using 50 m^3 batch ion-exchange vessels.[85] A similar product has been developed in France by Rhône-Poulenc using columns of silica–QMA ion exchange resins.[86] Protein binding capacities are given as 70 g/kg of resin, with total production costs for a 100 m^3/kg plant stated at 23 FF/kg of 90% protein purity isolate (zero value assumed for whey).

5.3. Proteins from Agricultural Crops

Agricultural crops are the largest source of proteins in the world; cereal crops for example often contain over 12% protein and soya bean flakes contain around 50% protein. These proteins are of low value, are recovered by relatively simple precipitation and physical separation processes, and are used mainly for animal feed supplements and as functional additives to human foods (e.g. gluten).

5.3.1. Soya Proteins

The major source of animal feed protein and low-cost functional proteins for the food industry is soya beans. These are cleaned, dried, cracked open and the hulls removed, conditioned with steam, flaked and then treated with solvent (usually hexane) to remove the majority of the 90% oil content of the whole bean. After solvent removal, the flakes are dried and ground to produce meal. For food grade proteins, the desolventised flakes are further processed successively to remove impurities; the block flowsheet for the production of protein isolate is shown in Fig. 7. This is the highest grade of concentrate, consisting of 96% protein. To produce a product with good functional properties, the protein in the flakes is dissolved in caustic (pH 8–9), and the waste carbohydrate solids are removed. Food grade acids are then used to alter the solution pH to the protein pI point of 4·5 where precipitation occurs. High yields of pure protein are possible because of the advantageous U-shape of the protein solubility curve about its pI point. Precipitate washing and neutralisation followed by spray drying produce the isolate.

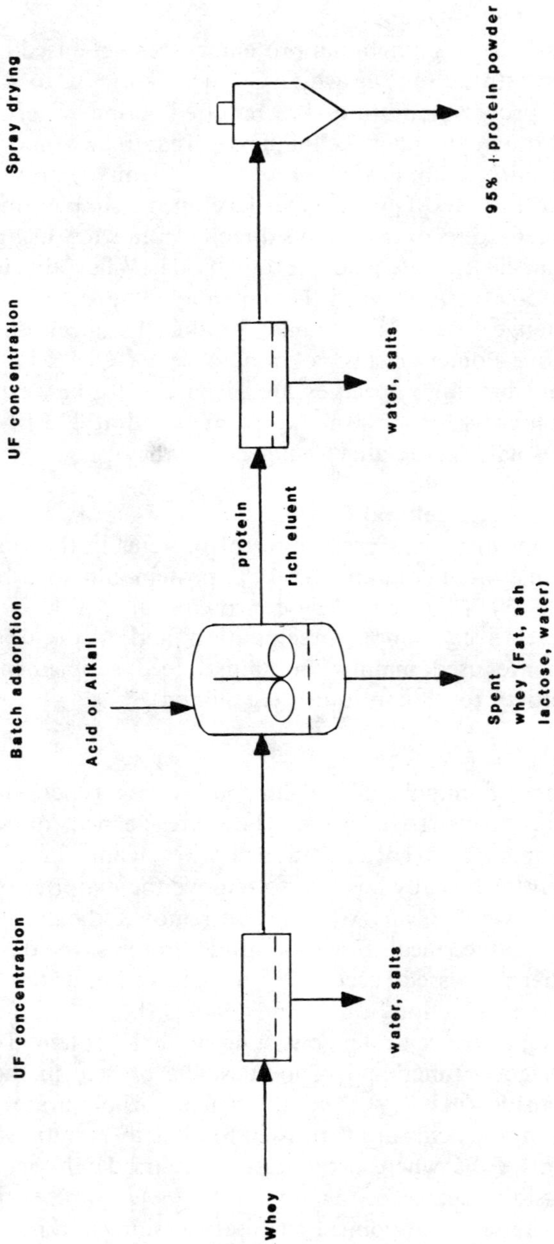

Fig. 6. Ion-exchange recovery of whey protein isolate.

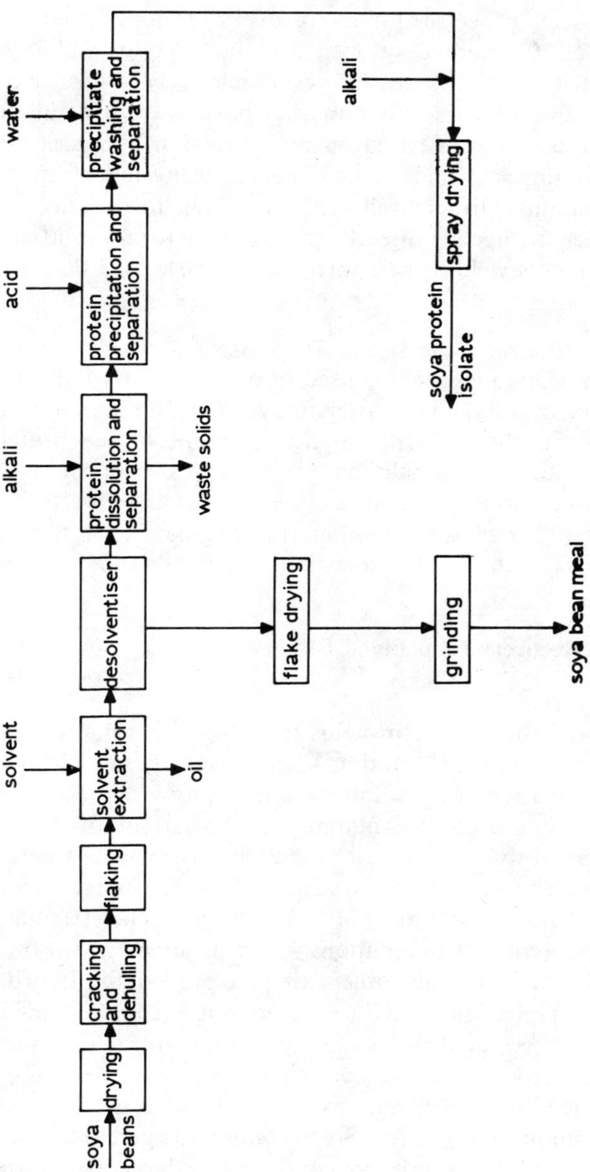

Fig. 7. Block flow-sheet for soya bean processing.

5.3.2. 'Leaf' Proteins

Commercial scale processes for the recovery of animal feed leaf protein concentrate (LPC) were developed and installed in California[87] and Hungary[88] in the early 1960s. However, relatively low soya costs, high energy costs and an unreceptive market have prevented full comercialisation of these processes based on the extraction by maceration of the protein-containing leaf juice. This is heated using live steam to around 80°C to precipitate protein followed by ageing to produce aggregates. These can be centrifuged or directly spray dried to form a solid concentrate, the overall process yield being approximately 50 kg of LPC from 1 tonne of leaf feed.

5.3.3. Vegetable Processing Waste Streams

Precipitation techniques may be used to recover nearly half of the 2–3% protein found in potatoes.[89] After maceration and starch removal, the waste juice is acidified to the isoelectric point of the protein, where precipitation occurs. Coagulation of the solids is encouraged by live steam injection to raise the temperature rapidly to 110–120°C. Decanter centrifuges may be used to dewater the coagulate to a 40% dry solids cake which is then dried. Dewatering may also be achieved using ultrafiltration.[90]

5.4. Protein Recovery from Blood Plasma

5.4.1. Human Blood

The recovery of therapeutic proteins from blood plasma is an important industrial scale operation[91] and as such provides a model for biological protein fractionation from solutions containing relatively few complex non-proteinaceous organic contaminants. Variations of the Cohn cold ethanol precipitation method[3] are used. In this, plasma fractionation is carried out in a five-parameter system (pH, ionic strength, temperature, protein concentration and alcohol concentration), the essential objective being to create sequential conditions in which some protein fractions are maintained in solution whilst others are precipitated out. In order for this and similar precipitation procedures to provide adequate yields, solubility differences between the dissolved and precipitated fractions should differ by at least two orders of magnitude. As an indication of this the solubility of albumin in 10% v/v ethanol is about 5 g/litre whereas in 30% v/v ethanol it is about 0·02 g/litre.[3] Since continuous precipitation processes give a better degree of process control than large batch vessels, the continuous small volume mixing process (CSVM) was developed[10,11]

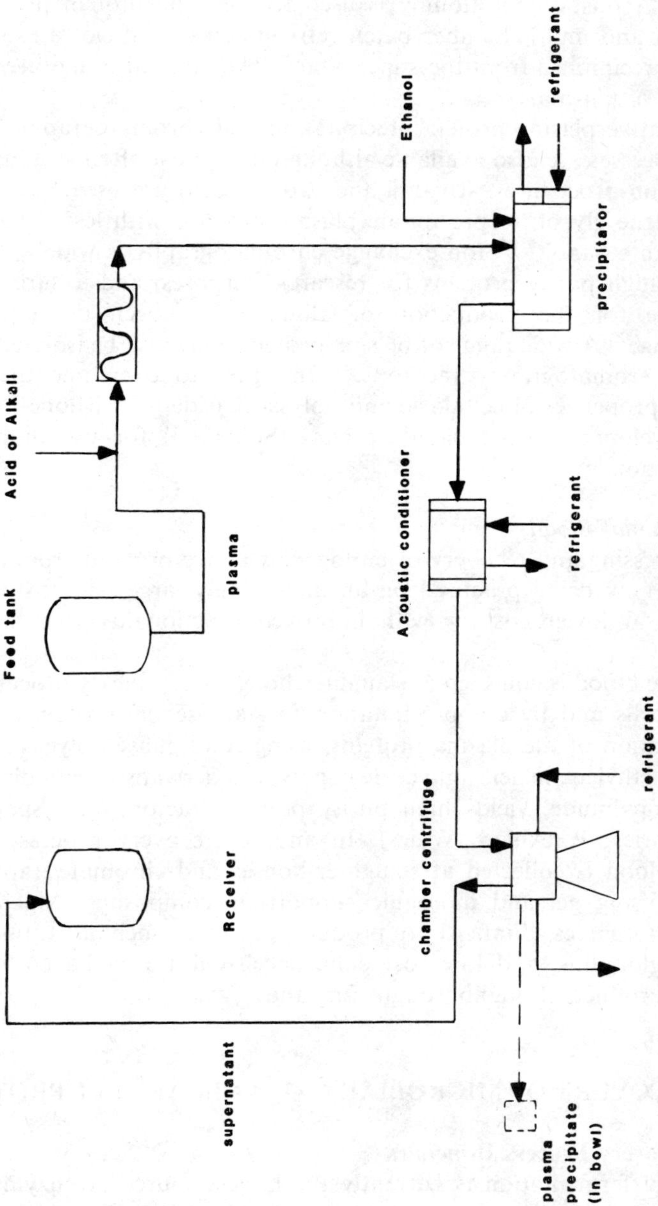

Fig. 8. The continuous small volume mixing (CSVM) process module.

(Fig. 8). Acoustic conditioning is used to 'age' the protein precipitate fractions, and multi-chamber batch refrigerated centrifuges are used to recover precipitates from the supernatant. The albumin fraction yield is 25 kg/m^3 of plasma.

Alternative plasma protein precipitation and chromatographic separation processes are also available although used most often as additional purification procedures. Rivanol and caprylic acid are used,[9] as well as polyethylene glycol, to precipitate plasma proteins with less denaturing effect than ethanol.[7,92] Ion exchange chromatography is widely used to produce high purity proteins for research purposes, and a 'large scale' procedure for the production of albumin has been developed by Pharmacia.[93] A wide range of plasma proteins may also be isolated using affinity chromatography methods.[94] In order to overcome the poor physical properties of cellulosic and polysaccharide gels, Rhône-Poulenc have developed a silica-based sorbent (Spherocil) for use in plasma fractionation.[95]

5.4.2. Animal Blood

The processing and recovery of biologically active proteins from animal blood is now being practised on an increasingly large scale to provide products at lower cost or with improved functionality than existing sources.

Porcine blood is collected at slaughter-houses and quickly processed to remove cells and by cryoprecipitation to stabilise the protein content. Fractionation of the plasma proteins, using solid-phase polyelectrolytes based on ethylene/maleic anhydride copolymers derivatised with dimethylaminopropylimide, yields high purity porcine Factor VIII (Speywood Laboratories, Wrexham, Wales). In another recovery process, sterile bovine blood is collected at slaughter-houses and chromatographically purified using gel and inorganic support/gel composites (Macrosorb, Sterling Organics Limited) to produce products such as transferrin, immunoglobulins, and low cost defined cell culture media (Advanced Protein Products, Loughborough, England).

6. RECOVERY OF MICROBIAL AND ANIMAL CELL PROTEINS

6.1. Recovery Process Principles

Microbial fermentation is currently the largest source of enzymes and increasingly becoming an important source of health care proteins. Tissue culture is also a source of biologically active proteins, particularly mono- and poly-clonal antibodies. Since virtually all microbial proteins are produced by suspended cell fermentations, the output stream leaving the fermenter consists of a slurry of whole cells (up to 3% dw) in aqueous solution containing inorganic salts, carbohydrate substrates, the excreted enzyme or protein product, and other enzymes, proteins and organic molecules produced by the microbe. Intracellular enzyme and protein products are retained by the microbe during fermentation and must be extracted from inside the microbe during the recovery process. When animal cells are used as sources for biologically active proteins then, depending on the cell type involved, both suspended cell or immobilised cell culture systems may be used. However, culture volumes are relatively small, around 10 m^3 being the maximum to date for suspended animal cell systems. Air-lift fermenter sizes of 1 m^3 are reported for hybridoma cultures producing kilogram quantities per annum of monoclonal antibodies.[96]

The generalised recovery sequence (commonly termed downstream processing) for extracellular and intracellular soluble products from suspended cell microbial fermentation broths consists of cell processing and removal, primary separation to remove water, fermentation nutrients, the bulk of the non-proteinaceous materials such as nucleic acids, and where possible very dissimilar proteinaceous components. Then follows purification to remove the remaining amounts of dissimilar contaminants as well as components similar to the product. Following this, finished product treatment includes the use of stabilisers, encapsulation techniques or immobilisation for bulk enzymes, and sterile preparations for therapeutic proteins.

The general principles and problems of microbial protein recovery are fairly well documented in the literature,[97–102] especially with regard to the recovery of intracellular products and the scale-up of laboratory and pilot-scale separation techniques. A generalised scheme of the many sequential separation techniques used in soluble protein recovery processes is shown in Fig. 9. For clarity, inter-stage operations such as concentration and desalting by ultrafiltration have been omitted. The process sequences selected[99] will be a compromise between step yield, number of steps and product purity requirements, and care must be taken to ensure that the choice and operating conditions of upstream separation stages do not compromise the selection or efficiency of later purification techniques. Consequently one of the aims of existing developments in downstream processing[54] is to increase yields by using techniques such as direct

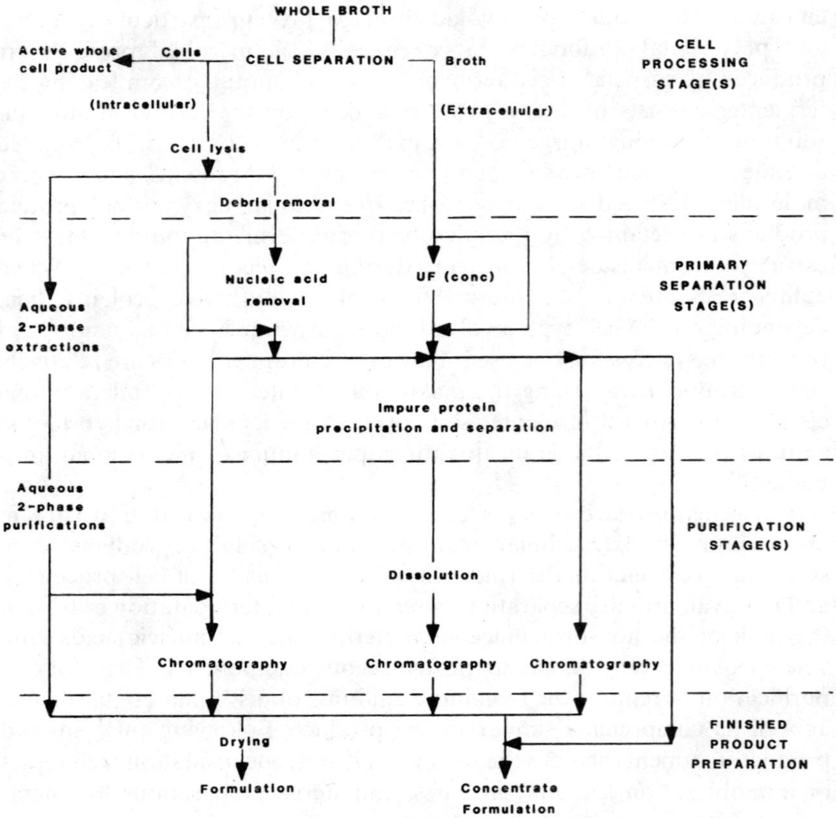

Fig. 9. Protein recovery procedures from fermentation broths.

chromatographic broth or homogenate extraction to reduce the number of stages, and by applying continuous recovery techniques using both conventional recovery steps[103] or newer techniques such as aqueous two-phase separations. These approaches may also reduce the costs of downstream processing which, for high purity intracellular enzymes, can approach three times the fermentation product cost.[100]

Microbial fermentations produce two general classes of polypeptide products, namely single-cell protein and soluble proteins and enzymes. It is convenient to consider the recovery processes for these separately.

6.2. Single-cell Protein

Single-cell protein (SCP) is the name given to the total dry matter of

Table 9
Microbial SCP Recovery Methods

SCP type	*Whole broth treatment*	*Cell–broth separation*	*Cell treatment*	*Drying*	*Product treatment*
Yeast	—	Centrifuges Filtration	Water washing	Evaporation Drum drying Spray drying	Grinding
Filamentous	—	Filtration Press dewatering	—	Drum drying	Pelletising
Bacterial (ICI)	Thermal disruption Acid treatment	Centrifugation	—	Flash drying	Grinding
Filamentous (food grade)	Heat shock	Filtration	Texturisation	—	Freeze storage Flavour addition

bacteria, yeasts, moulds and higher fungi when used for animal feed or human food.[104] The essence of the production process is feedstock treatment, microbial fermentation in batch or continuous culture, cell–broth separation, drying and often milling. SCP animal feed is a low-value product traditionally produced from molasses or more recently from waste carbohydrates. However, large-scale production using chemical feedstocks such as methanol (ICI's bacterium *Methylophilus methylotrophus*) and hydrocarbon substrates (genus *Candida*) is practised, the latter particularly in the USSR where annual production levels of $>10^6$ tonnes are estimated.[105]

Typical fodder yeast recovery process steps are shown in Table 9. Nozzle bowl centrifuges continuously separate the yeast (approximate diameter 8 μm) from the broth or wash water. Two or three wash stages are used, with wash water volumes approximately equal to the sludge concentrate volume. Vacuum evaporation increases the dry solids concentration from 15% to around 30%, and this is followed by roller drum drying and, if required, milling to a powder. Fungal SCPs are also available, the major production plant to date being the 10 000 tpa Pekilo process in which *Paecilomyces varioti* metabolises in continuous culture the waste cellulosic carbohydrates in spent sulphite liquors.[106,107] Filtration instead of centrifugation is required to separate the filamentous mycelium from the broth. This is followed by press dewatering and drying. Other fungal strains metabolise carbohydrates such as lactose in whey. Small plants have been installed in France.[108]

The ICI Pruteen plant[109] can produce up to 50 000 tonnes/annum of bacterial SCP using a 1500 m^3 pressure cycle fermenter. In order to facilitate cost-effective cell separation and drying, heat treatment and acid conditions rupture the bacteria; after this, flocs of sufficient size develop to allow high-rate centrifugal separation and continuous flash drying (see Table 9). Since water throughputs are immense (around 200 m^3/h for a 50 000 tonnes/annum plant), spent broth recycle is necessary and this imposes a severe sterility requirement on the separation process. The problems and cost of water removal in the downstream process, as typified by the ICI process, have been highlighted recently.[110]

The production of fungal SCP (*Fusarium graminearum*) for use as a high-quality protein food is now an important development in the UK.[111] Nucleic acid reduction is an essential feature of a food-grade SCP recovery process. Hydrolysis, rupture and precipitation, or heat treatment, all reduce cell nucleic acids[112,113] to mononucleotides which diffuse out of the cell and can be removed in a filtration stage (Table 9). Following this, proprietary texturising techniques are used to prepare a palatable finished product.

6.3. Extracellular Enzyme Recovery Processes

Of the few thousand enzymes known, only 30 or so are commercially important, with the 13 most common types, namely α-amylase, glucoamylase, protease, β-glucanase, cellulase, dextranase, lipase, pectinase, rennet and pullanase, all being extracellular. Annual production of the first three reaches hundreds of tons[114] and fermentations of 80–100 m^3 batch size are commonly used. It is a very cost-competitive market and, although downstream processing techniques are all very similar, process confidentiality prevents disclosure of the detailed technology and proprietary techniques used commercially. A recent review[115] provides a good introduction to industrial enzyme recovery process technology. Since user purity requirements are often very low,[116] minimal purification procedures are required. These basically consist of either ethanol or ammonium sulphate precipitation for solid enzyme products or water and impurity removal by ultrafiltration (10 000–20 000 MW cut-off membranes) for liquid concentrate products.

6.4. Intracellular Enzyme Recovery Processes

A number of factors make the recovery of intracellular enzymes more difficult than that of extracellular products, including:

(1) Product degradation by proteolytic enzymes necessitating relatively fast cell separation, rupture and primary separation stages.
(2) Recovery and purification is more difficult because of the very

low product concentrations and much higher levels of intracellular contaminants released during rupture.

(3) Physical separations are more difficult because of the relatively high density of the cell homogenate after rupture and the small cell debris size.

(4) Product purity requirements are often high.

Industrial fermentation batch sizes range from 3 to 30 m^3 for intracellular enzymes.[115] The major intracellular soluble contaminants are nucleic acids (up to 34% of bacterial cell dry weight (dw) and up to 10% of yeast cell dw) and polysaccharides and precursors (up to 10% of bacterial dw and 15% of yeast dw). Insoluble cell wall and membrane lipids account for up to 15% of bacterial cell dw and 6% of yeast cell dw. Multi-step procedures result in low recoveries; in the process example shown in Fig. 10,[117] the overall yield of the process was only 6%. In this example, 23 × and 3 × purification factors for the ion exchange chromatography and gel filtration steps respectively were achieved. On a larger scale, Higgins *et al.*[103] have demonstrated a continuous process for the recovery of a crude β-galactosidase precipitate from an *E. coli* strain, with a 43 × purification factor for nucleic acid removal and enzyme precipitation.

6.5. Recovery of Biologically Active Proteins

A range of proteins such as hormones, insulin, interferons, albumin and Factor VIII may be obtained from either animal cell cultures or fermentations of genetically engineered recombinant microorganisms.[118–120] Of these, human insulin is commercialised and interferons, lympokines and growth hormones are at advanced stages of clinical trials. *E. coli* has been the host organism for much of the early work, with the expressed foreign protein remaining within the cell as an intracellular product.[121] The recovery techniques used for these proteins, whilst very similar to those indicated above for intracellular enzymes, need to be more precise,[122] since:

(1) *E. coli* quickly and selectively degrades those polypeptides with abnormal (i.e. foreign) conformations via at least 8 soluble proteolytic activities.

(2) Fermentation times are often short (6–24 h), so harvest times for maximum yield are critical.

(3) *E. coli* releases lipopolysaccharide fever-forming pyrogens which must be removed from the final product.

Whole broth → Cell separation → Disruption → Centrifugal debris removal → $(NH_4)_2SO_4$ precipitation → Centrifugal precipitate collection → Dissolution and dialysis → DEAE-cellulose ion exchange → Gel filtration

Fig. 10. *β*-Glucosidase recovery from *Candida wickerhaus*.

Whole broth → Cross-flow microfiltration (↓ Spent broth) → Cell disruption → Cross-flow microfiltration (↓ Cell debris) → Concentration by UF → Nucleic acid precipitation by polyethyleneimine (↓ Waste precipitate) → Monoclonal antibody affinity chromatography

Fig. 11. Recovery of interferon from rDNA *E. coli*.[31]

(4) Biological proteins are very labile and so require careful processing.
(5) Purification levels exceeding 95–98% are necessary.

Consequently, the need for short processing times is one of the reasons why the downstream processing cost of rDNA products is relatively high, since this necessitates high flow-rates, hence high capital costs, for process equipment, particularly for cell separation and rupture, debris removal and soluble protein recovery. Overall process times for these operations of between 30 min and 1 h may be desirable, considerably shorter than the >12 h often used for extracellular products.

Primary separation stages are generally easier for expressed intracellular foreign proteins than for natural intracellular proteins because of their higher concentrations. For example, intracellular enzyme concentrations are typically only 10^{-3} g/litre of fermenter whole broth, and so, to isolate them, large quantities of both intracellular protein and nucleic acids must be removed. However, expressed foreign protein in *E. coli* may accumulate in insoluble granule or amorphous forms at up to 50% of the total cell protein.[123] Hence significantly less contamination has to be removed, the major component being nucleic acids which are commonly removed using precipitating agents such as polyethyleneimines or by enzyme lysis. Following this, affinity chromatography purification techniques are generally utilised to recover, at acceptable purity levels, the desired biological protein. As an example, a flow-chart for the recovery and purification of human leukocyte interferon is given in Fig. 11. The process indicated is similar to, though shorter than, other multi-step batchwise processes, reported for human insulin and human growth hormone.[65,122]

6.6. Impact of rDNA Techniques on Protein Recovery

Recombinant DNA techniques are being used to increase the microbial expression levels of foreign protein or native enzyme products, to decrease the expression levels of unwanted metabolites, to create long-life cell lines[124] and to decrease the proteolytic enzyme systems responsible for degrading abnormal proteins possibly by the use of cleavable fusion products or by the encouragement of protein aggregates to form.[123] In addition, the properties of cell walls may be altered in order to increase release yield and reduce the energy and time for cell disruption steps. However, many of these approaches will tend to improve the overall process yield rather than increase the specific effectiveness of DSP (downstream processing) separation techniques. The latter may be increased by making the desired protein product more 'reactive' to separation techniques. In one approach adopted,[125] a reversible attachment of the basic amino acid arginine to human urogastrone significantly increased its binding to ion exchange resins. Purification is enhanced because most bacterial proteins are acidic. An alternative approach[126] is the reversible attachment of an 'affinity' tail to the desired protein molecule.

The importance of reduced proteolytic enzyme attack is highlighted if process scale-up to tonnage outputs is reached for existing intracellular products. For example, a 10 tonnes/annum plant utilising an 18 h recombinant fermentation of an intracellular product to 0·5 g/litre, with 70% DSP recovery yield, will require 64 m^3 of fermenter capacity. Overall capital costs will be minimised by reducing the numbers of fermenters and cell–broth separators. Since the largest bacteria–broth disc stack centrifuges available have capacities in the range 3000–4000 litres/h and cost £100 000 to £150 000, prohibitively large capital expenditure will be required if a cell processing time of 1 h is required. Equally, to increase the number of fermenters to reduce batch volume will also increase plant capital.

Universal affinity chromatography with or without the 'tailing' of protein molecules may be, if successful, the ultimate solution to the rapid recovery at high purity of microbial proteins. However, one can presume that considerable research effort would be required to attach the ligand tail genetically and to develop a high-yield 'cleavage' process to separate the tail from the protein. Obviously the parallel requirement for this approach will be a specific affinity ligand/support system, and for low process costs this must be effective over many cycles and ideally extract the product as early as possible in the process stream in order to minimise the number of stages handling large volumes of broth. To date, no solid-support affinity chromatography medium is available. One possibility may be to utilise two-phase partition affinity chromatography to extract the product–ligand complex continuously from the cell homogenate and so avoid an expensive cell debris separation step. These concepts may be universally possible, look attractive and provide for the first time a true link between the microbe and the downstream process. However, one important objective and result of fermentation R&D on existing bio-process products such as enzymes and antibiotics has been to reduce overall process costs by significantly increasing the fermenter titres rather than attempt to integrate the microbe or fermentation step with the DSP. It is reasonable to expect therefore that increased titres will continue to be the prime objective of the recombinant R&D, particularly when protein outputs reach tonnage quantities. However, these titres will be limited to around 1·5–2·5 g/litre, assuming that a maximum of 50% 'foreign' protein can be expressed and stored within bacteria. Therefore, to raise titres significantly above this, recombinant R&D in the longer term must be directed towards long-life excretory microbial systems using microorganisms such as bacillus which excretes protein molecules, or to yeasts which have good product excretion mechanisms, are able to glycosylate protein molecules and, being much larger than bacteria, can be separated at far lower cost. Furthermore, release of pyrogens is very low. Genetic engineering companies such as Chiron of Oakland, California, claim to be well advanced in the use of yeasts for biological protein production.[120]

7. SAFETY ASPECTS OF PROTEIN RECOVERY PROCESSES

In addition to the 'conventional' hazards associated with industrial plants, there are a number of other specific safety factors to be considered in the design and operation of protein production and recovery processes. These may be categorised as follows:

(1) Explosion protected and flameproof plant designs.
(2) Minimisation (or elimination) of contact between the product (or production system, e.g. fermentation) and the operator and environment.

Hence category (1) safety features are required where large quantities of protein or enzyme dusts are dried, and/or where flammable solvents such as ethanol and hexane are used in large volumes in the recovery processes.

The data in Table 10 indicate that protein dust clouds can be ignited with explosive force. Explosion protection for dryers, hoppers and ducts usually consists of blow-out panels ducted to outside the process building and sized to prevent the pressure inside the equipment from exceeding its

Table 10
Explosion Parameters of Two Protein Dusts

	α-Casein	*Soya protein*
Minimum ignition temp. of cloud	460°C	540°C
Lower explosive limit	—	0·05 g/litre
Flame propagation in dust cloud	Yes	Yes

rupture pressure. As an alternative, or addition, explosion suppression devices may be used. Also, it is good practice to separate solids conveying equipment by mechanical devices, e.g. rotary valves, in order to prevent the spread of a dust explosion from one item to another. Electrical apparatus within equipment and dusty areas must conform to meet the appropriate classification codes for surface temperature and dust explosion characteristics. Flameproof or intrinsically safe electrical and instrumentation systems are conventionally used to prevent electrical spark ignition of solvent vapours and pose no special problems to the plant designer or operator. The prevention of dust clouds outside the drying, hopper and storage facilities is usually achieved by installing well engineered plants with a minimum of 'loose' flexible connections and with dust extraction at critical points.

Category (2) safety features concern operator or product system contamination. The most important examples are given here.

(1) The prevention of operator contact with tissue-destroying proteins, e.g. proteases, by containment of product powder, by using protective clothing and breathing systems for the operator, and/or by preparing these enzyme products in a non-dusting form, e.g. wax-coated granules.
(2) The use of sterile engineering techniques to prevent contamination of protein and enzyme fermentations with undesirable environmental microorganisms.
(3) The use of containment techniques to prevent the escape to the environment of genetically engineered microorganisms when used in protein and enzyme fermentations.
(4) The preparation, in a sterile environment, of parental dosage forms of health care proteins.

REFERENCES

1. Perutz, M., The birth of protein engineering, *New Scientist*, 13th June (1985).
2. Wood, W. B., Wilson, J. H., Benbow, R. M. and Hood, L. E., *Biochemistry: A Problems Approach*, Benjamin/Cummings Publishing Company, Menlo Park, California, 1981.
3. Kirstler, P. and Friedli, H., Ethanol precipitation, in *Methods of Plasma Protein Fractionation*, Curling, J. M. (Ed.), Academic Press, London, 1980.
4. Bell, D. J., Hoare, M. and Dunnill, P., The formation of protein precipitates and their centrifugal recovery, *Adv. Biochem. Eng. Biotechnol.*, **26**, 1–72 (1983).
5. Hoare, M. and Dunnill, P., Precipitation of food proteins and their recovery

by centrifuging and ultrafiltration, *J. Chem. Technol. Biotechnol.*, **34B**, 199–205 (1984).

6. Twineham, M., Hoare, M. and Bell, D. J., The effect of protein concentration on the break-up of protein precipitate by exposure to shear, *Chem. Eng. Sci.*, **39**, 509–13 (1984).
7. Hao, T. L., Ingham, K. C. and Wickerhauser, M., Fractional precipitation of proteins with polyethylene glycol, in *Methods of Plasma Protein Fractionation*, Curling, J. M. (Ed.), Academic Press, London, 1980.
8. Johnson, A. J., Macdonald, V. E., Semer, M., Fields, J. E., Schick, J., Lewis, C. and Brind, J., Plasma protein fractionation using solid phase polyelectrolytes, in *Methods of Plasma Protein Fractionation*, Curling, J. M. (Ed.), Academic Press, London, 1980.
9. Steinbeck, M., Protein fractionation by ammonium sulphate, rivanol and caprylic acid precipitation, in *Methods of Plasma Protein Fractionation*, Curling, J. M. (Ed.), Academic Press, London, 1980.
10. Foster, P. R., Dickson, A. J., Stenhouse, A. and Walker, E. P., *Sensors and Control in Protein Fractionation Processes*, 1st Edn, Inst. Chem. Eng. Symposium Series No. 94, Multi-Stream '85, Pergamon Press, London, 1985.
11. Foster, P. R. and Watt, J. G., The CSVM fractionation process, in *Methods of Plasma Protein Fractionation*, Curling, J. M. (Ed.), Academic Press, London, 1980.
12. Watson, J. S., Street, G. and Cumming, R. H., *Autolysis: A Feasible Method for Intracellular Protein Release*, private communication, Teeside Polytechnic, Middlesbrough, 1985.
13. Limon-Lason, J., Hoare, M., Orsborn, C. B. and Doyle, D. J., Reactor properties of a high speed bead mill for microbial cell rupture, *Biotechnol. Bioeng.*, **21**, 745–74 (1979).
14. Marffy, F. and Kula, M. R., Enzyme yields from cells of brewers yeast disrupted by treatment in a horizontal disintegrator, *Biotechnol. Bioeng.*, **16**, 623–34 (1974).
15. Follows, M., Hetherington, P. J., Dunnill, P. and Lilly, M. D., Release of protein from Baker's Yeast (*saccharomyces cerevisiae*) by disruption in an industrial homogeniser, *Trans. Instn Chem. Engrs*, **49**, 142–8 (1971).
16. Erickson, R. A., Disk stack centrifuges in biotechnology, *Chemical Engineering Progress*, 51–4, December (1984).
17. Brunner, K. H., *Separators in Biotechnology*, Westfalia Company Publication, Oelde, West Germany.
18. Brunner, K. H., Van Hemert, P., Kohlstette, W. and Tiesjema, R. H., *Continuous Centrifugation in Large Scale Vaccine Production*, Westfalia Separator Publication, Oelde, West Germany, 1984.
19. Westfalia Centrifuge Company, *New Centrifuge for Separating Human Blood Plasma*, press release, Interchemie, Paris, 1983.
20. Hemfort, H. and Kohlestette, W., *Centrifugal Clarifiers and Decanters for Biotechnology*, Westfalia Separator Publication, Oelde, West Germany, 1984.
21. Svarovsky, L. (Ed.), *Solid–Liquid Separation*, 2nd Edn, Butterworths, London, 1981.

22. Coulson, J. M. and Richardson, J. F., *Chemical Engineering*, Vol. 1, Pergamon Press, Oxford, 1963.
23. Crane, A. P., *Mycelia Removal in Penicillin Production*, Pall Filtration Ltd, Portsmouth, UK, 1984.
24. O'Sullivan, T. J., Epstein, A. C., Korchin, S. R. and Beaton, N. C., Applications of ultrafiltration in biotechnology, *Chemical Engineering Progress*, 68–75, January (1984).
25. Brocklebank, M. P., Usage and role of membranes in bio-processes, *Processing*, 27–31, April (1985).
26. Gutman, R. and Leaver, G., Membranes separate big from small in biotechnology, *Process Eng.*, 37–40, June (1984).
27. Literature from APV Co., UK and SFEC Co., France.
28. Membrane literature from Wafilin BV, Hardenberg, The Netherlands.
29. Le, M. S., Spark, L. B. and Ward, P. S., The separation of aryl acylamidase by crossflow microfiltration and the significance of enzyme/cell debris interaction, Paper presented at IMTEC (International Membrane Technology Conference), Sydney, 1983.
30. Quirk, A. V. and Woodrow, J. R., Tangential flow filtration: a new method for the separation of bacterial enzymes from cell debris, *Biotechnol. Lett.*, **5**, 277–82 (1983).
31. Vaks, B., Mony, Y., Pederson, J. U. and Horowitz, O., A semi-continuous process for the production of human interferon C from *E. coli* using tangential-flow microfiltration and immuno-affinity chromatography, *Biotechnol. Lett.*, **6**(10), 621–6 (1984).
32. Damiano, D., Shin, C. S., Ju, N. H. and Wang, S. S., Performance, kinetics and substrate utilisation in a continuous yeast fermentation with cell recycle by ultrafiltration membranes, *Appl. Microbial Technol.*, **21**, 69–77 (1985).
33. Atkinson, B. and Mavituna, F., *Biotechnology Handbook*, Macmillan, London, 1983, pp. 993–5.
34. Mattiasson, B., Applications of aqueous two-phase extraction systems in biotechnology, *Trends Biotechnol.*, **1**, 16–20 (1983).
35. Kula, M. R., Hustedt, H., Kroner, K. H. and Schütte, H., New developments in large scale protein isolation processes, *Proceedings of Biotech. 1983*, On-Line Publications, London, 359–72, 1983.
36. Kroner, K. H., Schütte, H., Stach, W. and Kula, M. R., Scale-up of formate dehydrogenase by partition, *J. Chem. Technol. Biotechnol.*, **32**, 130–7 (1982).
37. Fauquex, P.-F., Hustedt, H. and Kula, M. R., Phase equilibration in agitated vessels during extractive enzyme recovery, *J. Chem. Technol. Biotechnol.*, **35B**, 51–9 (1985).
38. Janson, J. C., Large scale affinity purification: state of the art and future prospects, *Trends Biotechnol.*, **2**, 1–8 (1984).
39. Flashel, E., Wandrey, C. and Kula, M. R., Ultrafiltration for the separation of biocatalysts, *Adv. Biochem. Eng. Biotechnol.*, **26**, 73–142 (1983).
40. Sweeney, M. J., Membrane based liquid separation systems, *Chemical Engineering Progress*, 32–5, January (1985).
41. Beaton, N. C. and Klinkowski, P. R., Industrial ultrafiltration design and application of diafiltration processes, *J. Separ. Proc. Technol.*, **4**(2), 1–10 (1983).

42. Skagins, K., Retrieving a needle from a haystack: isolation and purification of biological substances, *Proceedings of Biotech 1984*, On-Line Publications, London, 119–38.
43. Curling, J. M., Low, D. and Cooney, J. M., Downstream processing of fermentation products by chromatography, *Proceedings Biotech 1984*, On-Line Publications, London, 225–34.
44. Cooney, J. M., Chromatographic gel media for large scale protein purification, *BioTechnology*, 41–55, January (1984).
45. Janson, J. C. and Hedman, P., Large scale chromatography of proteins, *Adv. Biochem. Eng.*, **25**, 43–99 (1982).
46. Frej, A. K., Gustafsson, J. G. and Hedman, P., FPLC for monitoring microbial and mammalian cell cultures, *BioTechnology*, 777–81, September (1984).
47. Quirk, A. V., The work-up of fermentation products, in *Bioprocessing in the '80s*, Inst. Chem. Eng. Conf., Southampton, 1983.
48. Berglof, J. and Cooney, J. M., Scale-up of laboratory systems for large scale chromatography, *Proceedings of Biotech '83*, On-Line Publications, London, 833–42.
49. Miles, B. J. and Thompson, A. R., New inorganic materials for protein separation and recovery, Process Biochemistry, 11–16, May (1974).
50. Kent, C. A., Miles, B. J., Flaws, J. and Thompson, A. R., Biological process separations using inorganic adsorbents, in *Bioprocessing in the '80s*, Inst. Chem. Eng. Conf., Southampton, 1983.
51. Macrosorb product range from Sterling Organics, Cramlington, Newcastle, England.
52. Stanker, L. H., VanderLaan, M. and Juarez-Sdlinas, H., One step purification of mouse monoclonal antibodies from ascites fluid by hydroxylapatite chromatography, *J. Immunol. Methods*, **76**, 157–69 (1985).
53. Atkinson, A., Bradford, P. A. and Selmes, I. P., Large scale preparation of chromatographic grade hydroxylapatite and its application in protein separation procedures, *J. Appl. Chem. Biotechnol.*, **23**, 517–29 (1973).
54. Bruton, C. J., Large scale purification of enzymes, *Phil. Trans. Roy. Soc.*, **B300**, 249–61 (1983).
55. Janson, J. C., Large scale affinity purification: state of the art and future prospects, *Trends Biotechnol.*, **2**(2), 31–8 (1984).
56. Van der Wiel, J. P., Van Brakel, J., Duine, J. A. and Wesselingh, J. A., Continuous purification of proteins in fluidised sorption columns, poster presentation, 1st International Meeting on Downstream Processing, Toulouse, February 1985.
57. Anon, *Hydrophobic Chromatography for Hydrophilic Proteins*, Pharmacia Fine Chemicals, Uppsala, Sweden, 1979.
58. Kato, Y., Kitamura, T. and Hashimoto, T., Operational variables in high performance hydrophobic interaction chromatography of proteins on TSK gel phenyl-5PW, *J. Chromatog.*, **298**, 407–18 (1984).
59. Hedman, P., Gustafsson, J. G. and Frej, A. K., Protein recovery from unclarified cell homogenates, in *Discovery and Isolation of Microbial Products*, Verrall, M. S. (Ed.), SCI, Ellis Horwood, Chichester, 1985.
60. Scott, F., Stepping up to kg/hour production rates with HPLC, *Process Eng.*, 26–31, February (1984).

61. Dwyer, J. L., Techniques for scale-up to process HPLC, *Kemia-Kemi*, **3**, 247–51 (1985).
62. Dwyer, J. L., Scaling-up bio-product separation with high performance liquid chromatography, *BioTechnology*, 957–64, November (1984).
63. Di Bussolo, J. M., A practical introduction to reversed phase liquid chromatography of proteins and peptides, *Int. Biotechnol. Lab.*, **2**(3), 14–27 (1984).
64. Lowe, C. R., New developments in downstream processing, *J. Biotechnol.*, **1**, 3–12 (1984).
65. Hill, E. and Hirtenstein, M., Industrial affinity chromatography, in *Advances in Biotechnology Processing*, Vol. 1, Mizrah, A. and Van Wezel, A. (Eds), Allan R. Liss, New York, 1983, pp. 31–66.
66. Yang, Che-Ming and Tsao, G. T., Affinity chromatography, *Adv. Biochem. Eng.*, **25**, 19–42 (1982).
67. Yanchinski, S., Fermentec developing protein A process, *BioTechnology*, 667–8, August (1984).
68. Kopperschläger, G., Böhme, H. J. and Hofman, E., Cibacron Blue F3G-A and related dyes as ligands in affinity chromatography, *Adv. Biochem. Eng.*, **25**, 101–38 (1982).
69. Lowe, C. R. and Stead, C. V., The use of reactive dyestuffs in the isolation of proteins, in *Discovery and Isolation of Microbial Products*, Verrall, M. S. (Ed.), SCI, Ellis Horwood, Chichester, 1985.
70. Reed, P. B., Electrodialysis for the purification of protein solutions, *Chemical Engineering Progress*, 47–50, December (1984).
71. Commercial name 'Biostream', marketed by CJB (Developments), Portsmouth, UK.
72. Dickerson, C. H., Birch, J. R. and Cartwright, T., A novel rapid and continuous method for the resolution of cell dispersal activities in crude trypsin preparations, 3rd General Meeting of ESACT, Oxford, 1979.
73. Recycling Iso-Electric Focusing (RIEF) system, Ionics Inc., Watertown, Massachusetts, USA.
74. Gehle, R. D. and Schügerl, K., Protein recovery by continuous flotation, *App. Microbial Biotechnol.*, **20**, 133–8 (1984).
75. Dunlop, E. H., Feiler, W. A. and Mattione, M. J., Magnetic separations in biotechnology, *Biotechnol. Adv.*, **2**, 63–74 (1984).
76. Anon, Downstream protein recovery process stirs Sea Island bioengineers, *Biotechnol. Newswatch*, **4**, February 20 (1984).
77. Ledward, D. A. and Lawrie, R. A., Recovery and utilisation of by-product proteins of the meat industry, *J. Chem. Technol. Biotechnol.*, **34B**, 223–8 (1984).
78. Hopwood, A. P. and Rosen, G. D., Protein and fat recovery from effluents, *Process Biochem.*, 15–17, March (1972).
79. Grant, R. A., Protein recovery as an effluent treatment process, *Effluent and Water Treatment J.*, 616–21, December (1975).
80. Grant, R. A., Protein recovery from process effluents using ion exchange resins, *Process Biochem.*, 11–14, February/March (1974).
81. Anon, Protein from waste, *Food Manuf.*, 39, December (1984).
82. US Pat. 4,462,932, July 31 (1984).

83. *Cheese Whey Concentration by Reverse Osmosis*, PCI, Laverstoke Mill, Hampshire, England.
84. Darles, J. C., Molecular sieve separates protein from cheese whey, *Chem. Eng.*, 114–15, July 24 (1972).
85. Howell, J. A., Ion exchange recovery of proteins, paper presented at Inst. Chem. Eng. meeting, Glasgow, November 21, 1985.
86. Anon, Costs for purification of sweet whey proteins, *ProBiotech* (suppl. to *Process Biochem.*), December (1984).
87. Bray, W. J., Leaf protein recovery, *Chem. Eng.*, 76–7, January 22 (1973).
88. Anon, New protein route goes commercial in Hungary, *Chem. Eng.*, 68C–D, December 10 (1973).
89. Anon, Recovery of protein from potato juice, *Westfalia Separator Magazine*, Oelde, West Germany.
90. Wafilin BV, Membrane filtration literature, Hardenberg, The Netherlands.
91. Klausner, A., Adjustment in the blood fraction market, *BioTechnology*, **3**(2), 119–25 (1985).
92. Winkler, M. A., Wiseman, A. and Sadler, A. M., Polyethylene glycol precipitation of a microsomal enzyme, in *Discovery and Isolation of Microbial Products*, Verrall, M. S. (Ed.), SCI, Ellis Horwood, Chichester, 1985.
93. Curling, J. M., Albumin purification by ion exchange chromatography, in *Methods of Plasma Protein Fractionation*, Curling, J. M. (Ed.), Academic Press, London, 1980.
94. Eketorp, R., Affinity chromatography in industrial fractionation of human plasma, in *Methods of Plasma Protein Fractionation*, Curling, J. M. (Ed.), Academic Press, London, 1980.
95. Tayot, J. L., Tardy, M. and Gattel, P., Ion exchange and affinity chromatography on silica derivatives, in *Methods of Plasma Protein Fractionation*, Curling, J. M. (Ed.), Academic Press, London, 1980.
96. Randerson, D., Hybridoma technology and the process engineer, *Chem. Engnr*, 12–15, December (1984).
97. Thompson, A. R., Recent developments in protein recovery and purification, *J. Chem. Technol. Biotechnol.*, **34B**, 190–8 (1984).
98. Dunnill, P., Trends in downstream processing of proteins and enzymes, *Process Biochem.*, 9–13, October (1983).
99. Fish, N. M. and Lilly, M. D., The interactions between fermentation and protein recovery, *BioTechnology*, 623–7, July (1984).
100. Rosen, C. G., Primary separation steps in fermentation processes, *Proceedings of Biotech '83*, On-Line Publications, London, pp. 201–23, 1983.
101. Hedman, P., Interfacing fermentation with downstream processing, *Proceedings of Biotech '83*, On-Line Publications, London, pp. 343–58, 1983.
102. Brunt, J. V., Scale-up: the next hurdle, *BioTechnology*, **3**(2), 419–23 (1985).
103. Higgins, J., Lewis, D. J., Daly, W. H., Morqueira, F. G., Dunnill, P. and Lilly, M. D., Investigation of the unit operations involved in the continuous flow isolation of β-galactosidase from *Escherichia coli*, *Biotechnol. Bioeng.*, **20**, 159–82 (1978).
104. Foods, non-conventional, in Kirk-Othmer, *Encyclopedia of Chemical Technology*, 3rd Edn, Vol. 11, Wiley, 1980.

105. Rimmington, A., Single cell protein: the Soviet revolution, *New Scientist*, 27 June (1985).
106. Romantschuk, H. and Lehtomaki, M., Operational experience of first full scale Pekilo process SCP: mill application, *Process Biochem.*, **13**(3), 16–17 (1978).
107. Romanschuk, H., The Pekilo process: protein from spent sulphite liquor, in *Single Cell Protein*, II, Tannenbaum, S. R. and Wang, D. I. C. (Eds), MIT Press, Cambridge, Massachusetts.
108. Anon, A better way to make proteins from whey?, *Chem. Engng.*, 36C–D, March (1978).
109. McNairney, J., Modification of a novel protein product, *J. Chem. Technol. Biotechnol.*, **34B**, 206–14 (1984).
110. Atkinson, B. and Sainter, P., *Technology Forecasting for Downstream Processing*, final report, FAST Programme Contract FST/C/020/80/UK/H, 1982.
111. Edelman, J., Fewel, A. and Solomons, G. L., Myco-protein: the food of the future, *Food Manuf.*, 49–56, October (1984).
112. Viikari, L. and Linko, M., Reduction of nucleic acid content of SCP, *Process Biochem.*, 17–35, May (1977).
113. Emery, A. N., Barker, A. J. and Hargrave, A. L., Processing of microbial protein for food use, *Chem. Engnr*, 506–9, July (1977).
114. Godfrey, T. and Reichelt, J., *Industrial Enzymology: The Application of Enzymes in Industry*, Macmillan, London, 1983.
115. Street, G., Large scale industrial enzyme production, *CRC Crit. Rev. Biotechnol.*, **1**(1), 59–85 (1984).
116. Keay, L., Moseley, M. H., Anderson, R. G., O'Conner, R. J. and Wild, B. S., Production and isolation of microbial proteases, *Biotechnol. Bioeng.*, Symp. No. 3, 63–92 (1972).
117. Gilian, S., Prior, B., Venter, J. and Lategan, P., Production, purification and properties of *β*-glucosidase from *Candida wickerhaus*, *Appl. Microbial Biotechnol.*, **21**, 148–53 (1985).
118. Hammond, P., The role of recombinant DNA technology in the production of therapeutic human polypeptides, *Ind. Biotechnol.* (Wales), **3**(1), January (1984).
119. Anon, Biotechnology and the healthcare revolution, *Manuf. Chemist*, 49–51, January (1984).
120. ECN Chemscope, *BioTechnology*, special issue, May (1984).
121. ECN, *New Hosts Developed for Gene Spliced Proteins*, special report, ECN, Sept. 24 (1984).
122. McGregor, W. C., Large scale isolation and purification of proteins from recombinant *E. coli*, *Ann. NY Acad. Sci.*, 231–7.
123. Schoner, R. G., Ellis, L. F. and Schoner, B. E., Isolation and purification of protein granules from *Escherichia coli* cells over-producing bovine growth hormone, *Bio/Technology*, 151–4, February (1985).
124. Anon, Immobilised rDNA *E. coli* system secretes 40% enzyme for four months, *Bioproc. Technol.*, **7**(3), March (1985).
125. Sassenfeld, H. M. and Brewer, S. J., A polypeptide fusion designed for the purification of recombinant proteins, *BioTechnology*, 76–81 (1984).
126. Spark, L., Goodbye to 'bio-hype' but not to secrecy, *Process Engng*, 35–8, July (1985).

5

New and Modified Polysaccharides

V. J. MORRIS

AFRC Food Research Institute, Norwich, UK

1. INTRODUCTION

The majority of polysaccharide additives used by the food industry are extracted from higher plant tissue and seaweeds. Plants synthesise such polymers to serve as structural components of cell walls or as a source of stored energy. Historically mankind has learnt to recognise, extract and purify those polysaccharides with useful functional properties. In the context of this article useful functional properties will be taken to be the ability to thicken, stabilise or gel a food product. Examples of common structural polysaccharides are agar, carrageenans and alginate from marine algae and cellulose and pectin from land plants. Starch is an example of a storage polysaccharide and the galactomannans are probably hybrids having natural functions as both structural and storage components of seed endosperms. Each polymer has unique chemical features and unique functional properties. Traditionally each additive tends to be associated with a particular process or product. The functionality of the polymer will vary with variations in the physical and chemical structure, which in turn will depend on the method of extraction and the age, source and type of plant tissue. The cost and supply of polymers extracted from plants are subject to variations in climatic conditions.

To extend or improve the use of natural polysaccharides mankind has learnt to modify these polymers to suit its needs. Such modifications have been developed to offset the natural variability of the polymers, to optimise or extend the use in a particular product application or to generate new applications or products. In view of the 'uniqueness' of polysaccharides attempts have been made to modify given polymers to simulate the properties of other polysaccharides.

Section 2 will consider the modifications which have been made to conventional gelling, thickening or stabilising agents with particular emphasis on the potential of biotechnological methods. An alternative to natural or modified polymers is the development of new polysaccharides. Major sources of new polysaccharides are the mucilages or slimes secreted by mucoid microorganisms. The successful exploitation of xanthan gum by the food industry is an example of the potential of such polysaccharides. Microbial fermentation products offer the advantages of reproducible physical and chemical properties and a stable cost and supply. Section 3 will consider examples of the microbial production of 'traditional plant polysaccharides' and the production of new polymers with unique rheological properties.

Developments in biotechnology are designed to enhance the additives used by the food industry and thus to improve the cost, quality and acceptability of food products. New or modified polysaccharides must satisfy stringent toxicological tests before being approved for use in foods, and the cost in time or money is daunting. The manufacturer will need to be certain that there is a readily available market for a new gum, probably backed by wider non-food industrial applications, before seeking food clearance. To date it appears that there are no examples of enzymically modified polysaccharides used as additives by the food industry although there are examples of the selective breeding of plants to modify the functionality of the extracted polysaccharides. Xanthan gum is the only new microbial polysaccharide approved for use as a food additive.

In this article an attempt will be made to review with hindsight what has been done, what can be done at present, and to look into the future in order to see what might be done to improve the functionality of food polysaccharides.

2. MODIFIED POLYSACCHARIDES

This section will consider the modifications which have been made and can be made to conventional gelling, thickening or stabilising agents in order to improve their functional behaviour. The relative merits of enzymic and chemical modifications will be reviewed. Although the enzymic treatment of pectins[1] and starch[2] is a major processing aid in the food industry these applications involve extensive degradation of the polysaccharides. Such a large decrease in molecular weight may be considered to be a change in functional properties. However, in this article attention will be focused on subtle changes in chemical structure which drastically alter the function of the polymer as a gelling, thickening or stabilising agent.

2.1. Agar, Carrageenans and Furcellaran

These polysaccharides are extracted from the cell walls of red marine algae[3,4] (*Rhodophyceae*). The chemical structure of the polysaccharides in a given extract will be sensitive to the algal source and the method of extraction. Agar, κ-carrageenan, ι-carrageenan and furcellaran are approved as gelling and thickening agents for use in the food industry.[5] The idealised repeat units of these polysaccharides[3,4,6] are shown in Fig. 1. Agar differs from the carrageenans and furcellaran in containing 3,6-anhydro-L-galactose instead of 3,6-anhydro-D-galactose residues or their substituents. The carrageenans and furcellaran differ in their mode and extent of sulphation.

The gelation of these polysaccharides is discussed in a number of review articles.[7–10] Basically the idealised repeat units are capable of adopting an ordered (helical) conformation which is important for intermolecular association upon gelation. The polymer chains contain irregularities or departures from the idealised repeat unit which hinder helix formation and reduce the gelling ability of the polymer. In certain cases it may be possible to convert chemically certain structural irregularities to the required regular structure.[8] Thus L-galactose or L-galactose-6-sulphate irregularities in agar may be converted to 3,6-anhydro-L-galactose. Similarly D-galactose-6-sulphate or D-galactose-2,6-sulphate irregularities in carrageenans may be converted to 3,6-anhydro-D-galactose or 3,6-anhydro-D-galactose-2-sulphate (Fig. 2). Such modification forms the molecular basis[8] of the lime process for upgrading carrageenans.[11,12] Some of these structural irregularities are believed[13] to be the biological precursors of the anhydrogalactose residues. For example μ-carrageenan, an alternate copolymer of $(1\rightarrow3)$ D-galactose-4-sulphate and $(1\rightarrow4)$ D-galactose-6-sulphate, is believed to be the precursor to κ-carrageenan. Thus there are probably enzymes present in algae which are capable of carrying out such conversions. However, it seems unlikely that such enzymic treatments could at present compete economically with chemical methods.

The presence of hybrid structures containing irregular arrangements of 'gelling' repeat units and/or 'non-gelling' repeat units may inhibit or even completely prevent helix formation and gelation. Such structures are believed[8] to account for the poor gelling ability of carrageenans extracted

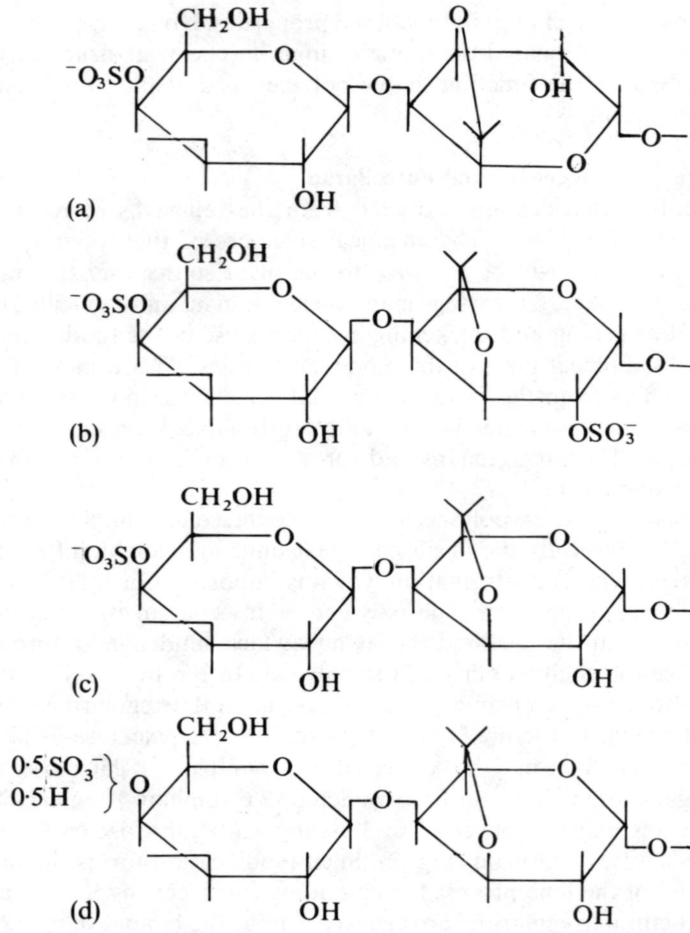

Fig. 1. Idealised repeat units for (a) agar, (b) ι-carrageenan, (c) κ-carrageenan and (d) furcellaran.

from certain species of *Gigartina*. This restricts the choice of seaweed available for commercial processing. The danger of the dependence on certain species of seaweed was emphasised in the early 1970s by the disappearance of the floating seaweed *Furcellaria fastigiate* and the consequent loss of furcellaran as a gelling agent. Fortunately κ-carrageenan was available to replace furcellaran in most food applications. To offset such a dependence on naturally occurring materials, whose

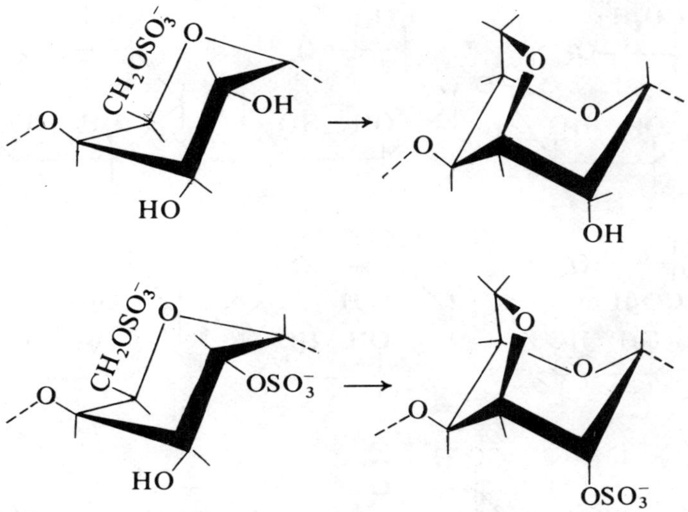

Fig. 2. Example of a chemical modification used to upgrade carrageenans. Note that the introduction of the anhydride bridge changes the shape of the sugar ring.

availability is subject to uncertain environmental conditions, carrageenans are now being farmed[14] in Nova Scotia and the Philippines. The farming of seaweed offers the possibility of optimising the use of the 'natural' enzyme systems within the algae by means of selective breeding or possibly genetic manipulation.

2.2. Alginates

The alginates are extracted from the cell walls of brown algae (*Phaeophyceae*).[3] Alginates are salts of alginic acid. Alginic acid has been shown to be a copolymer.[15] It contains (1→4) linked poly-β-D-mannuronic acid (M blocks), poly-α-L-guluronic acid (G blocks) and mixed MG blocks containing both residues (Fig. 3). The solution properties and the gelation of alginates have been discussed in several review articles[7-10] and are summarised in Section 3.2.

The functional properties of alginates as gelling and thickening agents are determined by the MG ratio and the block distribution within the polymer. These properties are dependent on the age, location and type of algae and the method of extraction. The fruiting bodies of *Ascophyllum nodusum* contain only ~15% G residues whereas older tissue may contain up to ~40% G residues present mainly in mixed (MG)

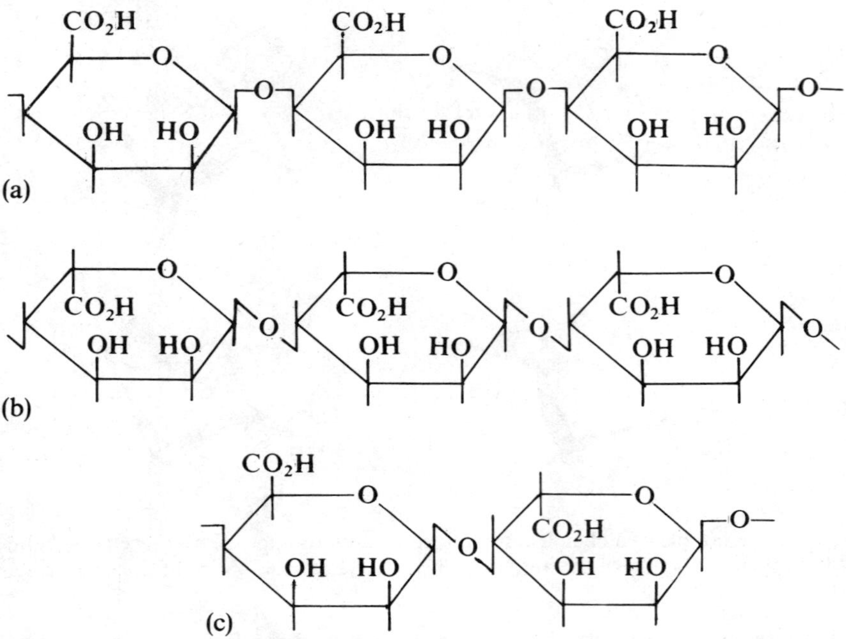

Fig. 3. Block structure of alginates: (a) mannuronic acid (M) blocks; (b) guluronic acid (G) blocks; (c) mixed (MG) blocks.

sequences. The stipes of *Laminaria hyperborea* are a source of high G alginates containing up to ~65% G residues with a predominance of G blocks. Thus the MG ratio and the block distribution may be very sensitive to the algal source. Alginates for specific applications tend to be isolated from selected algal species. The dependence upon naturally occurring supplies, subject to variable environmental conditions, may lead to problems with cost and supply. This was illustrated in 1977 when unusually high temperatures off the coast of California damaged the harvest of the giant kelp *Macrocystis pyrifera*.

There are at least two potential biotechnological solutions to the problem of an unpredictable supply of alginate. The production of alginate by microbial fermentation (Section 3.2) offers a readily controllable source and supply. The control of the activity of the C-5 epimerase, which converts D-mannuronic acid to L-guluronic acid as demonstrated by Larsen and Haug,[16] would permit control of the MG ratio of bacterial alginate. A second alternative would be to isolate the epimerase and then use the enzyme to upgrade low G alginates. Mannuronan C-5 epimerases are produced by the microorganism *Azotobacter vinelandii.*[16] The enzyme is extracellular and may be isolated and purified from liquid cultures.[17]

The enzyme will epimerise both algal and bacterial alginate provided the degree of polymerisation is greater than 10. Enzyme activity is sensitive to the calcium concentration of the medium.[16–18] At low calcium concentrations (0·86 mM) the enzyme preferentially inserts G residues next to G residues in the chain, thus creating homopolymeric G blocks. At higher calcium concentrations the G residues tend to be introduced randomly, creating heteropolymeric mixed MG blocks. Detailed analysis of enzyme modifications of alginates extracted from *Laminaria digitata*, *Dictyosiphon foeniculaceus* and *Ascophyllum nodusum* are discussed by Skjak-Braek.[18] Acetylated mannuronic acid residues found in bacterial alginate have been shown to inhibit epimerisation.[19]

Bacterial alginates or enzymically modified alginates are respectively new and modified polysaccharides. As far as the author is aware neither has been approved for use in foods. Farming of brown algae and the modification of epimerase activity either by varying growth conditions, as a consequence of selective breeding, or by genetic manipulation, could presumably minimise the problems of obtaining food clearance because such products would be similar to 'natural alginates'.

2.3. Pectins

The pectins are a class of complex polysaccharides which are extracted from the cell walls of land plants.[4,20,21] The macromolecules used by the food industry as gelling and thickening agents have structures based on a poly $(1 \rightarrow 4)$ linked α-D-galacturonan backbone substituted with a few neutral sugar side-chains. $(1 \rightarrow 2)$ linked L-rhamnose is the only known neutral sugar found in the polymer backbone.[21]

The remaining neutral sugars, chiefly galactose and arabinose, are believed to occur as side-chains attached to position 4 of $(1 \rightarrow 2)$ linked rhamnose residues.[22] Such substituted polygalacturonic acids are called pectinic acids and their salts are called pectates. Pectins are complete or partial methyl esters of pectinic acid. The salts of the partial methyl esters are called pectinates. Normal commercial extraction procedures will cleave some of the neutral sugar side-chains.

The gelation of pectins is discussed in several review articles.[4,7–10,23] Pectins may form one of two different types of gel depending upon the degree of esterification (DE) of the molecule. Pectins with DE > 55% are called high-methoxy (HM) pectins and are gelled[24] in the presence of sugar at acidic pH. Typical polymer concentrations would be ~1% and

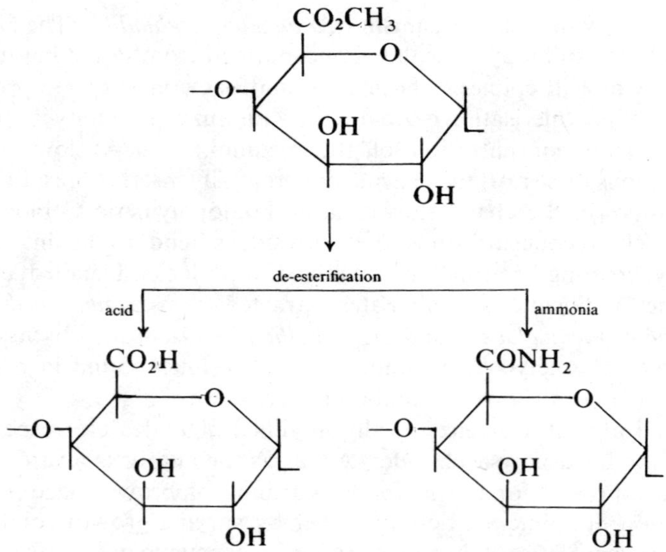

Fig. 4. Structural modification of pectins by chemical treatments.

the mix is prepared at high temperatures and the gel forms on cooling. The level of esterification controls the setting rate of the gel or, effectively, the temperature at which gelation commences on cooling. Pectins with DE ~70–80% are called 'rapid-set' pectins whereas those with DE ~55–65% are termed 'slow-set' pectins.[24] Low methoxy (LM) pectins are produced by a chemical de-esterification of HM pectin (Fig. 4). LM pectins gel[24] in the presence of divalent cations (usually calcium) and commercial gels usually contain sugar and are prepared at low pH. The gelation mechanism is similar to that of alginates (Section 3.2) and requires the presence of blocks of polygalacturonic acid. The DE value dictates the optimum calcium concentration for gelation (Fig. 5). Treatment with ammonia of alcoholic suspensions of pectins produces pectinic acid amides which can gel over a wide range of calcium concentrations (Fig. 5) and are thus less sensitive to the variable calcium concentrations in milk and water.

Pectinesterases which release methanol from methyl esterified galacturonic acid are available and could be used for de-esterification. Enzymic and chemical de-esterification leads to different acid and ester distributions for a given DE[25] (Fig. 5). The enzyme de-esterifies by acting at the reducing end of the polymer, or adjacent to a free carboxylic acid group,

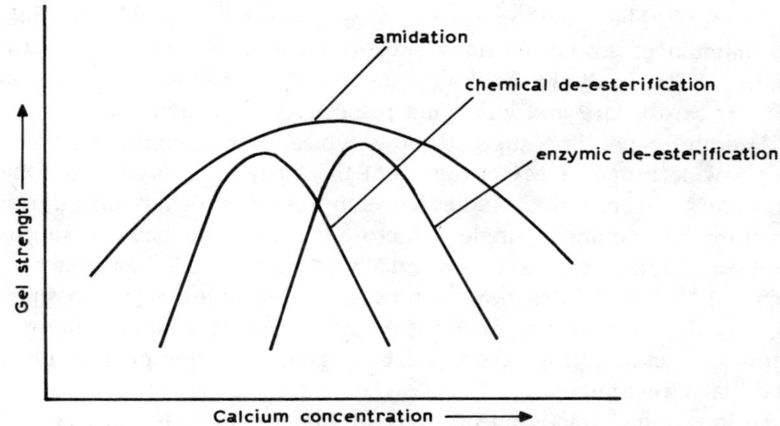

Fig. 5. Schematic diagram illustrating the effects of different methods of de-esterification. Amidation yields a broad calcium sensitivity. At the same level of de-esterification chemical and enzymic methods produce different optimum calcium levels.

and then proceeds stepwise along the polymer chain.[26,27] Thus enzymic methods are more effective at producing homopolymer carboxylic acid blocks than chemical methods which randomly de-esterify the chain. Pectinesterases can be found in many plants[28–31] and are produced by fungi and bacteria such as *Fusarium oxysporum*[32] and *Clostridium multifermentans.*[33,34] Pectinesterases are components of many commercial pectinase preparations which are usually produced from *Aspergillus niger*. Such preparations are widely used[1] for clarification of fruit juices, extraction of juice and the maceration and liquification of pulps. Despite the widespread use of pectinases, an enzymic de-esterification would produce a modified pectin with dubious legal status as a food additive. Food clearance would probably be relatively straightforward but the economics of chemical versus enzymic methods would at present appear to favour chemical methods.

2.4. Galactomannans

Galactomannans are extracted from the endosperms of a number of legume seeds. The three galactomannans which are available commercially are carob (locust bean) gum obtained from *Ceratonia siliqua*, guar gum obtained from *Cyamopsis tetragonolobus* and tara gum which is extracted from *Caesalpinia spinosa.*[35] Both guar and carob gum are approved for food use.[5] Tara gum has food approval but does not have an EEC

E number.[5] The galactomannans are composed of a (1→4) linked β-D-mannan backbone partially substituted at C-6 with α-D-galactosyl residues.[35] Typically the mannose/galactose (MG) ratios are 3·55, 3·0 and 1·56 for carob, tara and guar gum respectively.[35] Fractionation studies by Hui and Neukom[36] suggest that a typical extract contains polymers with a wide range of MG values. On the basis of X-ray diffraction,[37] chemical[38] and enzymic[39] studies it was suggested that guar had a uniform structure containing a single galactose residue attached to alternate mannose residues and that carob had a block structure. Recent enzymic,[40] chemical,[41] X-ray diffraction[42] and nmr[43] studies are taken to suggest an irregular to random distribution of galactose residues along the mannose backbone for carob and guar gum and thus presumably for other galactomannans.

An interesting functional property of galactomannans is the so-called synergistic interaction which occurs when they are mixed with certain specific polysaccharides.[44–54] Carob gum forms mixed gels with agarose, κ-carrageenan, xanthan, segmented agarose and segmented κ-carrageenan. The nature of the interaction is sensitive to the MG ratio and thus the type of galactomannan. Tara gum forms mixed gels with the above named polysaccharides but higher concentrations of tara gum are required to reproduce the gel strength and elastic modulus obtained for mixtures containing carob gum.[52] Guar gum does not gel but does show an enhanced viscosity when mixed with the above polysaccharides.[54] Such unusual rheological properties have been attributed[35,48,49,51] to a direct molecular interaction between the galactomannan and the helix-forming polysaccharide (Fig. 6). In the model proposed by Dea and coworkers,[35,48,49,51] it has been suggested that segments of the helical polymer bind to unsubstituted regions of the mannan backbone of the galactomannan (Fig. 6). Two variations of this model have also been suggested.[40,53,54] Although such models are widely accepted there is only circumstantial experimental evidence to support an intermolecular interaction. The stoichiometry and size of the suggested mixed junction zone are unspecified and the models cannot predict new synergistic interactions. X-ray fibre diffraction studies of mixed gels have as yet failed to reveal any evidence for such interactions in carob κ-carrageenan[55,56] and tara κ-carrageenan[57] mixtures. Kovacs[47] discusses an alternative model involving competition for solvent between xanthan and carob in order to explain gelation of mixed polymers. Certainly the solubility of the galactomannans follows the order guar > tara > carob and could explain the sensitivity to the MG ratio. However, a genuine molecular interaction has been demonstrated

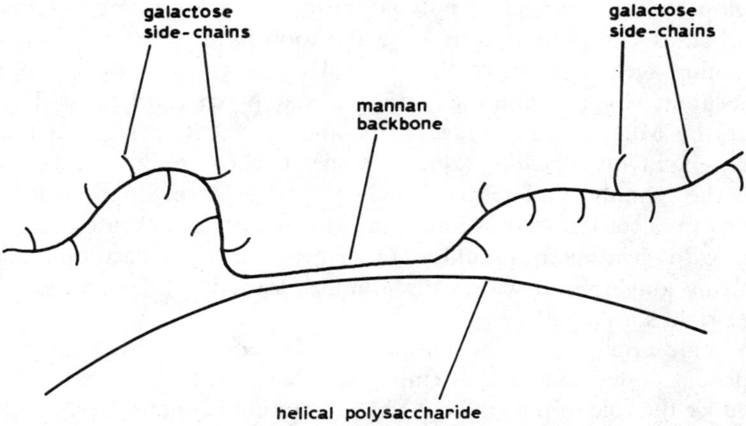

Fig. 6. Schematic model proposed[35,48,49,51] for the molecular interaction between a galactomannan and certain helix-forming polysaccharides.

for xanthan-carob and xanthan-tara mixed gels.[58,59] These data do not support the model shown in Fig. 6. Rather they suggest a cocrystallisation of the galactomannan with segments of the stereo-compatible denatured xanthan molecule.

The synergistic interactions are sensitive to the MG ratio and possibly the detailed distribution of galactose within, or between, galactomannans in a given extract.[60] Shortages in supply and fluctuations in the cost of carob have made the search for alternatives an attractive proposition. Tara gum may be used to replace carob gum but higher gum concentrations are needed to achieve the same results.[52] Tara gum has food approval but it would be necessary to obtain an E-number for use in Europe. An alternative is to modify the commercially available guar gum enzymically. α-D-Galactosidases are available and can be used for random cleavage of galactose side-chains.[40] Enzymic modification of guar gum produces a new polysaccharide with a higher MG ratio which can effectively duplicate the interaction properties characteristic of carob gum.[60–62] At present enzymically modified guar gum is not approved as a food additive.

2.5. Starch

Starch is the major reserve polysaccharide of most higher plants where it exists as water-insoluble, roughly spherical, partially ordered and crystalline starch granules.[63] The granules contain two polysaccharides called amylose and amylopectin.[63] Amylose is a linear (1 → 4) linked α-D-glucan.

Amylopectin is a branched polysaccharide containing amylosic chains branched at C-6. Starch is used by the food industry as a gelling and thickening agent. Details of the molecular changes which occur during the gelation of starch and the storage of starch gels can be found in the papers by Miles and coworkers[64–67] and in recent review articles.[9,10] Processing involves heating aqueous dispersions of starch to gelatinise or swell the granules and release amylose. The gelation of the starch on cooling involves the association of the amylose molecules into a network filled with gelatinised granules. On storage a crystallisation process involving amylopectin within the granules leads to stiffer granules and stiffer rigid starch gels.

In some applications the release and subsequent association of the amylose is undesirable and attempts have been made to modify starches to reduce the role of the amylose.[68] Most starches contain 20–25 wt% of amylose. By selective plant breeding it has been possible to develop high amylose starches such as amylomaize ($\sim$80% amylose) and low amylose starches such as waxy maize starch (0–10% amylose). Two basic types of chemical modification have been used to inhibit the release of amylose or its intermolecular association.[68] Cross-linking agents have been used to restrict swelling of the granules and release of amylose. Esterification and etherification, primarily of the alcohol residue at C-6, have been used to improve the solubility of the amylose and to inhibit intermolecular association. At present it seems unlikely that enzymic methods could be used to improve such treatments particularly in view of the cost involved in obtaining food approval.

3. NEW POLYSACCHARIDES

Major sources of new polysaccharides are the extracellular polysaccharides secreted by mucoid microorganisms. There are several review articles[69–73] which discuss bacterial polysaccharides with properties of proven and potential commercial importance. In this article only a few selected polysaccharides are discussed in detail. Such a choice is difficult and three factors have been taken into account in order to make the selection. Firstly the relevance to food use has been considered. Secondly an attempt has been made to illustrate how screening, mutation and genetic manipulation can and have been used to obtain polysaccharides with desirable properties. Finally, because the relationship between the primary structure and functionality of microbial polysaccharides is largely unknown, an attempt has been made to choose examples for which sufficient research has been carried out in order to link changes in functionality with changes in primary structure.

3.1. Xanthan Gum

Xanthan gum is the name given to the extracellular polysaccharides secreted by bacteria of the genus *Xanthomonas*. Most species of *Xanthomonas* are pathogenic to specific plant hosts.[74] *Xanthomonas campestris* is the common source[75] of xanthan gum although studies have been reported on the polysaccharides produced by other *Xanthomonas* bacteria such as *Xanthomonas juglandis*[76] and *Xanthomonas phaseoli*.[77] *Xanthomonas campestris* causes vascular disease of cauliflower, cabbages and rutabagas.[74] The exact origin of the original culture *Xanthomonas campestris* NRRL-B-1459 is uncertain. It is believed[75] that Prof. J. C. Walker, University of Wisconsin, sent the sample to Dr W. J. Zaumeyer, USDA, who in turn deposited the sample with the ARS Culture Collection at the North Regional Research Laboratory (NRRL) in 1953. The polysaccharide was isolated in the USDA Northern Regional Laboratory in 1959. The studies formed part of a programme designed to develop new indigenous polysaccharides synthesised by microorganisms utilising sugars derived from cereal starches as a source of carbon. The properties of the polymer were first described by Jeanes and coworkers[78] and it was first produced[79] successfully in commercial quantities by Kelco in 1961. Detailed toxicological and safety studies have been carried out[80] and the polymer was approved for food use by the FDA in 1969. Xanthan is approved for food use in the USA, UK and Europe.[5] Commercially xanthan is produced by aerobic fermentation in batch culture.[75] The feasibility of continuous culture production has been demonstrated by Evans *et al.*[76] The unique rheological properties of aqueous xanthan dispersions have led to a wide range of industrial applications.[46,50,70,72,73,80–85] The useful rheological properties of xanthan dispersions may be summarised as follows:

(1) *Thickening and suspending ability*. In the absence of shear and at sufficiently high polymer concentrations aqueous dispersions of xanthan exhibit weak gel-like properties.[86–88] The gel structure may be disrupted upon application of shear.[80] At low shear rates the samples have high viscosities and the viscosity decreases with increasing shear rate as the samples shear thin.[80] By appropriate choice of polymer concentration and ionic strength the viscosity can be made to be relatively insensitive to pH, temperature and ionic strength.[80]

(2) *Gelation.* Xanthan forms thermoreversible gels in the presence of the galactomannans carob[35,46,58] and tara gum.[49,52,59] Mixtures of xanthan and guar show an unexpectedly high viscosity.[54,80] Enzymic treatment of guar gum with an α-D-galactosidase raises the MG ratio and produces a new modified polysaccharide which also forms thermoreversible gels[60-62] with xanthan gum (Section 2.4). Mixtures of xanthan with certain glucomannans such as Konjac mannan also form thermoreversible gels.[35] Thickening and gelation of xanthan dispersions is enhanced in the presence of di- and tri-valent cations or borate anions.[80]

Xanthan may be dispersed in hot or cold water to produce thixotropic dispersions. This has led to the development of a number of dry mix food products such as desserts, sauces or gravies which may be heated or refrigerated without loss of desirable product characteristics. The weak gel-like behaviour permits stabilisation of emulsions, particulate suspensions and foams. Addition of xanthan has led to significant extensions in shelf-life for food emulsions. Acid stability is important in salad dressings, sauces, relishes or acidified milk products. Thixotropy permits pumping, pouring or spraying of samples. Xanthan is added to improve the processing and storage of batters and doughs and it has been suggested that it may be used to replace gluten in the production of gluten-free breads.[89] Alternatively xanthan can be used[90] to modify the properties of gluten and improve baking quality. The thixotropic properties have led to innovations in packaging such as the use of retort pouches. Xanthan–galactomannan mixes are used in gelled foods, to control particulate sedimentation in drinks and juices, and to impart freeze–thaw stability to products such as ice-cream.

Food related applications of xanthan include stabilisation of cattle feed supplements, calf milk substitutes, agricultural herbicides, pesticides, fungicides and fertilisers and to impart thixotropy to toothpastes. It has been approved for use in paper and cardboard packaging for foods. Xanthan can be used to control droplet size and hence drift of agricultural sprays. It is ironic that the polymer secreted by a plant pathogen now finds such wide applications in food and agriculture.

Wider industrial uses include stabilisation of suspensions, emulsions and foams. Examples include ceramic glazes, polishes, mining ores and paint pigments. Thixotropy is important for non-drip paints and jet printing. Foam stabilisation is important for agricultural foams, fire-fighting fluids and foam printing. Development of high viscosity and retention of this viscosity at high temperature permits uses for viscosity control of oil-well drilling fluids. Uses in enhanced oil recovery are summarised by Sandvik and Maerker.[85] Gelation in the presence of trivalent cations is used to formulate fracturing fluids used to provide high viscosity and to suspend propping agents during pumping operations. Xanthan–borax gels find uses in explosives and xanthan–iron mixtures are employed in board manufacture. Xanthan–carob mixtures are used for foam stabilisation, in the preparation of deodorant gels and in photographic processing.

Xanthan is one of the most exhaustively studied polysaccharides. The primary structure is a (1→4) linked β-D-glucan backbone (cellulose) substituted through position 3 on alternate glucose residues with a charged trisaccharide side-chain (Fig. 7).[91,92] Various substrains of *Xanthomonas campestris* have been identified which produce xanthans that differ in the level of substitution of the side-chain with pyruvate or acetate.[93–95] The secondary structure of the polymer has been studied[96–98] by X-ray fibre diffraction and molecular modelling. The addition of the trisaccharide side-chain is believed to cause the backbone to change from a two-fold ribbon-like structure of cellulose to a five-fold helical structure (pitch ~4·7 nm). Analysis of the patterns has been limited by the poor crystallinity of the samples. A single five-fold helix stabilised by hydrogen bonding of the side-chain to its backbone has been proposed[96,97] but double helical structures cannot be excluded.[98]

Retention of the helical structure in solution would stiffen the macromolecule and might be sufficient to account for the unusual rheological properties particularly when compared with simple cellulose derivatives.[99] The early studies of Dintzis *et al.*[100] distinguished between aggregated dispersions and true solutions. A variety of physical techniques have been used to probe the conformation of the polymer in solution. ^{1}H nmr shows two clearly resolved peaks assigned to the methyl protons of the pyruvate and acetyl groups.[77,101] The line broadening on cooling has been attributed to binding of the side-chain to the helical backbone.[77,101] Circular dichroism spectra show two pronounced bands attributable to the *O*-acetyl residues and the carboxyl group. The change in the acetyl band mirrors the change observed by nmr on heating and cooling. This has been attributed[77,101] to a change in the environment of the ester group, which is located on an inner residue of the side-chain, and would be consistent with a binding of the side-chain to the backbone on cooling. The terminally located carboxyl groups do not show a marked change in environment on heating and cooling.[77,101] Optical rotation studies[77,102–104] show a sharp change in optical rotation on heating and cooling. The contributions of the acetyl and carboxyl circular dichroism

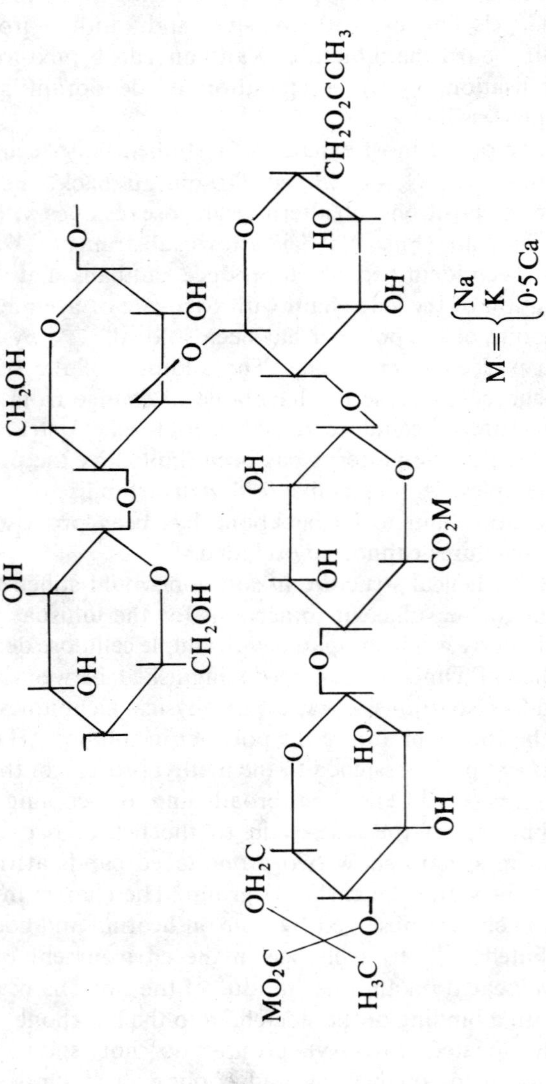

Fig. 7. Chemical repeat unit of xanthan gum. The pyruvate and acetate contents are variable.

bands to the optical rotation are opposite in sign to the observed change on heating or cooling[77,101] and this has been taken to suggest[77,101] that the optical rotation change is dominated by a change in the configuration of the backbone. The optical rotation data are believed to monitor an order(helix)–disorder(coil) transition with the nmr and circular dichroism data supporting binding of the side-chains to the helical backbone.[77] The transition mid-point temperature (T_m) is sensitive to cation type and concentration.[77,101–105] The reciprocal of T_m varies linearly with the logarithm of the cation concentration. This is consistent with a screening effect with cations stabilising the charged helix and raising the melting temperature of the helix. Divalent cations are particularly effective at stabilising the helix.[105] Dilution of xanthan solutions with water will reduce the ionic strength and can dilute out the ordered structure. The order–disorder transition has also been monitored by light scattering,[103,104] viscosity[102,103] and differential scanning calorimetry.[104]

Having established the existence of the ordered structure and the conditions under which this structure is formed in solution, attention has been directed towards determining the shape of the xanthan macromolecule in solution. To determine whether xanthan behaves as a worm-like coil or a rigid rod it is necessary to determine the stiffness of the polymer. Elongational flow birefringence measurements suggest an extended rigid structure.[88] An extended rigid structure for xanthan is supported by electric birefringence data on aqueous xanthan samples[88] and solutions in 4 M urea.[106] The rotatory relaxation times[88,106] are $\sim 10^3$ greater than the values obtained for typical cellulose derivatives.[107,108] This suggests[88] a Kratky–Porod persistence length (q) $\approx$ 100–300 nm. Burchard has suggested,[109] on the basis of a re-analysis of the light scattering data of Paradossi and Brant,[110] and an analysis of light scattering data on pyruvate-free xanthan,[109] that $q \approx 100 \pm 10$ nm. The light scattering data of Sato and coworkers[111–113] suggest $q \approx 120 \pm 20$ nm. Thus xanthan may be regarded as one of the stiffest natural biopolymers, lying between DNA[114,115] ($q \approx$ 50–150) and the triple stranded collagen[116] and schizophyllan.[117,118] Muller and coworkers have reported[119,120] a lower value of $q \approx 50$ nm for xanthan.

The stiffness of DNA, collagen and schizophyllan may be attributed to their multi-stranded structures. The question then arises as to whether xanthan is a single helix stabilised by hydrogen-bonded side-chains *or* a double helical structure? This problem has not been resolved and the evidence for single and double helical structures is presented below. The proponents of a single helix use the following arguments. There appears to be no change in molecular weight during the optical rotation change on cooling[77,103,104] although the molecular weight does increase at lower temperatures and this has been attributed to aggregation.[104] Estimates of the mass per unit length from light scattering[86,88,104,120–122] and the charge per unit length have been taken to favour a single helix. The transition monitored by nmr and optical rotation is broad and insensitive to polymer concentration favouring[77,101] an intra-unimolecular process. Stopped flow optical rotation[123] studies exploit the sensitivity of the order–disorder transition to ionic strength to probe the kinetics of the transition and favour a single helical model.[104,124,125] The evidence favouring a double helix is as follows. A double helix has been proposed on the basis of electron microscopy,[126] estimates of mass per unit length[110,127] and combined sedimentation, viscosity and light scattering data.[111–113] The data of Sato and coworkers[111–113] show that xanthan behaves as a flexible coil in cadoxen [tris(ethylenediamine)cadmium dihydroxide], as a rigid rod in 0·1 M NaCl for molecular weights $\lesssim 3 \times 10^5$ and as a semi-flexible rod at higher molecular weights. The molecular weight in 0·1 M NaCl is twice the value obtained in cadoxen. It is proposed[111–113] that xanthan is a double helix which may be dissociated into single strands in cadoxen. The discrepancies in mass per unit length and charge per unit length could be explained but, if xanthan possesses a unique structure, then the disagreement between the stopped flow data[104,124,125] and the studies by Sato *et al.*[111–113] represent an unresolved and major conflict of experimental data. The main experimental problem in studies on xanthan is the need to ensure that the samples are true solutions and do not contain aggregates. It has been suggested that xanthan may adopt different ordered structures under different conditions.[128]

There are two solution regimes for xanthan. Rheological studies[129,130] on the disordered polymer at low ionic strength shear thin due to molecular distortion. Studies on solutions of the ordered polymer would shear thin due to molecular orientation as evidenced by flow birefringence studies.[88] However, the commercially important thixotropic behaviour cannot be explained simply in terms of solutions of stiff macromolecules. Southwick and coworkers[121,131–134] have proposed that the unusually high viscosities are due to intermolecular entanglements. Quasi-elastic light scattering[121,131–134] data was interpreted, in the light of theories of particle motion in 'semi-dilute' solutions,[135,136] as due to restricted diffusion leading to enhanced viscosity. The thixotropy disappeared after heat treatment in the presence of 4 M urea and, on the basis of optical

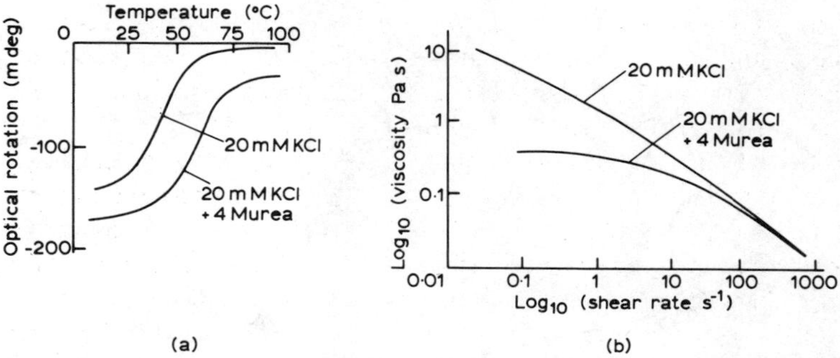

Fig. 8. Effect of heat treatment in the presence of 4M urea on the behaviour of a xanthan (potassium salt) sample. (a) Optical rotation studies suggest that urea stabilises the xanthan helix. (b) Rheological studies show that urea reduces the viscosity and reduces the extent of shear thinning.

rotation data, this was attributed to denaturation of the helix. The Southwick model is not supported by the following experimental observations. Theories of particle motion in 'semi-dilute' solutions[135,136] suggest that the rotational diffusion coefficient should be very concentration dependent whereas experimentally no such concentration dependence is observed.[86,88,106] At sufficiently high ionic strength the stiffness of xanthan as monitored by electric birefringence[86,88,106] remains essentially unchanged even after heat treatment in the presence of 4 M urea. Recent optical rotation[87,88] data suggest that urea stabilises the helical structure (Fig. 8a). The denaturation observed by Southwick *et al.* was due to a decrease in ionic strength.[86–88,121] Even at high ionic strength heat treatment with 4 M urea still reduced the low shear rate viscosity[87,88] (Fig. 8b). It has been demonstrated that the slow modes observed by quasi-elastic light scattering, and attributed to restricted diffusion,[121,131–134] may be removed by restricting molecular association of the ordered (helical) structure.[86,88] This is consistent with the early observations of Dintzis and coworkers.[100] The weak gel-like behaviour and shear thinning properties are believed to be due to association of the ordered (helical) molecules.[86–88,104] Norton *et al.*[104] propose that the conformational ordering is incomplete and that the xanthan molecules contain helical sequences separated by disordered sequences. The helical sequences are considered[104] to associate into a molecular network (Fig. 9). Morris *et al.*[86] suggest a cruder 'dispersion model' for commercial xanthan preparations. When powdered xanthan is dispersed in water it is considered that the particles swell but do not completely dissolve. Moderate shearing

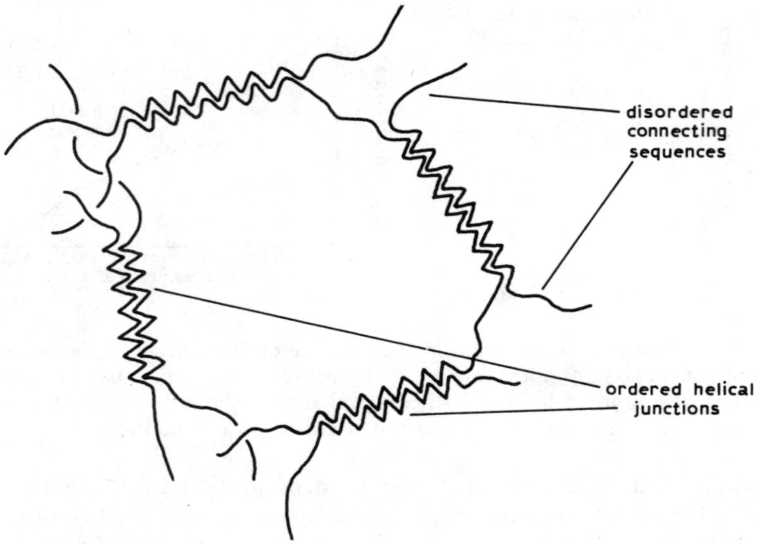

Fig. 9. Molecular network model proposed[104] to explain the weak gel-like structure of xanthan samples.

would distort these microgel particles and may lead to flow of particles past each other resulting in some plastic deformation. This would account for the irreversibility of the transition from 'disperison-like' to 'solution-like' properties following centrifugation, filtration or heat treatment.[86,88] This could explain the cation sensitivity of the rheological properties observed by Ross-Murphy *et al.*[88] The calcium salts gave the highest viscosities at low shear rates and calcium appeared to stabilise the molecular association even following heat treatment in urea. This stabilising effect seems to depend on xanthan concentration and is less effective at lower xanthan concentrations.[137] Divalent cations are particularly effective at stabilising the xanthan helix[105] and even at moderate polymer concentrations the transition temperature may be too high to permit disruption of the helix and dissolution of the aggregate.[137] The degree of molecular association will also control the clarity of commercial samples.

The effect of variable acetate and pyruvate substitution has been discussed.[78,93,95,138–140] It is useful in such studies to demonstrate that such changes in acetyl or pyruvate content are not accompanied by changes in molecular weight and cation content. The stiffness of the macromolecule leads to liquid crystalline behaviour at high polymer concentrations.[141] The rheology of xanthan dispersions is drastically modified by the addition of certain plant galactomannans.[35,47–51,60–62] Addition of guar leads to an unexpected increase in viscosity.[54] Mixtures of xanthan with carob, tara or enzyme modified guar form rubbery thermoreversible gels. Such synergistic interactions have already been discussed in Section 2.4. Mixtures of xanthan and the glucomannan Konjac mannan also gel[35] but the molecular basis for gelation is unknown. The recent studies reported by Cairns *et al.*[58,59] suggest a basis for xanthan-glucomannan binding and gelation.

3.2. Bacterial Alginate

Alginate was first discovered[142,143] in 1880. Alginates are the salts of alginic acid which is a structural component of the cell walls of brown algae (*Phaeophyceae*).[3] A form of alginic acid is also produced by certain strains of the microorganisms *Pseudomonas aeruginosa*[144–146] and *Azotobacter vinelandii*.[147–149]

Pseudomonas aeruginosa is not naturally mucoid.[150] However, mucoid forms of *P. aeruginosa* have been isolated[146,151] and laboratory cultures of non-mucoid forms may be induced[152–154] to produce extracellular polysaccharide. The mucoid strains of *P. aeruginosa* which have been isolated[146,151] have been found in association with chronic respiratory tract infections accompanying cystic fibrosis or in connection with secondary infections accompanying major surgery, burns, leukaemia or following treatment with immunosuppressive drugs. Mucoid forms of *P. aeruginosa* have been isolated from non-mucoid laboratory cultures on the basis of their resistance to bacteriocin,[153] carbenicillin,[154] fluctoxacillin,[152] tobramycin[152] and bacteriophages.[154] Govan[153] has reported the transfer of a mucoid determinant by conjugation from a mucoid to a non-mucoid strain leading to extracellular polysaccharide production. The natural and induced mucoid strains are believed[155,156] to revert to the favoured non-mucoid form on subculturing. This transformation may be inhibited by favouring selection of the mucoid form by addition of surfactant to the maintenance medium[155] or, in continuous culture production, by limiting[157] the availability of sulphate, magnesium and iron. For possible commercial production of bacterial alginate *A. vinelandii* is preferred to *P. aeruginosa* because of the association of the latter bacterium with pathogenic infections in mankind.

Bacteria of the genus *Azotobacter* are normally mucoid soil microorganisms. Several strains of *A. vinelandii* secrete bacterial alginate[147–149] in their natural habitat and retain this ability under laboratory culture conditions. Large-scale production of bacterial alginate using *A. vinelandii* was considered by Tate and Lyle Ltd, UK.[158–161] Continuous culture production has the advantage over batch culture production in that the availability of oxygen can be closely controlled,[162] as is necessary if alginate is to be produced. The bacteria are fastidious. If the oxygen level becomes too high then the bacteria increase their rate of respiration leading to conversion of substrate to carbon dioxide.[163,164] If the oxygen level drops too low then the bacteria start to accumulate poly-3-hydroxybutyrate rather than produce bacterial alginate.[165]

Alginic acid has been shown to be a (1→4) linked copolymer of β-D-mannuronic acid and its C-5 epimer α-L-guluronic acid.[15] Haug *et al.*[166] showed that the polymer was a block copolymer (Fig. 3) containing polyguluronic acid (G) blocks, polymannuronic acid (M) blocks and mixed MG blocks containing both uronic acids. It was originally thought that the MG blocks contained alternating M and G monomers but more recent enzymic[167,168] and ^{13}C nmr[169] studies suggest a more random distribution of monomers in these mixed blocks. The isolation of mannuran C-5 epimerases from marine algae[170] and bacterial culture media[16] has led to the suggestion that alginic acid is biosynthesised as polymannuronic acid and then enzymically converted to a block copolymer. The MG ratio and the block distribution within the polymer are variable.[171] Cation binding and gelation studies on alginates suggest[25,172,173] the presence of G blocks containing more than 20 monomers. Enzymic studies have been taken to suggest[174] the presence of M blocks of length ~24 residues.

Algal alginic acid and its ammonium, sodium, potassium and calcium salts are approved for food use in the USA, UK and Europe.[5] A variety of alginic acid derivatives have been prepared[3] but perhaps the most important derivative is the propylene glycol ester. Propylene glycol (propane-1,2-diol) alginate is prepared[175] by reacting moist alginic acid with propylene oxide. The propylene glycol ester was first introduced in 1941 and is also approved as a food additive.[5] As far as the author is aware bacterial alginate is not approved for food use.

Bacterial alginate from *P. aeruginosa*[144–146] and *A. vinelandii*[147–149] has been shown to be a block copolymer containing (1→4) linked β-D-mannuronic acid and α-L-guluronic acid present as G, M and MG blocks. The bacterial alginates are generally *O*-acetylated at C-2 and/or C-3 and enzymic studies[176] suggest that acetylation may be restricted to mannuronic acid residues. This has recently been confirmed.[19] The degree of acetylation is variable with values between 0 and 1 acetyl group per

1·5 residues reported for *P. aeruginosa*[146] and values between 1 per 5·2 residues and 1 per 66 residues for *A. vinelandii*[147,148] depending on culture conditions. Acetylation affects and controls epimerisation.[19] It may be possible to control or eliminate the degree of acetylation. It would be easy to include a deacetylation step in the production process. The functional properties of alginates are sensitive to the MG ratio and the block distribution within the polymer. The MG ratio is variable[146,148] but naturally occurring bacterial alginates are not rich in G blocks. Following the discovery of an extracellular epimerase Larsen and Haug[16] showed that the epimerase activity was sensitive to the calcium concentration in the medium. By controlling the calcium concentration in the growth medium Larsen and Haug[16] demonstrated that they could raise the G block content from ~18% to ~67%. Higher G block contents could perhaps be achieved by enzymic modification after extraction (Section 2.2).

The major industrial applications of alginate are determined by the molecular weight and the detailed chemical composition of the polymer. Dispersions of alginate in water may be used as thickening or stabilising agents. Examples include stabilisation of foams and emulsions and/or suspensions of solid particles. Typical food applications are given in Table 1 and non-food applications are shown in Table 2. Cold-setting thermo-irreversible gels may be prepared by introducing divalent cations (usually calcium) into alginate solutions or dispersions. The mechanism by which the calcium is added may be used to vary the quality and texture of the final product. Food and non-food applications are reported in Tables 1 and 2. The gelation of alginates forms the basis for the new technological[177] production of reconstituted or fabricated foods. The nature of the gelation process lends itself to such factory production rather than use in household products where the variable calcium concentration in milk and/or water makes the setting process uncertain and difficult to control. The gelation process may be adapted to the production of films and fibres.[178] Alginates are also added to foods to improve 'water retention' (Table 1) and to inhibit ice crystallisation and thus impart freeze–thaw stability. Propylene glycol ester is claimed[179] to be one of the few polysaccharides to function as a genuine emulsifier.

The thickening and stabilising behaviour will depend on the viscosity of the sample. In dilute solution monovalent salts of alginic acid behave as typical polyelectrolytes.[180,181] The molecular weight and hence the viscosity of the solution can be controlled by varying the extraction conditions. Molecular weights may be monitored osmotically,[182] by

Table 1
Applications of Alginates in Foods

Gelling agent	Cooked and instant puddings Pie and pastry fillings Milk desserts Dessert jellies Animal foods Fabricated foodstuffs
Thickening and stabilising agent	Emulsion stabilisation in salad dressings, sauces and relishes Foam stabilisation in beer and whipped products Thickening of sauces, gravies and fillings Suspension of particles in fruit juices, drinks and toppings
Film formation	Sausage skins
Water retention agent	Freeze–thaw stability of frozen foods Enhanced solubility of dry mixes

Table 2
Non-food Applications of Alginates

Gelling agent	Explosives (borate–alginate gels) Air-freshener gels Impression-dentistry and toy manufacturing Pharmaceutical preparations Grouting
Thickening and stabilising agent	Emulsion stability in paints, polishes, latexes Thickening of creams, lotions, liquid detergents and shampoos, textile printing liquids Suspending solids in ceramics, cleaners, mica in wallpaper printing Foam stabilisation in detergents, shampoos
Film formation	Dissolvable fibres
Water retention	Paper coating and sizings Primer for porous surfaces Root dipping Textile printing and dyeing
Other uses	Hardening gelatin in photography Water treatment: flocculating impurities

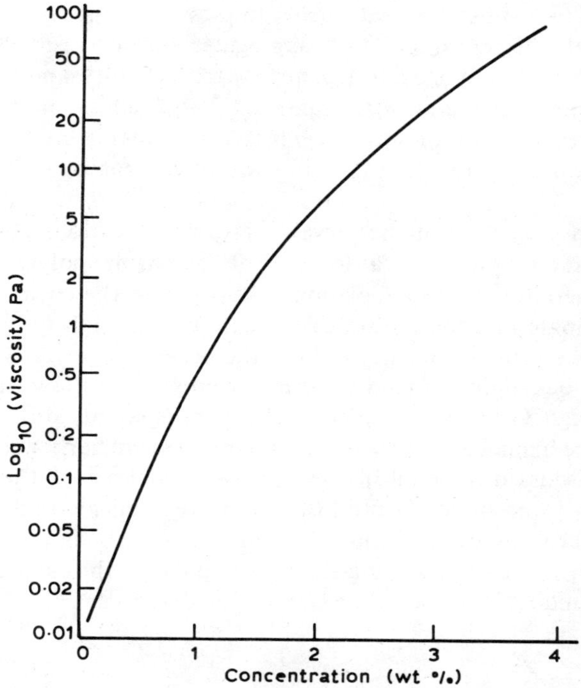

Fig. 10. Typical viscosity–concentration profile for an alginate sample.

sedimentation[183] analysis or by light scattering.[180] Typical weight average molecular weights for 'native' alginate are[180] $\sim 5 \times 10^5$. Light scattering studies[180,184] reveal the presence of aggregates and the difficulty in obtaining true solutions. Even in the presence of sufficient quantities of electrolyte to screen the molecular charge the alginate molecules behave as stiff and inflexible macromolecules.[180,181,184] Conformational analysis calculations[185,186] support this conclusion. The flexibility of the polymer, and thus its size and shape in solution, has been shown theoretically[185,186] and experimentally[184,187] to depend on the relative proportions of M, G and MG blocks. When alginates are employed as thickening agents the concentrations are[178] normally $>1\%$. At these concentrations the solutions are no longer dilute and molecular entanglement will enhance the viscosity and increase the dependence of viscosity on polymer concentration. The typical viscosity–concentration curve[178] shown in Fig. 10 is atypical of a simple entangled polymer solution. The rate of increase of viscosity with concentration decreases with increasing concentration

(Fig. 10). It is possible that such samples are dispersions containing incompletely dispersed polymer aggregates or microgels. Addition of divalent cations (e.g. calcium) increases the viscosity and enhances the shear thinning behaviour of the sample.[178] Such effects are likely to arise from the development of a weak gel network probably involving localised precipitation and interchain binding within the entangled polymer network.

Gelation occurs at higher levels of divalent cations. Gels may be prepared at concentrations as low as $\sim 0{\cdot}1\%$ but the polymer concentration is generally[178] $> 1\%$. Gelation is sensitive to the chemical structure of the alginate and the nature of the cation. For high G-block content alginates the gelling efficiency follows the order[188] $Ba^{2+} > Sr^{2+} > Ca^{2+}$. High M-block alginates tend to form turbid gels with low elastic moduli whereas high G-block alginates produce transparent, stiffer and brittle gels.[188] The main features of the gelation process are fairly well understood and are discussed in detail in recent review articles.[7–10] Cation binding studies[25,172] on isolated G and M blocks have implicated calcium binding within inter-G-block junction zones in calcium alginate gels. This has been confirmed by monitoring the environment of the carboxyl group of the guluronic acid residues using circular dichroism.[173,189–193] The model

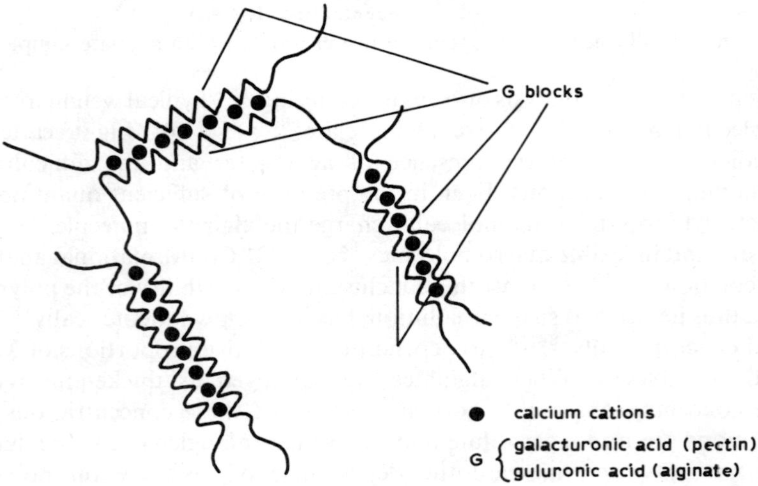

Fig. 11. 'Egg-box' model for the gelation of alginate and low-methoxy pectin. Recent results suggest that the junction zones may be simply two G-block regions linked by bound calcium cations.

for the junction zones was based on evidence gleaned from conformational analysis[185,186,190,194,195] calculations and polarised infrared and X-ray fibre diffraction data.[196–201] The two-fold buckled ribbon structure contains cavities in which calcium ions are bound to guluronic acid residues (Fig. 11). The model for gelation has been called the 'egg-box' model.[190] Originally the junction zones were considered[190] to contain several G-blocks. Recent studies[193,202–204] are consistent with the formation of just dimers. This could explain the poor crystallisation of calcium alginate fibres.[200]

3.3. Curdlan

Curdlan is a homopolymer of (1 → 3) linked β-D-glucose which was originally isolated from the microorganism *Alcaligenes faecalis* var. *myxogenes*. The polymer was discovered in 1964 by Harada and coworkers[205–210] whilst they were engaged in a research programme designed to screen for soil microorganisms which are capable of metabolising petroleum fractions. *Alcaligenes faecalis* var. *myxogenes* (strain 10C3) was found to grow in a medium containing 10% ethylene glycol as the sole carbon source. The bacterium produced two structurally distinct polysaccharides: a water-soluble acidic heteroglycan and a water-insoluble neutral homoglycan. The heteroglycan was found to contain succinyl groups and was called succinoglycan.[209] Aqueous suspensions of the homoglycan gelled on heating[205] and the polymer was called curdlan.[211] Curdlan forms complexes with the dyes aniline blue[212] and congo red.[213,214] The development of a blue colour with aniline blue has been shown[212] to be dependent upon polymer concentration, molecular weight and gel-forming ability. Incorporation of aniline blue into agar plates provided a screening test for curdlan production. Microorganisms were screened for the development of blue colonies. A series of mutants of *A. faecalis* var. *myxogenes* (10C3) were obtained by subjecting strain 10C3 to a variety of mutagenic agents. Mutant 22 produced[215] mainly succinoglycan, mutants 11 and 60 produced[215,216] only curdlan but in low yield and a spontaneous mutant 10C3K produced[205,217] only curdlan in high yield. A further mutant 10C3k-u (ATCC 21680, IFO 13140) was finally selected as a high-yielding curdlan producer. This is a uracil-requiring auxotroph isolated[218] by the Takeda Chemical Company and specified in their 1973 US patent application[219] as PS-13140. The aniline blue method was used to screen the IFO Culture Collection (Osaka, Japan) for curdlan producers. Four out of the nine stock cultures of *Agrobacterium* strains were selected for pure curdlan production. The 1973 US patent[219] specifies *Agrobacterium radiobacter* (ATCC 21679, IFO 13126) as micro-organism PS-13127. Mutant forms of *Rhizobium* have also been found[220] which produce curdlan in high yield.

The Takeda Chemical Company produced curdlan using a batch process.[218] The potential of continuous culture production has been discussed by Railton-Phillips and Lawford.[221] Isolation of curdlan involved the following procedure. The fermentation broth was treated with alkali. Under alkaline conditions curdlan is soluble and the bacterial cells are removed by centrifugation. The supernatant was then neutralised and the insoluble polymer removed by centrifugation, washed, and dried. The Takeda Chemical Company developed curdlan to pilot plant production but then abandoned commercial exploitation.[221] The Wako Chemical Company marketed curdlan in 1981. Curdlan has not been granted food approval but may be acceptable in certain countries as a 'natural product'.

Potential food applications for curdlan were considered.[222] Suggested applications include use as: a 'water retention' agent in sausages, starch-based jellies; a binding agent in spaghetti, noodles, hamburgers or starch-based jellies; or a gelling agent for sweet jellies and desserts. As a thickening agent curdlan is unlikely to compete successfully with xanthan gum. However, curdlan does show unique gelation behaviour. Curdlan forms a 'high-set' gel[223,224] when an aqueous suspension is heated to ~70–100°C. This gel has been said to be thermally irreversible although melting of the gel has been reported[225] at temperatures above 100°C. An aqueous suspension when heated to ~60–65°C and then cooled forms a 'low-set' thermally reversible gel.[223,224] Curdlan gels show syneresis but this can be inhibited by the presence of tannins, sugars or starch.[225,226] The addition of up to ~5% starch is claimed to be sufficient to completely inhibit syneresis.[226] Canned multilayered jellies containing both 'high-set' and 'low-set' gels have been described.[224] Curdlan is apparently not degraded by the human digestive enzymes but will be degraded by colonic bacteria suggesting possible applications in low-calorie or calorie-reduced foods. Other industrial uses include[221] artificial foods for fish or silkworms, tertiary oil recovery, a support for immobilised enzymes and as a binding agent in tobacco products. Curdlan may be used to produce water-insoluble, biodegradable films impermeable to oxygen.[221,224] There are reports that curdlan may have anti-tumour activity.[227,228]

Curdlan has been shown to be a homopolymer containing (1 → 3) linked β-D-glucose[211,229,230] (Fig. 12). The degree of polymerisation (dp) is believed[229] to vary with fermentation conditions with values ranging from

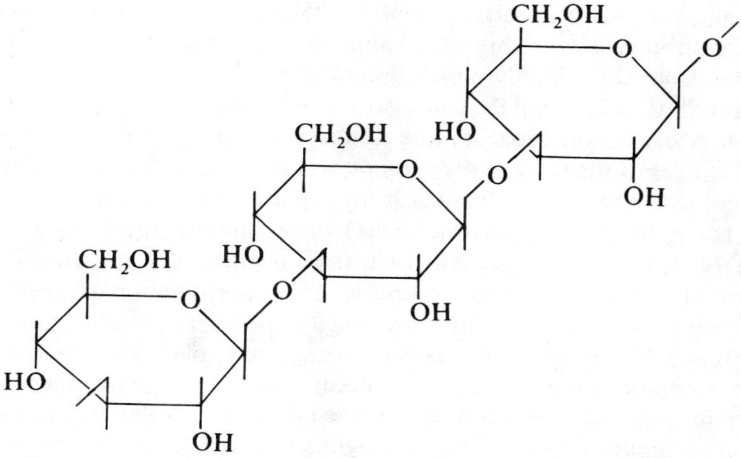

Fig. 12. Chemical structure of curdlan.

dp ~ 135–455, although values as high as dp ~ 540 have been reported.[231] The solution properties of curdlan are difficult to determine because curdlan forms suspensions rather than solutions when dispersed in water. It is useful to compare and contrast curdlan with the related polymer schizophyllan. Schizophyllan is a branched polysaccharide composed of a (1 → 3) linked β-D-glucan backbone substituted through position 6 of every third residue with a β-linked D-glucose.[232,233] The branched structure appears to inhibit molecular association. Sedimentation, viscosity and light-scattering studies[117,118,234–236] of schizophyllan in water or 0·01M NaOH suggest that the polymer behaves[117,118] as a semi-flexible rod with a persistence length of ~ 180 nm. The samples in water show some evidence of slight aggregation. The data[117,118,234–236] have been taken to be compatible with a triple-helical structure similar to that proposed[237,238] for a homopolymer of (1 → 3) linked β-D-xylan. Comparative studies[117,118,234–236] using water–dimethyl sulphoxide (DMSO) mixtures as solvents suggest that the triple helix dissociates irreversibly into coil-like strands when the DMSO concentration is greater than 87%. Light scattering and viscosity studies on molecular weight fractions of curdlan in cadoxen–water (1:1) as solvent suggest[239] that the polymer behaves as a flexible coil. Comparative studies on curdlan[239] and cellulose[240,241] in cadoxen–water mixed solvents suggest that curdlan is more flexible than cellulose. ^{13}C nmr studies[231] on aqueous suspensions of curdlan (dp ~ 540) and on alkaline solutions show a sharp decrease in line width for concentrations of NaOH > 0·22M. The high-resolution spectra obtained[241] at higher alkaline conditions (>0·22M NaOH) are characteristic of a highly mobile 'random coil' structure. The transition observed at 0·22M NaOH can also be monitored[212,213,242] by optical rotatory dispersion, viscosity and flow birefringence studies and absorption studies of the curdlan–dye complexes obtained with aniline blue and congo red. By analogy with schizophyllan such studies[212,213,241,242] are taken to indicate an order(helix)–disorder(coil) transition at 0·22M NaOH. Unlike schizophyllan the transitions observed for curdlan are reported[212,213,241,242] to be reversible. This interpretation appears to be imcompatible with deuterium exchange experiments[243] on curdlan in NaOD–D_2O solvents and it is not possible to explain this discrepancy. The transition is sensitive to the molecular weight of the polymer.[213,224] For dp <25 no transition is observed. For dp >200 the transition becomes insensitive to molecular weight. A similar transition is observed when non-solvents such as dioxane, water or 2-chloroethanol are added to solutions of curdlan in DMSO.[244] The consensus of opinion is that aqueous suspensions of curdlan may be regarded as poorly dispersed aggregates of an ordered and, on the basis of work discussed below, possibly a triple helical structure. On the basis of nmr data[231] it has been proposed that the suspension may also contain single helical species although there is no direct unambiguous evidence for such structures.

Curdlan gels may be prepared by several methods. Aqueous suspensions of curdlan become less turbid on heating, yielding clear sols at 50–55°C. If these sols are heated to ~60°C and then cooled they produce 'low-set' thermally reversible gels.[223,224] If the sols are heated to higher temperatures >95°C a 'high-set' thermally irreversible translucent gel is produced.[223,224] It has been reported[225] that 'high-set' gels may be melted when heated to 140–160°C. Gelation may also be induced by dialysis of an alkaline solution against a non-solvent water, by cooling a suspension in 0·2–0·63M DMSO[245] or 3–7M urea,[246] or by adding calcium ions to a weakly alkaline solution.[224]

Gels have been probed by ^{13}C nmr,[231,247] small-angle X-ray scattering[248] (SAXS), X-ray fibre diffraction and deuterium exchange experiments monitored by infrared spectroscopy.[243–252] The morphology of heat-treated curdlan suspensions has been studied by electron microscopy.[253] These studies suggest the presence of fibrillar structures in the suspensions which assemble into networks on heat treatment.[253] SAXS studies have been taken[248] to support the existence of fibrillar structures. The X-ray fibre diffraction patterns obtained for curdlan samples depend on the method of sample preparation. Marchessault and coworkers[250–252] prepared oriented fibres by extrusion of curdlan solutions in DMSO into methanol. The fibres were washed with water and they showed[250] weakly crystalline patterns. After annealing two crystalline polymorphs were obtained.[250,252] Crystalline form A was obtained at high relative humidity and crystalline B was obtained at low relative humidity. Fulton and Atkins[243] obtained B patterns for annealed 'high-set' gels. The B patterns are believed to represent a 6-fold triple helical structure[243,250,252] similar to that proposed[237,238] for (1 → 3) linked β-D-xylan and this structure has been shown[252] to persist in the hydrated state (pattern A). Takeda *et al.*[249] obtained a third type of pattern (type C) from curdlan films cast from aqueous suspensions. The pattern was taken[249] to represent a mixture of 7-fold single and 7-fold triple helices. Comparative ^{13}C nmr studies[231,247] on 'low-set' gels and alkaline solutions of curdlan have been interpreted as supporting the existence of such single helices. Whereas the chemical shift observed for C-1 and C-3 may be indicative of a conformational change it is difficult to see how the ordered state can be definitely assigned to the presence of a single helical species within the gel. The type C pattern was also obtained[243] from stretched films cast from curdlan solutions in 1 M NaOH. The data have been interpreted[243] in terms of a 7-fold triple helix which may be regarded as a loose distorted state of the 6-fold triple helix. The 7-fold triple helix converts[243] to the 6-fold triple helix on annealing. X-ray data[243,250–252] and deuterium exchange experiments[243] have led to a micellar model[243] (Fig. 13) for the 'high-set' gel consisting of ordered triple helices associated into micellar domains. The domains contain helices hydrogen-bonded via interstitial water and association of micelles forms the junction zones of the gel. Presumably, according to Fulton and Atkins,[243] aqueous samples of curdlan are dispersions composed of incompletely dispersed aggregates of micelles containing triple helices. Heating appears to disperse the aggregates leading eventually to a clear micellar solution. The distinction between a micellar solution and a true solution of triple helices relies heavily on the interpretation of deuterium exchange experiments.[243] The 'high-set' gel results from an association of these micelles into ordered domains (Fig. 13). It would appear that the distinction between 'high-set' and 'low-set' would then rest on the extent of aggregation and/or crystallisation of the micellar domains. A conflicting model has been suggested by Saito and coworkers[229,247,254] and Kasai and Harada.[248] They envisage a mixture of single and triple helices present as micelles within the aqueous dispersion. The 'low-set' gel is pictured as an

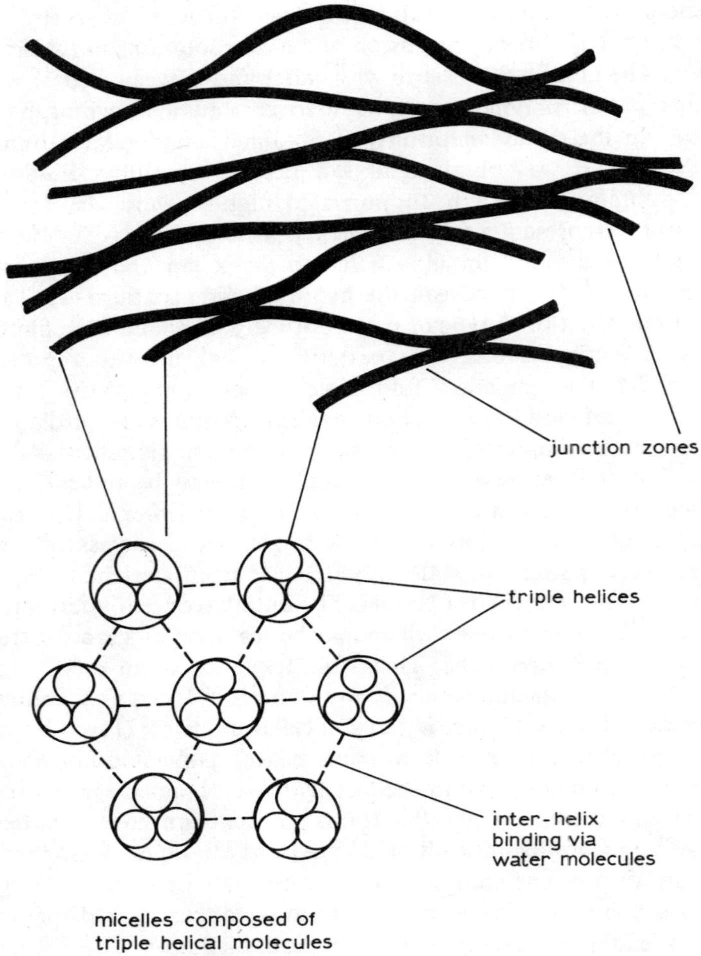

Fig. 13. Micellar model for curdlan gels.

assemblage of such mixed micelles into a network. Heat treatment to form a 'high-set' gel is considered to lead to a conversion of single to triple helices. It is not clear whether the triple helices are considered as separate entities which then associate or whether each strand is considered to participate in several triple helical junction zones. The low crystallinity in curdlan gels, when compared to the biosynthetically produced crystalline paramylon, has been discussed by Chuah *et al.*[252]

3.4. Gellan Gum

Gellan gum is the generic name for the extracellular polysaccharide secreted by the previously unrecognised microorganism *Pseudomonas elodea*.[255,256] The polymer has been referred to by the code names S-60 and PS-60 in the earlier literature.[255–259] The polymer was discovered by Kelco as part of a screening programme. The bacterium was isolated from plant (*Elodea*) tissue. Gellan gum is produced by aerobic submerged fermentation of *Pseudomonas elodea*.[260] Batch culture conditions are described by Kang and coworkers.[258–261] A deacetylated form of gellan gum is produced commercially by Kelco under the tradename Gelrite. By varying the extraction conditions it is possible to obtain three 'types' of gellan gum.[259] Gellan gum is isolated from the fermentation broth by alcohol precipitation. The polysaccharide can also be precipitated from the broth by reducing the pH but this is not used commercially. The native form of gellan gum is *O*-acetylated. A deacetylated product may be obtained by alkali treatment (80°C, pH ~ 10, 10 min) followed by alcohol precipitation. Clarified products, freed from bacterial cell debris, may be obtained by filtration of the hot fermentation liquor. The clarified deacetylated product is sold as Gelrite. Gellan gum has not been approved as a food additive. However, extensive toxicological trials have been carried out and Kelco have announced[262,263] that they intend to seek FDA approval for use as a broad-spectrum gelling agent.

Potential food applications of gellan gum are discussed by Sanderson and Clark.[262] Cold-setting gels may be prepared by adding monovalent or divalent cations to sufficiently concentrated aqueous dispersions of gellan gum. At appropriate ionic strength and polymer concentration gellan gum forms thermoreversible gels. The melting temperature of the gel can be increased by raising the ionic strength. Thus there is the possibility for preparing cold-setting or thermosetting, thermoreversible or thermoirreversible gels. This indicates the versatility of gellan gum as a broad-spectrum gelling agent. Gellan gum may be used to replace or partially replace all of the commonly used plant polysaccharide gelling agents often at significantly lower polymer concentrations.[262] Recipes for the use of gellan gum for preparing confectionery, jams, jellies, fabricated foods, water-based gels, pet foods, pie fillings and puddings, icings and frostings, and in dairy products are described by Sanderson and Clark.[262] Several non-food applications have been suggested for gellan gum. Gelrite is currently being marketed as an agar replacement for the preparation of growth media for microorganisms or plant tissue culture.[259,264–267] Gelrite gels can be subjected to several autoclaving

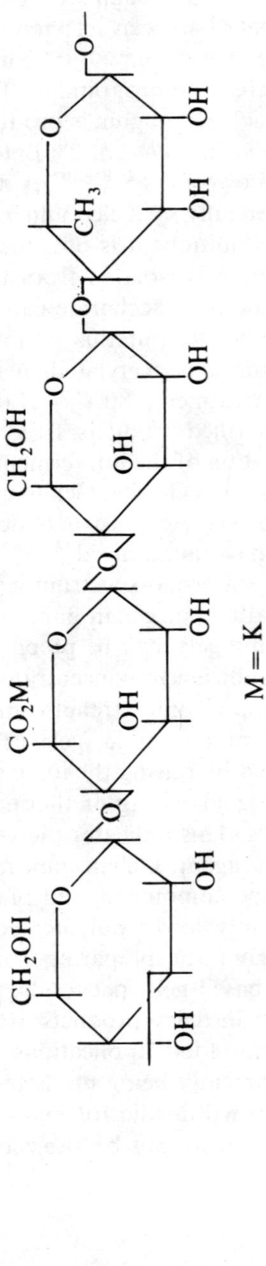

Fig. 14. Chemical repeat unit of gellan gum. The commercial product Gelrite is generally the potassium salt of the deacetylated polymer.

cycles without degradation, they are resistant to enzymic degradation, chemically inert to most biological growth media additives and melt and set like agar gels.[261] The advantages of Gelrite when compared to agar are cited[261] as improved clarity, high purity and hence reduced toxicity to sensitive bacteria or plant cells, and the ability to develop the same gel strength at lower polymer concentrations. Gelrite has been shown to compare favourably with agar for the growth of clinically important[265] and thermophilic eubacteria[264] and for the growth of mesophilic *Methanobacterium* and *Methanobrevibacter* species.[267] High concentrations of certain sugars can inhibit gelation.[263] It has been suggested that gellan gum could be used as a matrix[261] for immobilised enzymes or whole cells, and in the production[70] of air-freshener gels. Gelatin–gellan gum mixes may be used to produce rigid coacervate gels with potential uses for microencapsulation.[70] It is not known whether gellan gum can be digested or whether it can be fermented by gut microorganisms. Should gellan gum be accepted for food use then it would provide an ideal system for studying changes in gut microflora in response to a new polysaccharide.

Gellan gum has been shown[268,269] to be a linear anionic heteropolysaccharide with the tetrasaccharide repeat unit shown in Fig. 14. Additional structural data is given in the articles by Miles *et al.*[270] and O'Neill *et al.*[271] Native gellan gum contains $\sim 5{\cdot}6$ wt% *O*-acetyl groups. This level of substitution would correspond to an average of one acetyl group per repeat unit. The location of the acetyl substituents is uncertain but nmr studies[269] suggest substitution at C-6 of one or both of the glucose residues.

Dilute aqueous samples of gellan gum are highly viscous and exhibit marked shear thinning behaviour. Creep compliance studies[272] suggest that the shear thinning behaviour involves breakdown of a weak gel-like structure. This behaviour is similar to that seen for xanthan gum (Section 3.1) and suggests that the aqueous gellan samples are dispersions rather than true solutions. DMSO appears to be a good solvent for polysaccharides and solutions of gellan gum in DMSO have a lower viscosity and markedly reduced shear thinning behaviour.[272,273] Light scattering studies in DMSO suggest a high molecular weight (~ 1–2×10^6) stiff polymer coil.[270,272,273] Comparative electric birefringence studies[272] of gellan gum in water and DMSO suggest an extended rigid structure in water. The rigidity could imply adoption of an ordered (helical) structure and/or may reflect molecular aggregation.

Gelation occurs at higher concentrations and is sensitive to the degree of acetylation of the polymer and the cation content and composition of

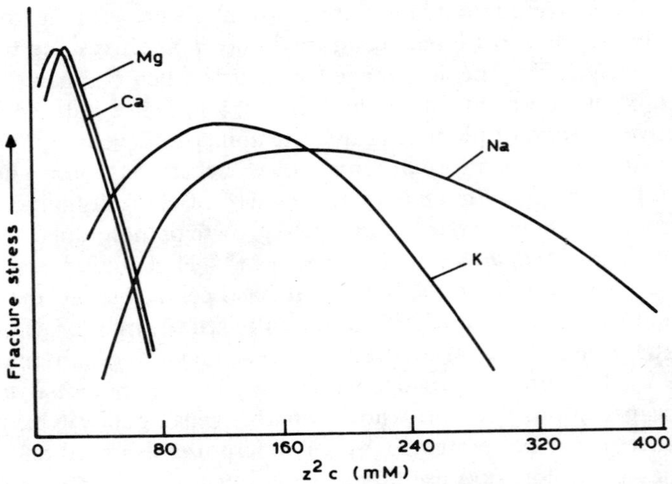

Fig. 15. Effect of cations upon the fracture stress of gellan gum gels. Cation concentrations are plotted as z^2c where z is the cation valency and c is the concentration in mM.

the medium.[259] The native material produces soft elastic gels whereas deacetylated gellan gum forms hard brittle gels.[259] Gelation is sensitive to the type and concentration of cations in the medium. Figures 15 and 16 show the effects of common monovalent and divalent cations on the fracture strength[259,263] and the shear modulus of the gels. At an equivalent ionic strength divalent cations produce stronger gels with higher elastic moduli. Detailed studies on the effects of monovalent and divalent cations on the mechanical properties of gellan gels suggest[274] that gelation may be sensitive to cation type as well as cation valency. With increasing ionic strength both the fracture strength and the elastic modulus pass through a maximum and then decrease at higher ionic strength. At high ionic strength the gels become noticeably turbid. It was originally reported[270] that there was no change in optical rotation upon melting of the gel. Recent studies[275] show a sharp change in optical rotation accompanying melting and setting of the gels. Such changes are normally taken to indicate order(helix)–disorder(coil) transitions. Further studies are needed to confirm these observations on gellan gum. It would be useful to probe the carboxyl and acetyl groups by circular dichroism to test for conformational changes on heating and cooling. The use of tetra methyl or tetra ethyl salts may provide a means of preventing inter-molecular

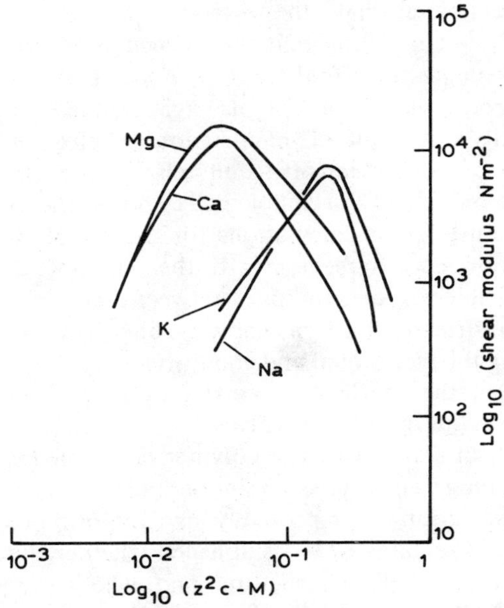

Fig. 16. Effect of cations on the shear modulus. Concentration of cations is plotted as z^2c where z is cation valency and c is the concentration in mol/litre. Polymer concentration 0·6%.

association and allowing the conformation and conformational transitions to be studied.

The nature of the junction zones within gellan gels has been probed by X-ray fibre diffraction.[270,272,276,277] Partial deacetylation of gellan gum markedly affects the mechanical properties of the gels and this change is also apparent in the X-ray diffraction patterns. The X-ray data suggest that deacetylation does not alter the secondary structure of the polymer but does improve intermolecular association and crystallinity. Increasing crystallinity accompanies the transition to a harder more brittle gel. Dehydration of the partially deacetylated fibres does not alter the secondary structure of the polymer or the molecular alignment but does reduce the level of crystallinity.[270] Thus the crystalline patterns obtained are not artifacts of dehydration and probably reflect the structure of the junction zones of the hydrated gels. X-ray fibre diffraction studies[277] on monovalent salt forms of deacetylated gellan gum suggest that the helical conformation and the unit cell dimensions appear to be insensitive to the nature of the cation. The axial advance per chemical repeat unit is approximately half the extended length of chemical repeat unit[270,272,276,277] suggesting that the molecular conformation could be a compressed single three-fold helix or an extended three-fold double helix. A stereochemically acceptable right-handed single helix and a stereochemically acceptable left-handed double helix can be built.[277] Both structures can be packed into the unit cell.[277] The carboxyl groups are located on the outside of the double helix and on the inside of the single helix.[276] Structure factor calculations for the double helix and a single helix model show poor agreement with the experimental data and it has not been possible to discriminate between the two models.[277] The experimental diffraction patterns show a concentration of the diffracted radiation along the meridional and equatorial axes[270,272,276,277] and this, plus the sites of the charged carboxyl groups, may favour the double helical model. Thus gelation involves the formation of junction zones containing helical sections of the polymer chain packed into a trigonal unit cell containing two polymer chains per unit cell. The cation sensitivity suggests an association which probably involves local precipitation and/or crystallisation. It remains to be established whether the polymer adopts a double or single helical structure, and whether acetyl groups act stereochemically to prevent helix formation or association, or whether they control the polymer solubility.

3.5. XM6

XM6 is the codename assigned to the extracellular polysaccharide secreted by *Enterobacter* (NC1B 11870). This microorganism was isolated and purified from a laboratory culture discovered in the Microbiology Department, University of Edinburgh, by Sutherland and coworkers.[278] A sample of the microorganism has been deposited with the National Collection of Industrial Bacteria, UK. Conditions for batch culture are described by O'Neill *et al.*[279] The polymer may be isolated by alcohol precipitation and clarified by dispersion in water, centrifugation, dialysis against water and freeze-drying.

XM6 is an example of a gelling polysaccharide.[278] At concentrations $\sim 0{\cdot}3\%$ the polymer may be dispersed in water to produce a viscous liquid.[278] Addition of cations changes the sample from a viscous liquid to a thermoreversible gel.[278] Cations may be added at room temperature to produce a cold-setting gel. Alternatively the polymer and electrolyte may be mixed at high temperature and cooled to produce thermosetting gels. The gel 'melts' over a narrow temperature interval and the 'melting point' increases with increasing ionic strength.[278] The author is unaware

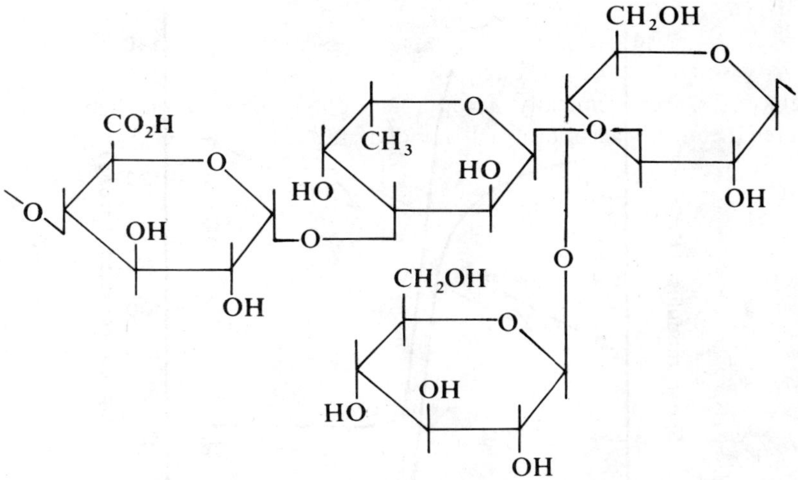

Fig. 17. Chemical repeat unit of XM6.

of any toxicity studies on XM6 and the polymer is not approved for food use.

The primary structure of XM6 has recently been determined[279] and the chemical repeat unit is an anionic branched tetrasaccharide shown in Fig. 17. XM6 may be regarded as a deacetylated form of the extracellular polysaccharide produced by the unrelated microorganism *Klebsiella aerogenes* serotype K54. The K54 structure[280] is identical to that of XM6 with the exception that every alternate repeat unit contains a single *O*-acetyl substituent at C-2 or C-4 of the fucosyl residue.[271,279] The K54 polysaccharide forms viscous liquids when dispersed in water but does not gel upon addition of cations. Aqueous dispersions of deacetylated K54 do gel upon addition of cations.

Comparative X-ray fibre diffraction studies on K54[281] and XM6[282] have provided an indication of the effect of acetylation upon gelation. The deacetylated K54 yields X-ray fibre diffraction patterns identical to those obtained for XM6 but very different from those obtained for native K54. Deacetylation leads to a marked increase in crystallinity. Recent studies[282] suggest that both XM6 and K54 may adopt the same helical structure (8_3 double helix) but that the acetyl groups inhibit crystallisation for K54. The X-ray patterns for XM6 suggest that the junction zones of the gel are highly crystalline assemblages of segments of the helical macromolecules. The X-ray patterns indicate the presence of at least two crystalline modifications.

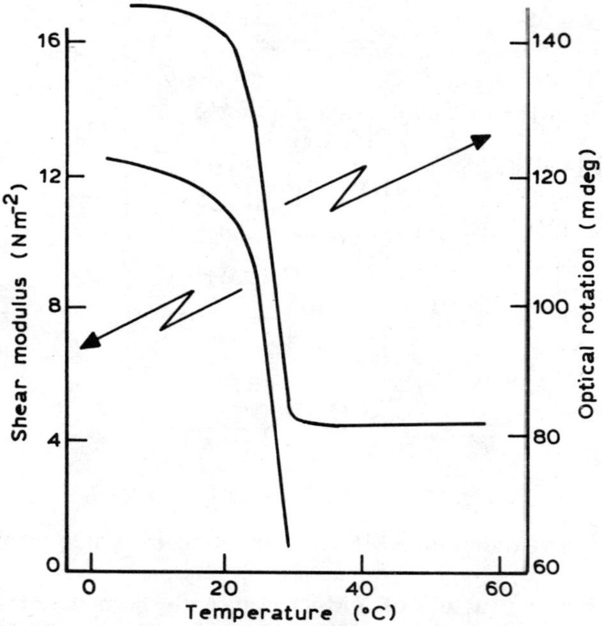

Fig. 18. Change in optical rotation accompanying the melting of an XM6 gel.

Above the gel–sol transition temperature XM6 samples behave as viscoelastic liquids.[278] The gel–sol transition is accompanied by an abrupt change in optical rotation[278] (Fig. 18). The transition mid-point temperature is insensitive to polymer concentration but highly dependent on ionic strength.[278] Circular dichroic studies show a sharp transition on melting and setting which has been taken to indicate[278] a change in the local environment of the carboxyl group of the glucuronic acid residue. Gelation and the 'melting temperature' are both sensitive to the presence of both divalent and monovalent cations.[278] Addition of electrolyte would screen molecular charge and could stabilise helical conformations and promote intermolecular association. The effects cannot be explained solely in terms of screening effects because there appears to be a dependence on cation type. This dependence does not follow a simple Hofmeister series and the cations sodium and calcium are claimed[278] to be most effective at gelling the polymer.

The present data are consistent with the formation of crystalline junction zones in which cation-selective effects could be explained in terms of cation size and binding within the junction zones. There is a need to establish whether cation binding occurs upon gelation. This would help to establish whether the circular dichroism and optical rotation data monitor an order(helix)–disorder(coil) transition or interchain binding via cations of the ordered helical structures.

3.6. Other Polysaccharides

Bacteria, fungi and yeasts are all potential sources of new polysaccharides. The following is intended as an introduction to the literature on such polysaccharides and references will be confined to review articles and recent or pertinent papers. In addition to xanthan gum (Section 3.1), bacterial alginate (Section 3.2), curdlan (Section 3.3) and gellan (Section 3.4) the following polysaccharides have been produced on a large scale and are available commercially. These include dextran[70,283] from *Leuconostoc mesenteroides* (NRRL512(F)), Biozan (S130)[72,73,79,261] from *Alcaligenes* (ATCC 31555), S-194[72,73,79,261] from *Alcaligenes* (ATCC 31961), Actigum C5 (scleroglucan)[69,70,283] from *Sclerotium rolfsii* and pullulan[272,283,284] from *Aureobasidium pullulans*. Scleroglucan has the calorific equivalence of starch[69,70,283] and has been shown to lower cholesterol levels in chicks and rats.[285] Suggested uses in food[69] are as a suspending, coating and gelling agent. Pullulan is non-toxic and non-digestible by human enzymes and possible food uses for preparing food films as emergency rations have been suggested.[286] The microorganisms *Bacillus polymyxa*[283] and *Rhizobium meliloti* (IFO 13336)[287] produce polysaccharides which gel. The microorganisms *Arthrobacter*[69,71,100,283] *viscus* (NRRL B-1973, NRRL B-1797), *Arthrobacter stabilis* (NRRL B-3225), *Azotobacter indicus* var. *myxogenes*,[69,71,283] *Erwina tahitica*,[69,71,283] *Alcaligenes* (ATCC 31853)[79,261] and *Acetobacter* (NBI 1022)[288] produce polysaccharides with interesting rheological properties. Bakers' yeast glucan[69,70] is the cell wall material obtained from *Saccharomyces cerevisiae*. It has been suggested[69,70] that this material and similar material extracted from spent brewers' yeast cell walls could be added to foods to simulate the mouthfeel of fats and/or oils. Several types of polysaccharides may be extracted from yeasts.[71] Emulsan[289,290] from *Acinetobacter calcoaceticus* RAG-1 is an example of a lipopolysaccharide which functions as a true emulsifier.

4. CONCLUSIONS

The use of biotechnology to produce or modify polysaccharides is clearly of industrial importance. Applications to the food industry are restricted by the cost in time and money needed to obtain approval for use in foods. It is possible that in the future it may become necessary not only to demonstrate that a new additive is safe but also that it is *desirable* and *necessary* to introduce a *new* additive.

The potential problems with traditional plant polysaccharides are variations in cost, supply and chemical structure. For land plants plant breeding and chemical modification have been successful at overcoming these problems. In the case of marine algae farming could help to secure supply. At present breeding programmes, farming of selected species and chemical modifications appear to be preferable on economic grounds to enzymic modification after extraction. The major hurdle is cost and, in particular, the need to obtain food approval. Major advances in the production and purification of enzymes could make enzymic processes economically viable. Alternatively, if an enzymic modification could be used to generate a high-value polysaccharide from an abundant or cheap source then the process may become economical.

At present the discovery of new polysaccharides relies on the screening of the extracellular polysaccharides produced by bacteria. Fungi and yeasts are also potential sources of new polysaccharides. The microbially produced polysaccharides have the advantages of controlled cost, supply and chemical structure. It is possible that the substrates may be chosen from waste or surplus products. Bacterial alginate is an example of a polysaccharide which is also produced by plants. This is not unique as the $\beta(1 \rightarrow 3)$ glucans are examples of polysaccharides produced by plants and bacteria. Screening procedures have led to a range of useful industrial polymers including xanthan, Biozan, S194, gellan, actigum C5, pullulan, dextrans and emulsan. Within the food industry only xanthan has been approved for food use and gellan may be the second such polymer approved as a food additive. Xanthan and gellan would cover the main areas of thickening, suspending and gelling ability. It would be desirable to have alternatives to xanthan and gellan but the cost of obtaining food approval would suggest that any new polymer would need to have demonstrable advantages over xanthan or gellan to warrant exploitation. New polysaccharides with new properties may generate new market opportunities. Perhaps a genuine emulsifier acceptable to the food industry can be found. Is it possible that the newly discovered cyclic polysaccharides produced by microorganisms such as *Rhizobium* can provide an alternative to cyclodextrins as complexing agents? Microbial polysaccharides are usually neutral or anionic. The discovery of cationic polysaccharides could open new markets.

Having found a potentially useful polysaccharide it would be nice to be able to optimise production or improve the functionality of the polymer. However, little is known about the genetic control of biosynthesis, about biosynthetic mechanisms and the influence of chemical structure on functionality. The only general observation is that gelation seems to be favoured by microbial polysaccharides which can crystallise upon association. It appears to be possible to increase yields by the deliberate mutation of selected strains of microorganisms. The work on *Pseudomonas aeruginosa* suggests the possibility of genetically altering bacteria to enhance yield. The molecular weight of the polymer is of importance industrially but little is known about how to control this parameter. An obvious advantage would be the ability to transfer production of a polysaccharide from one microorganism to another microorganism. Is it possible to engineer a bacterium to enable it to produce several polysaccharides? The ability to control or tune functionality by modifying the primary structure of polymers suffers from the difficulty in designing screening tests to detect such changes and from a lack of knowledge

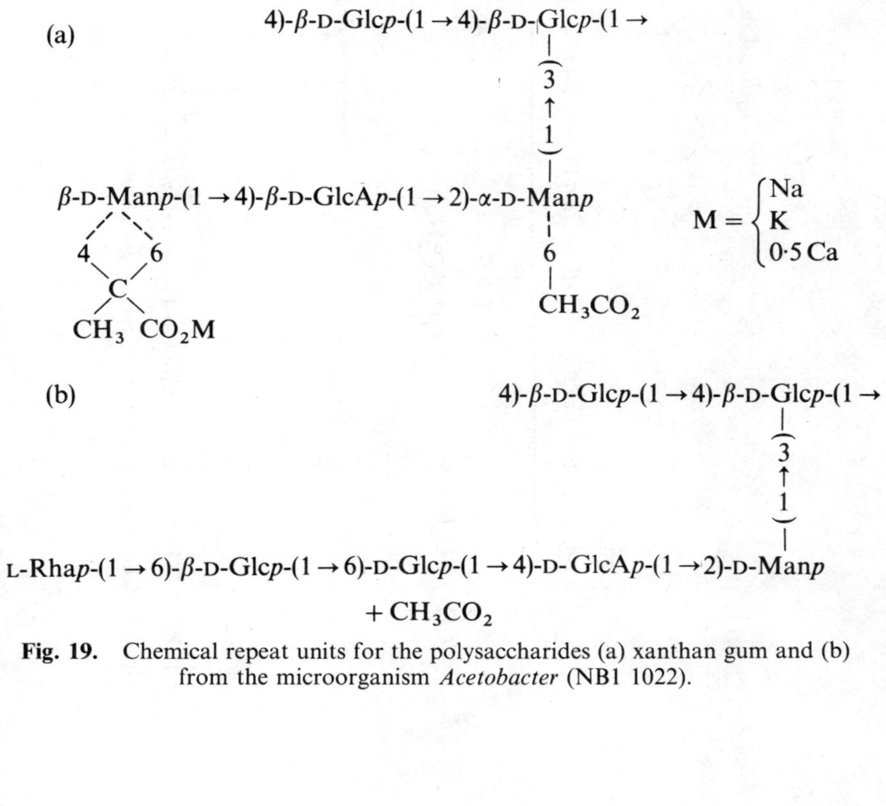

Fig. 19. Chemical repeat units for the polysaccharides (a) xanthan gum and (b) from the microorganism *Acetobacter* (NB1 1022).

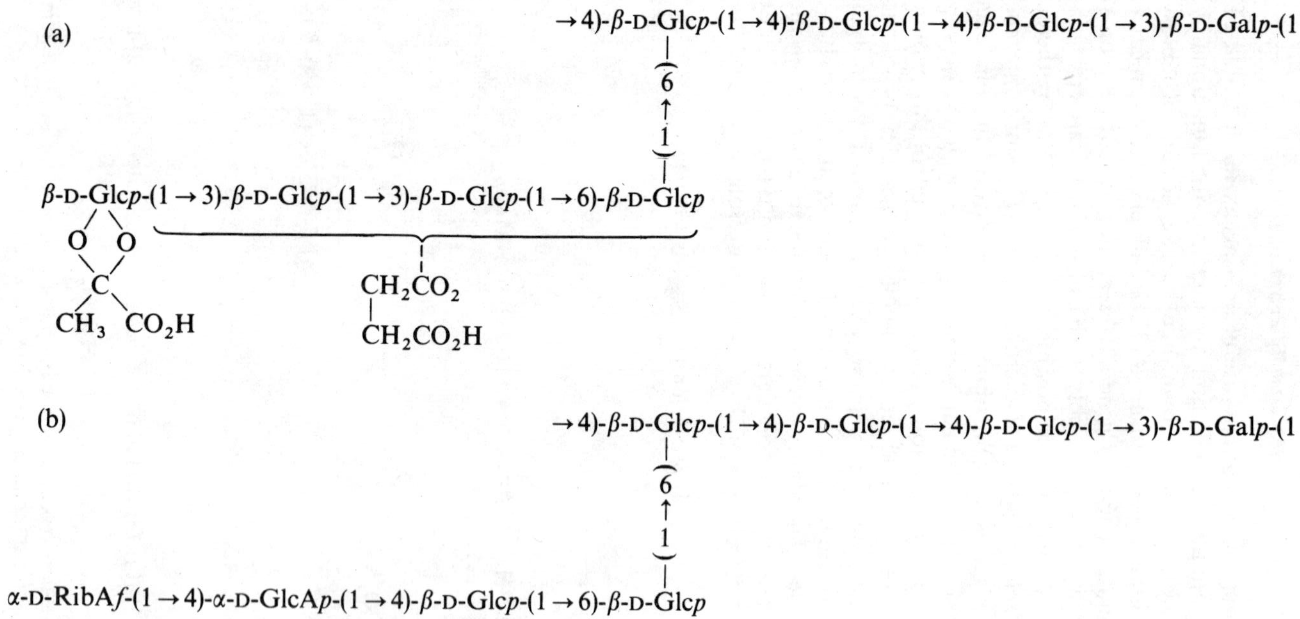

Fig. 20. Chemical repeat units for the polysaccharides (a) succinoglycan and (b) from the microorganism *Rhizobium meliloti* (IFO 13336).

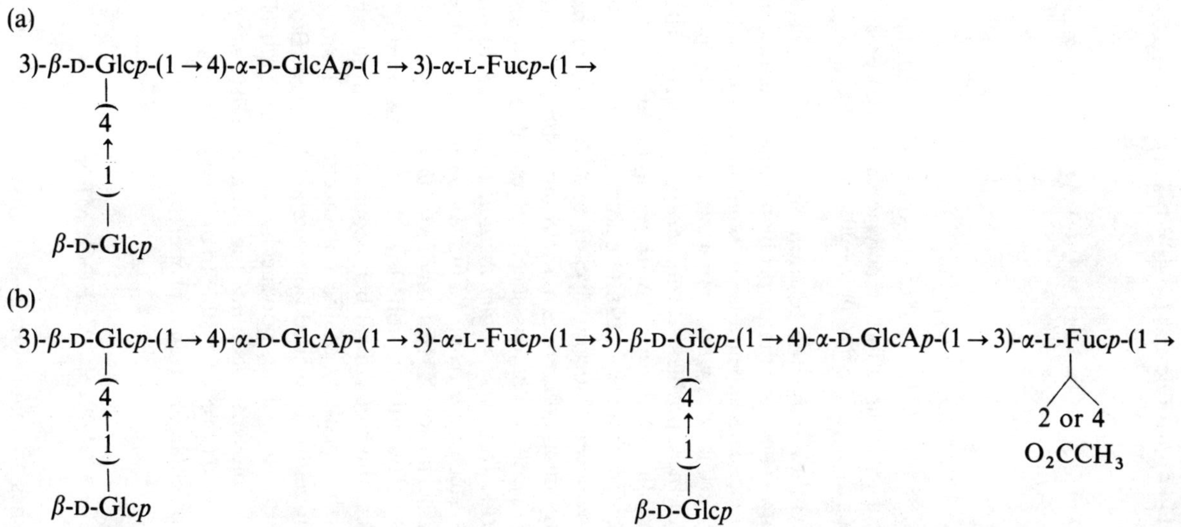

Fig. 21. Chemical repeat units for (a) XM6 and (b) the polysaccharide produced by *Klebsiella aerogenes* serotype K54.

(a)

3)-β-D-Glc*p*-(1 → 4)-β-D-GlcA*p*-(1 → 4)-β-D-Glc*p*-(1 → 4)-α-L-Rha*p*-(1 →

(b)

3)-β-D-Glc*p*-(1 → 4)-β-D-GlcA*p*-(1 → 4)-β-D-Glc*p*-(1 → 4)-α-L-Rha*p*-(1 →
3
↑
1
α-L-Rha*p* or α-L-Man*p*

Fig. 22. Chemical repeat units for the polysaccharides (a) gellan gum and (b) Biozan (S-130).

about the biosynthesis and the relationship between functionality and primary structure. These are major problems and little research has been done in these areas. The appearance of what may ultimately prove to be families of polysaccharides with similar structures may provide a route to the solution of these problems. Thus *Acetobacter* (NBI 1022) produces a polymer resembling xanthan but with an additional two sugars in the pentasaccharide side-chain (Fig. 19). The polysaccharides from *Alcaligenes faecalis* var. *myxogenes* (succinoglycan) and *Rhizobium meliloti* (IFO 13336) show only small differences in the composition of the tetrasaccharide side-chain (Fig. 20). Whereas succinoglycan does not appear to gel, the polysaccharide from *Rhizobium meliloti* (IFO 13336) does gel. The non-gelling polysaccharide from *Klebsiella aerogenes* serotype K54 and the gelling polysaccharide XM6 from *Enterobacter* (NCIB 11870) differ only in their *O*-acetyl substitution (Fig. 21 and Section 3.5). *O*-Acetyl substituents affect the mechanical behaviour of gellan gum gels (Section 3.4) and affect the rheology of the polysaccharides from *Arthrobacter* species. The non-gelling branched polysaccharide S-130 (Biozan) has the same backbone as the gelling polysaccharide gellan gum (Fig. 22). By investigating the biosynthesis and the functionality of such similar polysaccharides it may prove possible to understand how nature engineers polymers for particular functional properties.

ACKNOWLEDGEMENTS

The author wishes to thank Dr B. H. Kirsop and M. A. O'Neill for reading an early draft and suggesting changes which have improved the chapter.

REFERENCES

1. Pilnik, W. and Rombouts, F. M., in *Polysaccharides in Food*, Blanshard, J. M. V. and Mitchell, J. R. (Eds), Butterworths, London, 1979, p. 109.
2. Norman, B. E., in *Microbial Polysaccharides and Polysaccharases*, Berkeley, R. C. W., Gooday, G. W. and Ellwood, D. C. (Eds), Academic Press, London, 1979, p. 339.
3. Percival, E. and McDowell, R. H., *Chemistry and Enzymology of Marine Algal Polysaccharides*, Academic Press, London, 1967.
4. Rees, D. A., *Adv. Carb. Chem. Biochem.*, **24**, 267 (1969).
5. Overeem, A., in *Gums and Stabilisers for the Food Industry: 2, Application of Hydrocolloids*, Phillips, G. O., Wedlock, D. J. and Williams, P. A. (Eds), Pergamon Press, Oxford, 1984, p. 369.
6. Araki, C. and Arai, K., *Bull. Chem. Soc. Japan*, **40**, 1452 (1967).
7. Rees, D. A., Morris, E. R., Thom, D. and Madden, J. K., in *The Polysaccharides*, Aspinall, G. O. (Ed.), Vol. 1, Academic Press, New York, 1982, p. 195.
8. Rees, D. A., *Chem. Ind.*, 630 (1972).
9. Morris, V. J., *Chem. Ind.*, 159 (1985).
10. Morris, V. J., in *Functional Properties of Food Macromolecules*, Mitchell, J. R. and Ledward, D. A. (Eds), Elsevier Applied Science, London, 1986, pp. 121–70.
11. Stanley, N. F., US Pat. 3,094,517 (1963).
12. Anderson, N. S., Dolan, T. C. S., Penman, A., Rees, D. A., Mueller, G. P., Stainioff, D. J. and Stanley, N. F., *J. Chem. Soc. (C)*, 602 (1968).
13. Lawson, C. J. and Rees, D. A., *Nature*, **227**, 392 (1970).
14. Glicksman, M., *Progr. Food Nutr. Sci.*, **6**, 299 (1982).
15. Drummond, D. W., Hirst, E. L. and Percival, E., *J. Chem. Soc.*, 1208 (1962).
16. Larsen, B. and Haug, A., *Carb. Res.*, **17**, 287 (1971).
17. Skjak-Braek, G. and Larsen, B., *Carb. Res.*, **139**, 273 (1985).
18. Skjak-Braek, G., in *Gums and Stabilisers for the Food Industry: 2, Application of Hydrocolloids*, Phillips, G. O., Wedlock, D. J. and Williams, P. A. (Eds), Pergamon Press, Oxford, 1984, p. 523.
19. Skjak-Braek, G., Larsen, B. and Grasdalen, H., *Carb. Res.*, **145**, 169 (1985).
20. Kertesz, Z. I., *The Pectic Substances*, Interscience, New York, 1951.
21. Aspinall, G. O., *The Polysaccharides*, Vol. 1, Academic Press, New York, 1970.
22. Keegstra, K., Talmadge, K. W., Bauer, W. D. and Albershiem, P., *Plant Physiol.*, **51**, 188 (1973).
23. Rees, D. A., *Carb. Polym.*, **2**, 254 (1982).
24. Lockwood, B., *Food Process Ind.*, 47 (1972).
25. Kohn, R., *Pure Appl. Chem.*, **42**, 371 (1975).
26. Solms, J. and Deud, H., *Helv. Chim. Acta*, **38**, 321 (1955).
27. Kohn, R., Furda, J. and Kopec, Z., *Coll. Czech. Chem. Commun.* (Engl. Edn), **33**, 264 (1968).
28. Versteeg, C., Rombouts, F. M. and Pilnik, W., *Lebenum. Wiss. Technol.*, **11**, 267 (1978).
29. MacDonnell, J. R., Jansen, E. F. and Lineweaver, H., *Arch. Biochem.*, **6**, 389 (1945).

30. Delincée, H. and Radola, B. J., *Biochim. Biophys. Acta*, **214**, 178 (1970).
31. Brady, C. J., *Aust. J. Plant Physiol.*, **3**, 163 (1976).
32. Miller, L. and McMillan, J. D., *Biochemistry*, **10**, 570 (1971).
33. Miller, L. and McMillan, J. D., *J. Bact.*, **102**, 72 (1970).
34. Sheimann, M. I., McMillan, J. D., Miller, L. and Chase, T., *Eur. J. Biochem.*, **64**, 565 (1976).
35. Dea, I. C. M. and Morrison, A., *Adv. Carb. Chem. Biochem.*, **31**, 241 (1975).
36. Hui, P. A. and Neukom, H., *Tappi*, **47**, 39 (1964).
37. Palmer, K. J. and Ballantyne, M., *J. Amer. Chem. Soc.*, **72**, 736 (1950).
38. Baker, C. W. and Whistler, R. L., *Carb. Res.*, **45**, 237 (1975).
39. Courtois, J. E. and LeDizet, P., *Bull. Soc. Chim. Biol.*, **52**, 15 (1970).
40. McCleary, B. V., *Carb. Res.*, **71**, 205 (1979).
41. Hoffman, J., Lindberg, B. and Painter, T., *Acta Chem. Scand.*, **B29**, 137 (1975).
42. Marchessault, R. H., Buleon, A., Deslandes, Y. and Goto, T., *J. Coll. Interface Sci.*, **71**, 205 (1979).
43. Grasdalen, H. and Painter, T., *Carb. Res.*, **81**, 59 (1980).
44. Baker, G. L., Carrow, J. W. and Woodmansee, C. W., *Food Ind.*, **21**, 617 (1949).
45. Deud, J., Huber, G. and Solms, J., *Experientia*, **6**, 138 (1950).
46. Rocks, J. K., *Food Technol.*, **27**, 476 (1971).
47. Kovacs, P., *Food Technol.*, **27**, 26 (1973).
48. Dea, I. C. M., McKinnon, A. A. and Rees, D. A., *J. Mol. Biol.*, **68**, 153 (1972).
49. Dea, I. C. M. and Morris, E. R., in *Extracellular Microbial Polysaccharides*, Sandford, P. A. and Laskin, A. (Eds), ACS Symp. Ser. 45, Amer. Chem. Soc., Washington, DC, 1977, p. 174.
50. Sanderson, G. R., *Progr. Food Nutr. Sci.*, **6**, 77 (1982).
51. Dea, I. C. M., Morris, E. R., Rees, D. A., Welsh, E. J., Barnes, H. A. and Price, J., *Carb. Res.*, **57**, 249 (1977).
52. Cairns, P., Miles, M. J., Morris, V. J. and Brownsey, G. J., *Food Hydrocolloids*, **1**, 89 (1986).
53. Tako, M., Asato, A. and Nakamura, S., *Agric. Biol. Chem.*, **48**, 2995 (1984).
54. Tako, M. and Nakamura, S., *Carb. Res.*, **138**, 207 (1985).
55. Miles, M. J., Morris, V. J. and Carroll, V., *Macromolecules*, **17**, 2443 (1984).
56. Carroll, V., Miles, M. J. and Morris, V. J., in *Gums and Stabilisers for the Food Industry: 2, Application of Hydrocolloids*, Phillips, G. O., Wedlock, D. J. and Williams, P. A. (Eds), Pergamon Press, Oxford, 1984, p. 501.
57. Cairns, P., Miles, M. J. and Morris, V. J., *Int. J. Biol. Macromol.*, **8**, 124 (1986).
58. Cairns, P., Miles, M. J. and Morris, V. J., *Nature*, **322**, 89 (1986).
59. Cairns, P., Miles, M. J., Morris, V. J. and Brownsey, G. J., *Carb. Res.* (in press).
60. McCleary, B. V. and Neukom, H., *Progr. Food Nutr. Sci.*, **6**, 109 (1982).
61. McCleary, B. V., Amado, R., Waibel, R. and Neukom, H., *Carb. Res.*, **92**, 269 (1981).
62. McCleary, B. V., Dea, I. C. M., Windust, J. and Cooke, D., *Carb. Polym.*, **4**, 253 (1984).

63. Banks, W. and Greenwood, C. T., *Starch and its Components*, Edinburgh University Press, Edinburgh, 1975.
64. Miles, M. J., Morris, V. J. and Ring, S. G., *Carb. Polym.*, **4**, 73 (1984).
65. Miles, M. J., Morris, V. J. and Ring, S. G., *Carb. Res.*, **135**, 257 (1985).
66. Miles, M. J., Morris, V. J., Orford, P. D. and Ring, S. G., *Carb. Res.*, **135**, 271 (1985).
67. Miles, M. J., Morris, V. J., Orford, P. D. and Ring, S. G., in *New Approaches to Research on Cereal Carbohydrates*, Hill, R. D. and Munck, L. (Eds), Elsevier, Amsterdam, 1985, p. 109.
68. O'Dell, J., in *Polysaccharides in Food*, Blanshard, J. M. V. and Mitchell, J. R., Butterworths, London, 1979, p. 171.
69. Sandford, P. A., in *Polysaccharides in Food*, Blanshard, J. M. V. and Mitchell, J. R. (Eds), Butterworths, London, 1979, p. 251.
70. Sandford, P. A. and Baird, J., in *The Polysaccharides*, Aspinall, G. O. (Ed.), Vol. 2, Academic Press, New York, 1983, p. 411.
71. Slodki, M. E. and Cadmus, M. C., *Adv. Appl. Microbiol.*, **23**, 19 (1978).
72. Colegrave, G. T., *Ind. Eng. Chem. Prod. Res. Dev.*, **22**, 456 (1983).
73. Sandford, P. A., Cottrell, I. W. and Pettitt, D. J., *Pure Appl. Chem.*, **56**, 879 (1984).
74. Buchanan, R. E. and Gibbons, N. E., *Bergey's Manual of Determinative Bacteriology*, 8th Edn, Williams and Wilkins, Baltimore, 1974, p. 243.
75. Jeanes, A., Rogovin, P., Cadmus, M. C., Silman, R. W. and Knutson, C. A., Agricultural Research Service, USDA report ARS-NC-51, Nov. 1974.
76. Evans, C. G. T., Yeo, R. G. and Ellwood, D. C., in *Microbial Polysaccharides and Polysaccharases*, Berkeley, R. C. W., Gooday, G. W. and Ellwood, D. C. (Eds), Ch. 3, Academic Press, London, 1979, p. 51.
77. Morris, E. R., Rees, D. A., Young, G., Walkinson, M. D. and Darke, A., *J. Mol. Biol.*, **110**, 1 (1977).
78. Jeanes, A., Pittsley, J. E. and Senti, F. R., *J. Appl. Polym. Sci.*, **5**, 519 (1961).
79. Baird, J. K., Sandford, P. A. and Cottrell, I. W., *BioTechnology*, **1**, 778 (1983).
80. Sanderson, G. R., *Brit. Polym. J.*, **13**, 71 (1981).
81. Andrew, T. R., in *Extracellular Microbial Polysaccharides*, Sandford, P. A. and Laskin, A. (Eds), ACS Symp. Ser. 45, 1977, p. 231.
82. McNeely, W. H. and Kang, K. S., in *Industrial Gums*, 2nd Edn, Academic Press, London, 1973, p. 473.
83. Pettitt, D. J., in *Polysaccharides in Food*, Blanshard, J. M. V. and Mitchell, J. R. (Eds), Butterworths, London, 1979, p. 263.
84. Finnerty, W. R. and Singer, M. E., *BioTechnology*, **1**, 47 (1983).
85. Sandvik, E. I. and Maerker, J. M., in *Extracellular Microbial Polysaccharides*, Sandford, P. A. and Laskin, A. (Eds), ACS Symp. Ser. 45, 1977, p. 242.
86. Morris, V. J., Franklin, D. and I'Anson, K., *Carb. Res.*, **121**, 13 (1983).
87. Frangou, S. A., Morris, E. R., Rees, D. A., Richardson, R. K. and Ross-Murphy, S. B., *J. Polym. Sci., Polym. Lett.*, **20**, 531 (1982).
88. Ross-Murphy, S. B., Morris, V. J. and Morris, E. R., *Faraday Symp. Chem. Soc.*, **18**, 115 (1983).
89. Christianson, D. D., Gardner, H. W., Warner, K., Boundy, B. K. and Inglett, G. E., *Food Technol.*, **28**, 23 (1974).

90. Singer, N. S. and Murray, D. W., US Pat. 4,198,438 (1980).
91. Jansson, P. E., Kenne, L. and Lindberg, B., *Carb. Res.*, **45**, 275 (1975).
92. Melton, L. D., Mindt, L., Rees, D. A. and Sanderson, G. R., *Carb. Res.*, **46**, 245 (1976).
93. Sandford, P. A., Pittsley, J. E., Knutson, C. A., Watson, P. R., Cadmus, M. C. and Jeanes, A., in *Extracellular Microbial Polysaccharides*, Sandford, P. A. and Laskin, A. (Eds), ACS Symp. Ser. 45, 1977, p. 192.
94. Sandford, P. A., Watson, P. R. and Knutson, C. A., *Carb. Res.*, **63**, 253 (1978).
95. Smith, I. H., Symes, K. C., Lawson, C. J. and Morris, E. R., *Int. J. Biol. Macromol.*, **3**, 129 (1981).
96. Moorhouse, R., Walkinshaw, M. D. and Arnott, S., in *Extracellular Microbial Polysaccharides*, Sandford, P. A. and Laskin, A. (Eds), ACS Symp. Ser. 45, 1977, p. 90.
97. Moorhouse, R., Walkinshaw, M. D., Winter, W. T. and Arnott, S., in *Cellulose Chemistry and Technology*, Arthur, J. D. (Ed.), 1977, p. 133.
98. Okwyama, K., Arnott, S., Moorhouse, R., Walkinshaw, M. D., Atkins, E. D. T. and Wolf-Ullish, C. H., in *Fiber Diffraction Methods*. French, A. D. and Gardner, K. D. (Eds), ACS Symp. Ser. 141, 1980, p. 411.
99. Rinaudo, M. and Milas, M., *Biopolymers*, **17**, 2663 (1978).
100. Dintzis, F. R., Babcook, G. E. and Tobin, R., *Carb. Res.*, **13**, 257 (1970).
101. Powell, D. A., in *Microbial Polysaccharides and Polysaccharases*, Berkeley, R. C. W., Gooday, G. W. and Ellwood, D. C. (Eds), Academic Press, London, 1979, p. 117.
102. Holzworth, G., *Biochemistry*, **15**, 4333 (1976).
103. Milas, M. and Rinaudo, M., *Carb. Res.*, **76**, 189 (1979).
104. Norton, I. T., Goodall, D. M., Frangou, S. A., Morris, E. R. and Rees, D. A., *J. Mol. Biol.*, **175**, 371 (1984).
105. Lambert, F., Milas, M. and Rinaudo, M., *Int. J. Biol. Macromol.*, **7**, 49 (1985).
106. Morris, V. J., I'Anson, K. and Turner, C., *Int. J. Biol. Macromol.*, **4**, 362 (1982).
107. Foweraker, A. R. and Jennings, B. R., *Polymer*, **16**, 720 (1975).
108. Foweraker, A. R. and Jennings, B. R., *Makromol. Chem.*, **178**, 505 (1977).
109. Burchard, W., *Faraday Symp. Chem. Soc.*, **18**, 233 (1983).
110. Paradossi, G. and Brant, D. A., *Macromolecules*, **15**, 874 (1982).
111. Sato, T., Norisuye, T. and Fujita, H., *Polym. J.*, **16**, 341 (1984).
112. Sato, T., Kojima, S., Norisuye, T. and Fujita, H., *Polym. J.*, **16**, 423 (1984).
113. Sato, T., Norisuye, T. and Fujita, H., *Macromolecules*, **17**, 2696 (1984).
114. Hagerman, P. J., *Biopolymers*, **20**, 1481 (1981).
115. Kam, Z., Borochov, N. and Eisenberg, H., *Biopolymers*, **20**, 2671 (1981).
116. Sato, T., Iso, N., Mizuno, H., Onda, N., Yamato, H. and Odashima, H., *Biopolymers*, **21**, 715 (1982).
117. Yanaki, T., Norisuye, T. and Fujita, H., *Macromolecules*, **13**, 1462 (1980).
118. Kashiwagi, Y., Norisuye, T. and Fujita, H., *Macromolecules*, **14**, 1220 (1981).
119. Muller, G., Lecourtier, J., Chauveteau, G. and Allain, C., *Makromol. Chem., Rapid Commun.*, **5**, 203 (1984).

120. Muller, G., Anrhourrache, M., Lecourtier, J. and Chauveteau, G., in *International Workshop on Plant Polysaccharides, Structure and Function*, Nantes, France, 1984, p. 255.
121. Jamieson, A. M., Southwick, J. G. and Blackwell, J., *Faraday Symp. Chem. Soc.*, **18**, 131 (1983).
122. Dentini, M., Coviello, T., Crescenzi, V., Kajiwara, K. and Burchard, W., in *International Workshop on Plant Polysaccharides, Structure and Function*, Nantes, France, 1984, p. 242.
123. Goodall, D. M. and Cross, M. T., *Rev. Sci. Instrum.*, **46**, 391 (1975).
124. Rees, D. A., *Pure Appl. Chem.*, **53**, 1 (1981).
125. Norton, I. T., Goodall, D. M., Morris, E. R. and Rees, D. A., *J. Chem. Soc. Chem. Commun.*, 545 (1980).
126. Holzworth, G. and Prestridge, E. B., *Science*, **197**, 757 (1977).
127. Holzworth, G., *Carb. Res.*, **66**, 173 (1978).
128. Milas, M. and Rinaudo, M., *Polym. Bull.*, **12**, 507 (1984).
129. Thurston, G. B., *J. Non-Newtonian Fluid Mechanics*, **9**, 57 (1981).
130. Thurston, G. B. and Pope, G. A., *J. Non-Newtonian Fluid Mechanics*, **9**, 69 (1981).
131. Southwick, J. G., McDonnell, M. E., Jamieson, A. M. and Blackwell, J., *Macromolecules*, **12**, 305 (1979).
132. Southwick, J. G., Lee, H., Jamieson, A. M. and Blackwell, J., *Carb. Res.*, **87**, 287 (1980).
133. Southwick, J. G., Jamieson, A. M. and Blackwell, J., *Macromolecules*, **14**, 1728 (1981).
134. Jamieson, A. M., Southwick, J. G. and Blackwell, J., *J. Polym. Sci., Polym. Phys.*, **20**, 1513 (1982).
135. Doi, M. and Edwards, S. F., *J. Chem. Soc., Faraday Trans. II*, **74**, 560 (1978).
136. Doi, M. and Edwards, S. F., *J. Chem. Soc., Faraday Trans. II*, **74**, 918 (1978).
137. Morris, V. J. and Chilvers, G. R., unpublished data.
138. Bradshaw, I. J., Nisbet, B. A., Kerr, M. H. and Sutherland, I. W., *Carb. Polym.*, **3**, 23 (1983).
139. Smith, I. H., Symes, K. C., Lawson, C. J. and Morris, E. R., *Carb. Polym.*, **4**, 153 (1984).
140. Tako, M. and Nakamura, S., *Agric. Biol. Chem.*, **48**, 2987 (1984).
141. Rinaudo, M. and Milas, M., *Carb. Polym.*, **2**, 264 (1982).
142. Stanford, E. C. C., Brit. Pat. 142 (1881).
143. Standford, E. C. C., *J. Soc. Chem. Ind.*, **3**, 297 (1884).
144. Linker, A. and Jones, R. S., *Nature*, **204**, 187 (1964).
145. Carlson, D. M. and Matthews, L. W., *Biochemistry*, **5**, 2817 (1966).
146. Evans, L. R. and Linker, A., *J. Bacteriol.*, **116**, 915 (1973).
147. Gorin, P. A. J. and Spencer, J. F. T., *Can. J. Chem.*, **44**, 993 (1966).
148. Larsen, B. and Haug, A., *Carb. Res.*, **17**, 287 (1971).
149. Page, W. J. and Sadoff, H. L., *J. Bacteriol.*, **122**, 145 (1975).
150. Doggett, R. G., *Appl. Microbiol.*, **18**, 936 (1969).
151. Doggett, R. G. and Harrison, G. M., in *Proc. 5th Int. Cystic Fibrosis Conf.*, Lawsen, D. (Ed.), Cambridge University Press, London, 1969, p. 175.

152. Govan, J. R. W. and Fife, J. A. M., *J. Antimicrobial Chemotherapy*, **4**, 233 (1978).
153. Govan, J. R. W., *Proc. Soc. Gen. Microbiol.*, **3**, 187 (1976).
154. Martin, D. R., *J. Med. Microbiol.*, **6**, 111 (1973).
155. Govan, J. R., *J. Med. Microbiol.*, **8**, 513 (1975).
156. Main, F., Jarman, T. R. and Righelato, R. C., *J. Bacteriol.*, **134**, 418 (1978).
157. Jones, S. E., Brown, M. R. W. and Govan, J. R. W., *J. Pharm. Pharmacol.*, **29**, 69 (1977).
158. Deavin, L., Jarman, T. R., Lawson, C. J., Righelato, R. C. and Slocombe, S., in *Extracellular Microbial Polysaccharides*, Sandford, P. A. and Laskin, A. (Eds), ACS Symp. Ser. 45, 1974, p. 14.
159. Tate and Lyle Ltd, *The Microbial Production of Alginates*, Reading, UK, 1975.
160. Bucke, C., Deavin, L., Lawson, C. J. and Pinder, D. F., *Biochem. Soc. Trans.*, **3**, 844 (1975).
161. Pinder, D. F. and Bucke, C., *Biochem. J.*, **152**, 617 (1975).
162. Jarman, T. R., in *Microbial Polysaccharides and Polysaccharases*, Berkeley, R. C. W., Gooday, G. W. and Ellwood, D. C. (Eds), Academic Press, London, 1979, p. 35.
163. Still, G. C. and Wang, C. H., *Arch. Biochem. Biophys.*, **105**, 126 (1964).
164. Downs, A. J. and Jones, C. W., *FEBS Lett.*, **60**, 42 (1975).
165. Dawes, E. A., in *Proc. Int. Symp. Macromol.*, Mano, E. B. (Ed.), Elsevier, Amsterdam, 1975, p. 433.
166. Haug, A., Larsen, B. and Smidsrod, O., *Acta Chem. Scand.*, **20**, 183 (1966).
167. Min, H. E., Sasaki, S. F., Kashiwabara, Y., Umekawa, M. and Nisizawa, K., *J. Biochem.* (Tokyo), **81**, 555 (1977).
168. Simonescu, R. I., Popa, V. I., Liga, A. and Rusan, V., *Cellulose Chem. Technol.*, **9**, 547 (1975).
169. Grasdalen, H., Larsen, B. and Smidsrod, O., *Carb. Res.*, **89**, 179 (1981).
170. Madgwick, J., Haug, A. and Larsen, B., *Acta Chem. Scand.*, **27**, 3592 (1973).
171. Penman, A. and Sanderson, G. R., *Carb. Res.*, **25**, 273 (1972).
172. Kohn, R. and Sticzay, T., *Coll. Czech. Chem. Commun.*, **42**, 2372 (1977).
173. Morris, E. R., Rees, D. A. and Thom, D., *J. Chem. Soc., Chem. Commun.*, 245 (1973).
174. Boyd, J. and Turvey, J. R., *Carb. Res.*, **66**, 187 (1978).
175. Steiner, A. B. and McNeely, W. H., US Pat. 2,128,551 (1949).
176. Davidson, I. W., Sutherland, I. W. and Lawson, C. J., *J. Gen. Microbiol.*, **98**, 603 (1977).
177. Sime, W. J., in *Gums and Stabilisers for the Food Industry: 2, Application of Hydrocolloids*, Phillips, G. O., Wedlock, D. J. and Williams, P. A. (Eds), Pergamon Press, Oxford, 1984, p. 177.
178. McDowell, R. H., *Properties of Alginates*, Alginate Industries, London, 1955.
179. Sanderson, G. R., *Food Technol.*, **35**, 50 (1981).
180. Smidsrod, O. and Haug, A., *Acta Chem. Scand.*, **22**, 797 (1968).
181. Smidsrod, O., *Carb. Res.*, **13**, 359 (1970).
182. Donnan, F. G. and Rose, R. C., *Can. J. Res.*, **B28**, 105 (1950).

183. Cook, W. H. and Smith, D. B., *Can. J. Biochem. Physiol.*, **32**, 227 (1954).
184. Mackie, W., Noy, R. and Sellon, D. B., *Biopolymers*, **19**, 1839 (1980).
185. Whittington, S. G., *Biopolymers*, **10**, 1481 (1971).
186. Whittington, S. G., *Biopolymers*, **10**, 1617 (1971).
187. Smidsrod, O., Glover, R. M. and Whittington, S. G., *Carb. Res.*, **27**, 107 (1973).
188. Smidsrod, O., *Faraday Discuss. Chem. Soc.*, **57**, 263 (1974).
189. Bryce, T. A., McKinnon, A. A., Morris, E. R., Rees, D. A. and Thom, D., *Faraday Discuss. Chem. Soc.*, **57**, 221 (1974).
190. Grant, G. T., Morris, E. R., Rees, D. A., Smith, P. J. C. and Thom, D., *FEBS Lett.*, **32**, 195 (1973).
191. Liang, J. N., Stevens, E. S., Frangou, S. A., Morris, E. R. and Rees, D. A., *Int. J. Biol. Macromol.*, **2**, 204 (1980).
192. Morris, E. R., Rees, D. A., Sanderson, G. R. and Thom, D., *J. Chem. Soc., Perkin Trans. II*, 1418 (1975).
193. Morris, E. R., Rees, D. A., Thom, D. and Boyd, J., *Carb. Res.*, **66**, 145 (1978).
194. Smidsrod, O., Haug, A. and Whittington, S. G., *Acta Chem. Scand.*, **26**, 2563 (1972).
195. Yathindra, N. and Rao, V. S. R., *J. Polym. Sci. A2*, **10**, 1369 (1972).
196. Atkins, E. D. T., Mackie, W. and Smolko, E. E., *Nature*, **225**, 626 (1970).
197. Atkins, E. D. T., Mackie, W., Parker, K. D. and Smolko, E. E., *J. Polym. Sci. B*, **9**, 311 (1971).
198. Atkins, E. D. T., Nieduszynski, I. A., Mackie, W., Parker, K. D. and Smolko, E. E., *Biopolymers*, **12**, 1865 (1973).
199. Atkins, E. D. T., Nieduszynski, I. A., Mackie, W., Parker, K. D. and Smolko, E. E., *Biopolymers*, **12**, 1879 (1973).
200. Mackie, W., *Biochem. J.*, **125**, 89 (1971).
201. Mackie, W., Perez, S., Rizzo, R., Taraval, F. and Vignon, M., *Int. J. Biol. Macromol.*, **5**, 329 (1983).
202. Morris, E. R., Rees, D. A., Robinson, G. A. and Young, G. A., *J. Mol. Biol.*, **138**, 363 (1980).
203. Mackie, W., Sellon, D. B. and Sutcliffe, J., *Polymer*, **19**, 9 (1978).
204. Kay, P. Y. and Sellon, D. B., *J. Polym. Sci., Polym. Phys.*, **20**, 659 (1982).
205. Harada, T., Masada, M., Fujimori, K. and Maeda, I., *Agric. Biol. Chem.*, **30**, 196 (1966).
206. Harada, T. and Yoshimura, T., *Agric. Biol. Chem.*, **29**, 1027 (1965).
207. Harada, T. and Yoshimura, T., *Biochem. Biophys. Acta*, **83**, 374 (1964).
208. Harada, T. and Yoshimura, T., *J. Ferm. Technol.*, **42**, 615 (1964).
209. Harada, T., *Arch. Biochem. Biophys.*, **112**, 65 (1965).
210. Harada, T., Yoshimura, T., Hidaka, H. and Koreeda, A., *Agric. Biol. Chem.*, **29**, 757 (1965).
211. Harada, T., Misaki, A. and Saito, H., *Arch. Biochem. Biophys.*, **124**, 292 (1968).
212. Nakanishi, I., Kimura, K., Kusui, S. and Yamazaki, E., *Carb. Res.*, **32**, 47 (1974).
213. Ogawa, K., Tsurgi, J. and Watanabe, T., *Carb. Res.*, **29**, 397 (1973).
214. Ogawa, K. and Hatano, M., *Carb. Res.*, **67**, 527 (1978).

215. Harada, T., Amenura, A., Saito, H., Kanamara, S. and Misaki, A., *J. Ferm. Technol.*, **46**, 679 (1968).
216. Harada, T., in *Fermentation Technology Today*, Terui, G. (Ed.), Proc. 4th Int. Ferment. Symp. (Kyoto, Japan), 1972, p. 603.
217. Harada, T., Fujimori, K., Hirose, S. and Masada, M., *Agric. Biol. Chem.*, **30**, 764 (1966).
218. Nakanishi, I., Kanamara, K., Kimura, K., Matsukura, A., Asai, M., Suzuki, T. and Yamatodani, S., Proc. 284th Meeting of Kansai Branch of Agric. Chem. Soc. (Osaka, Japan), 1972.
219. Kimura, H., Sato, S., Nakagawa, T., Nakatani, H., Matsukura, A., Suzuki, T., Asai, M., Kanamaru, T., Shibata, M. and Yamatodani, S., US Pat. 3,754,925 (1973).
220. Ghai, S. K., Hisamatsu, M., Amenura, A. and Harada, T., *J. Gen. Microbiol.*, **122**, 33 (1981).
221. Railton-Phillips, K. and Lawford, H. G., *Progr. Ind. Microbiol.*, **18**, 201 (1983).
222. Kimura, H., Moritaka, S. and Misaki, M., *J. Food Sci.*, **38**, 668 (1973).
223. Konno, A., Kimura, H., Nakagawa, T. and Harada, T., *Nippon Nogeikagaku Kaishi*, **52**, 247 (1978).
224. Harada, T., in *Polysaccharides in Foods*, Blanshard, J. M. V. and Mitchell, J. R. (Eds), Butterworths, London, 1979, p. 283.
225. Kuge, T., Suetzugu, N. and Nishiyama, K., *Agric. Biol. Chem.*, **7**, 1315 (1977).
226. Ishida, K., Oshima, K. and Yokoo, T., *Nippon Shokuhin Kogyo Gakkaishi*, **25**, 677 (1978).
227. Sasaki, T., Abiko, N., Sugino, Y. and Nitta, K., *Cancer Res.*, **38**, 379 (1978).
228. Sasaki, T., Abiko, N., Nitta, K., Takasuka, N. and Sugino, Y., *Eur. J. Cancer*, **15**, 211 (1979).
229. Saito, H., Misaki, A. and Harada, T., *Agric. Biol. Chem.*, **32**, 1261 (1968).
230. Harada, T., *Process. Biochem.*, **9**, 21 (1974).
231. Saito, H., Ohki, T. and Sasaki, T., *Biochemistry*, **16**, 908 (1977).
232. Kikumoto, S., Miyajima, T., Kimura, K., Okubo, S. and Komatsu, N., *J. Agric. Chem. Soc.* (Japan), **45**, 162 (1971).
233. Tabata, K., Ito, W., Kojima, T., Kawabata, S. and Misaki, A., *Carb. Res.*, **89**, 121 (1981).
234. Norisuye, T., Yanaki, T. and Fujita, H., *J. Polym. Sci., Polym. Phys.*, **18**, 547 (1980).
235. Sato, T., Norisuye, T. and Fujita, H., *Carb. Res.*, **95**, 195 (1981).
236. Sato, T., Norisuye, T. and Fujita, H., *Macromolecules*, **16**, 185 (1983).
237. Atkins, E. D. T., Parker, K. D. and Preston, R. D., *Proc. Roy. Soc.*, **B173**, 209 (1969).
238. Atkins, E. D. T. and Parker, K. D., *J. Polym. Sci., C*, **28**, 69 (1969).
239. Hirano, I., Einaga, Y. and Fujita, H., *Polym. J.*, **11**, 901 (1979).
240. Henley, D., *Arkiv Kemi*, **18**, 327 (1961).
241. Brown, W. and Wilkinson, R., *Eur. Polym. J.*, **1**, 1 (1965).
242. Ogawa, K., Watanabe, T., Tsurgi, J. and Ono, S., *Carb. Res.*, **23**, 399 (1972).
243. Fulton, W. S. and Atkins, E. D. T., in *Fiber Diffraction Methods*, French, A. D. and Gardener, K. H. (Eds), ACS Symp. Ser. 141, 1980, p. 385.

244. Ogawa, K., Miyagi, M., Fukumoto, T. and Watanabe, T., *Chem. Lett.*, 943 (1973).
245. Aizawa, M., Takahashi, M. and Suzuki, S., *Chem. Lett.*, 193 (1977).
246. Maedo, I., Saito, H., Masada, M., Misaki, A. and Harada, T., *Agric. Biol. Chem.*, **31**, 1184 (1967).
247. Saito, H., in *Solution Properties of Polysaccharides*, Brant, D. A. (Ed.), ACS Symp. Ser. 150, 1981, p. 125.
248. Kasai, N. and Harada, T., in *Fiber Diffraction Methods*, French, D. A. and Gardener, K. H. (Eds), ACS Symp. Ser. 141, 1980, p. 363.
249. Takeda, H., Yasuoka, N., Kasai, N. and Harada, T., *Polym. J.*, **10**, 365 (1978).
250. Marchessault, R. H., Deslandes, Y., Ogawa, K. and Sundarajan, P. R., *Can. J. Chem.*, **55**, 300 (1977).
251. Marchessault, R. H. and Deslandes, Y., *Carb. Res.*, **75**, 231 (1979).
252. Chuah, C. T., Sarko, A., Deslandes, Y. and Marchessault, R. H., *Macromolecules*, **16**, 1375 (1983).
253. Harada, T., Koreeda, A., Sato, S. and Kasai, N., *J. Electron Microsc.*, **28**, 147 (1979).
254. Saito, H., Miyata, E. and Sasaki, T., *Macromolecules*, **11**, 1244 (1978).
255. Kaneko, T. and Kang, K. S., in *Abstr. Ann. Meeting Amer. Soc. Microbiol.*, 1979, p. 101.
256. Kang, K. S., Veeder, G. T., Mirrasoul, P. J., Kanecko, T. and Cottrell, I. W., *Appl. Environ. Microbiol.*, **43**, 1086 (1982).
257. Veeder, G. T., Kang, K. S., Mirrasoul, P. J., Koupal, L. and Stapley, E. O., in *Abstr. Ann. Meeting Amer. Soc. Microbiol.*, 1979, p. 200.
258. Kang, K. S., Veeder, G. T. and Cottrell, I. W., in *Abstr. Ann. Meeting Amer. Soc. Microbiol.*, 1979, p. 200.
259. Moorhouse, R., Colegrave, G. T., Sandford, P. A., Baird, J. and Kang, K. S., in *Solution Properties of Polysaccharides*, Brant, D. A. (Ed.), ACS Symp. Ser. 150, 1981, p. 111.
260. Kang, K. S. and Veeder, G. T., US Pat. 4,326,053 (1982).
261. Kang, K. S., Veeder, G. T. and Cottrell, I. W., *Progr. Ind. Microbiol.*, **18**, 231 (1983).
262. Sanderson, G. R. and Clark, R. C., *Food Technol.*, **37**, 63 (1983).
263. Sanderson, G. R. and Clark, R. C., in *Gums and Stabilisers for the Food Industry: 2, Application of Hydrocolloids*, Phillips, G. O., Wedlock, D. J. and Williams, P. A. (Eds), Pergamon Press, Oxford, 1984, p. 201.
264. Shungu, D., Valiant, M., Tutlane, V., Weinburg, E., Weisberger, B., Koupal, L., Gadebusch, H. and Stapley, E., *Appl. Environ. Microbiol.*, **46**, 840 (1983).
265. Lin, C. C. and Casida, L. E., Jr, *Appl. Environ. Microbiol.*, **47**, 427 (1984).
266. Cottrell, I. W., in *Industrial Potential of Fungal and Bacterial Polysaccharides*, ACS Symp. Ser. 126, Amer. Chem. Soc., Washington, DC, 1980, p. 251.
267. Harris, J. E., *Appl. Environ. Microbiol.*, **50**, 1107 (1985).
268. O'Neill, M. A., Selvendran, R. R. and Morris, V. J., *Carb. Res.*, **124**, 123 (1983).
269. Jannson, P. E., Lindberg, B. and Sandford, P. A., *Carb. Res.*, **124**, 135 (1983).
270. Miles, M. J., Morris, V. J. and O'Neill, M. A., in *Gums and Stabilisers for*

the Food Industry: 2, Application of Hydrocolloids, Phillips, G. O., Wedlock, D. J. and Williams, P. A. (Eds), Pergamon Press, Oxford, 1984, p. 485.
271. O'Neill, M. A., Morris, V. J. and Selvendran, R. R., in *Gums and Stabilisers for the Food Industry: 3*, Phillips, G. O., Wedlock, D. J. and Williams, P. A. (Eds), Elsevier Applied Science, 1986, p. 29.
272. Carroll, V., Chilvers, G. R., Franklin, D., Miles, M. J., Morris, V. J. and Ring, S. G., *Carb. Res.*, **114**, 181 (1983).
273. Brownsey, G. J., Chilvers, G. R., I'Anson, K. and Morris. V. J., *Int. J. Biol. Macromol.*, **6**, 211 (1984).
274. Attwool, P. T., Atkins, E. D. T., Miles, M. J. and Morris, V. J., unpublished data.
275. Crescenzi, V., Dentini, M., Coviello, T. and Rizzo, R., *Carb. Res.*, **149**, 425 (1986).
276. Carroll, V., Miles, M. J. and Morris, V. J., *Int. J. Biol. Macromol.*, **4**, 432 (1982).
277. Attwool, P. T., Atkins, E. D. T., Upstill, C., Miles, M. J. and Morris, V. J., in *Gums and Stabilisers for the Food Industry: 3*, Phillips, G. O., Wedlock, D. J. and Williams, P. A. (Eds), Elsevier Applied Science 1986, p. 135.
278. Nisbet, B. A., Sutherland, I. W., Bradshaw, I. J., Kerr, M., Morris, E. R. and Shepperson, W. A., *Carb. Polym.*, **4**, 377 (1984).
279. O'Neill, M. A., Morris, V. J., Selvendran, R. R., Sutherland, I. W. and Taylor, I. T., *Carb. Res.*, **148**, 63, (1986).
280. Dutton, G. G. S. and Merrifield, E. H., *Carb. Res.*, **105**, 189 (1982).
281. Atkins, E. D. T., Isaac, D. H. and Elloway, H. F., in *Microbial Polysaccharides and Polysaccharases*, Berkeley, R. C. W., Gooday, G. W. and Ellwood, D. C. (Eds), Academic Press, London, 1979, p. 161.
282. Attwool, P. T., Atkins, E. D. T., Miles, M. J., Morris, V. J., O'Neill, M. A. and Sutherland, I. W., *Int. J. Biol. Macromol.*, in press.
283. Sandford, P. A., *Adv. Carb. Chem. Biochem.*, **36**, 265 (1979).
284. Catley, B. J., in *Microbial Polysaccharides and Polysaccharases*, Berkeley, R. C. W., Gooday, G. W. and Ellwood, D. C. (Eds). Academic Press, London, 1979, p. 69.
285. Griminger, P. and Fisher, H., *Proc. Soc. Exp. Biol. Med.*, **122**, 551 (1966).
286. Kennedy, J. F., Griffiths, A. J. and Atkins, D. P., in *Gums and Stabilisers for the Food Industry: 2, Application of Hydrocolloids*, Phillips, G. O., Wedlock, D. J. and Williams, P. A. (Eds), Pergamon Press, Oxford, 1984, p. 417.
287. Harada, T., *Biochem. Soc. Symp.*, **48**, 97 (1983).
288. Tayami, K., Minakami, H., Entani, E., Fujiyama, S. and Masai, H., *Agric. Biol. Chem.*, **49**, 959 (1985).
289. Rosenberg, E., Perry, A., Gibson, D. T. and Gutrick, D. L., *Appl. Environ. Microbiol.*, **37**, 409 (1979).
290. Rosenberg, E., Zuckerberg, A., Rubinowitz, E. and Gutrick, D. L., *Appl. Environ. Microbiol.*, **37**, 402 (1979).

6

Detoxifying Enzymes

IRVIN E. LIENER
Department of Biochemistry, University of Minnesota, St Paul, Minnesota, USA

1. INTRODUCTION

Proteins of plant origin offer considerable promise for alleviating the shortage of food protein now facing many segments of the world's population—a situation which will certainly become more acute if the global expansion of the population continues unabated. It should be recognized, however, that many plants commonly consumed by man contain substances which may have an adverse effect on the nutritional properties of the protein unless inactivated or eliminated by suitable methods of processing or household preparation.[1,2] In many cases these substances are relatively innocuous in their native form but may be converted to a toxic state when acted upon by enzymes which frequently accompany these substances in the tissues of the same plant. Paradoxically, advantage can sometimes be taken of the action of these endogenous enzymes to effect the removal of these toxicants by appropriate methods of processing. Alternatively, exogenous enzymes may be used for detoxification, such as fermentation by molds or bacteria or the use of enzyme preparations derived from these microorganisms. In this chapter examples will be provided which illustrate the manner in which enzymes have been used, either endogenously or exogenously, to detoxify various sources of foods so that they can be used in the diet of man or animals.

2. GOITROGENIC GLYCOSIDES

A number of cruciferous plants such as the rapeseed (*Brassica napus*), mustard seed (*Brassica juncea*) and Abyssinian kale (*Crambe abyssinica*) provide a potentially valuable source of protein, not only for feeding animals but for use in the human diet as well.[3] The protein content of the meal remaining after extraction of the oil is relatively high (30–50%), and the amino acid composition compares favorably with that of soybeans.[4,5] The use of such oil seeds, however, is limited by the presence of high levels of sulfur-containing glycosides (glucosinolates). Upon hydrolysis these glycosides yield products that, unlike the parent glycoside, are goitrogenic and act as growth depressants.[6] These goitrogenic products, primarily isothiocyanates and oxazolidine-thiones, are liberated from the parent glycoside by an endogenous enzyme, myrosinase (thioglucoside glucohydrolase), which acts on the glucosinolate when the seed is crushed and moistened. Although a wide variety of glucosinolates are present in cruciferous seeds,[6] the principal one present in rapeseed and crambe is progoitrin. When progoitrin is acted upon by myrosinase it yields, among other products, an isothiocyanate which spontaneously cyclizes to goitrin (5-vinyloxazolidine-2-thione) as shown in Fig. 1.

The toxicity of rapeseed can be reduced in one or a combination of several different ways: (a) destruction of the enzyme by moist heat treatment;[7] (b) removal of the glucosinolates by extraction with hot water,[8,9] dilute alkali[10] or acetone,[11] or by decomposition with iron salts[12] or soda ash;[13] (c) removal of goitrogenic end-products following potentiation of the action of the endogenous enzyme.

Although procedure (a) prevents further breakdown of the glucosinolates in the meal, it leaves the glucosinolate intact and thus presents the possibility that it could be hydrolyzed to toxic end-products after ingestion by enzymes introduced by other dietary ingredients, or by the microflora

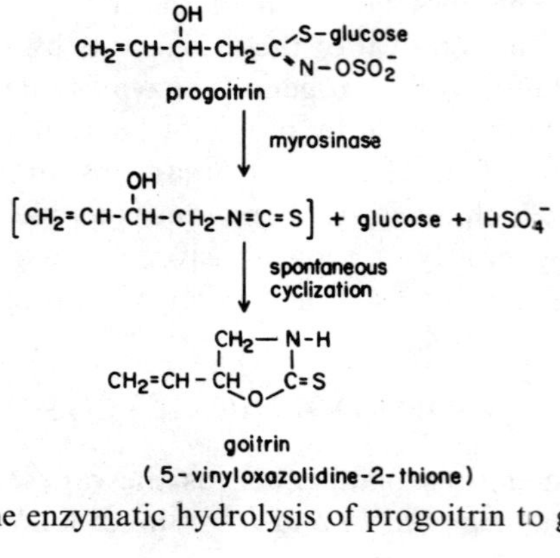

Fig. 1. The enzymatic hydrolysis of progoitrin to goitrin.

of the gut.[14–16] In one instance at least, enzyme-inactivated mustard seed cake was found to be toxic to cattle because kohlrabi, another cruciferous plant which contains myrosinase, was fed at the same time to the animals.[17] On the other hand, microorganisms in the digestive tract of the chick do not lead to a breakdown of glucosinolates to goitrogenic end-products.[18] Although procedure (b), especially in combination with heat treatment, effects a marked reduction in toxicity, it does not eliminate any goitrogenic compounds which may have been produced prior to processing.

Most relevant to the subject at hand is procedure (c) wherein the glucosinolates are deliberately permitted to undergo enzymatic conversion to goitrogenic end-products which are subsequently removed by various procedures. For example, Belzile and Bell[19] noted a reduction in toxicity towards mice of raw rapeseed meal which had been moistened, allowed to autolyze (1 h at 22°C or 50°C), and then extracted with buffer solutions (pH 3, 6 or 9) to remove any goitrogenic products which had formed. Goering[20] developed a process in which ground rapeseed was moistened with cold water and, to potentiate enzyme action, digestion was carried out at 45–55°C. The liberated isothiocyanates were then removed by steam-stripping. Although this procedure effectively removes the volatile isothiocyanates, the oxazolidine-thione and other non-volatile products of enzymatic hydrolysis remain in the meal. Mukherjee *et al.*[21,22] modified this procedure by using hexane extraction to remove over two-thirds of the goitrogenic compounds produced by the autolysis of ground rapeseed. This method, however, suffers from the fact that the oil becomes heavily contaminated with the hydrolytic products of the glucosinolates. Other techniques for removing the end-products of autolyzed rapeseed include extraction with water,[23] acetone[24] or salt solution.[25,26]

In the case of crambe seed, autolysis of the raw moistened meal was followed by extraction of the goitrin with acetone containing 2–12% water or a ternary mixture of 53% acetone, 44% hexane and 3% water.[27] The marked reduction in toxicity in rats fed such a product is shown in Fig. 2. It is evident from these results that enzymatic hydrolysis followed by extraction with aqueous acetone provides a very effective means for detoxifying crambe seed meal.

With mustard seed meal, the principal undesirable component is allyl isothiocyanate which arises from the hydrolysis of the thioglucoside sinigrin (Fig. 3). Unless removed from the meal, allyl isothiocyanate imparts an extremely pungent flavor which limits the amount of mustard seed meal that can be used for feed. In a process developed by Mustakas

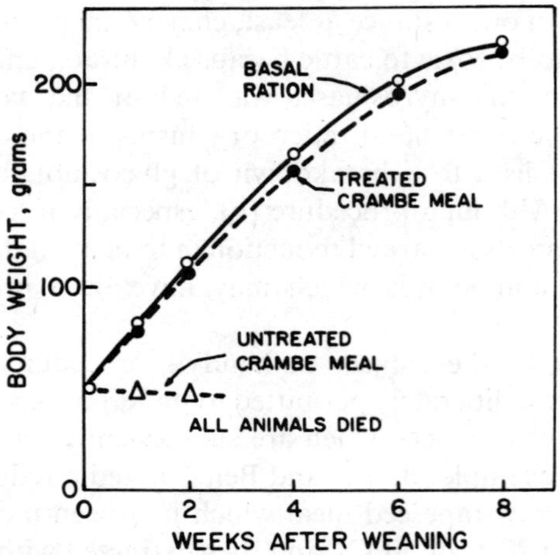

Fig. 2. Growth of weanling rats fed diets containing 28% crambe meal, either treated (autolysis followed by extraction with aqueous acetone) or untreated, and 72% basal ration. Comparison is made with a control group fed basal ration alone. Taken from Tookey *et al.*[27]

and co-workers,[28–30] advantage is taken of the fact that allyl isothiocyanate is not only water-soluble but volatile as well, so it can be removed by distillation after enzymatic hydrolysis of the glycoside. A flow-sheet of the complete process which incorporates this feature is shown in Fig. 4. In this process the raw ground meal is moistened to 30%, and the conversion of the thioglucoside to allyl thiocyanate by myrosinase is allowed to proceed for 15–45 min at 55°C, followed by a combined cooking and distillation step at 100°C. The latter step permits the recovery of allyl isothiocyanate (mustard oil) which can be utilized commercially as a condiment. Subsequent extraction with hexane removes the oil which contains erucic acid, to the extent of 25%. This step not only serves to eliminate most of the erucic acid from the meal, which if present would

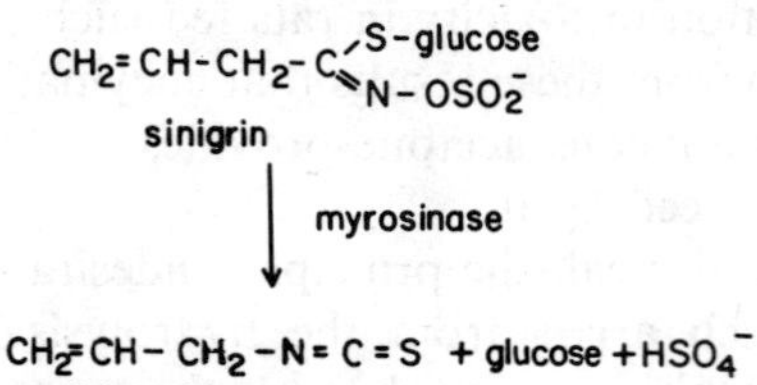

Fig. 3. Enzymatic hydrolysis of sinigrin, the principal glucosinolate of mustard seed.

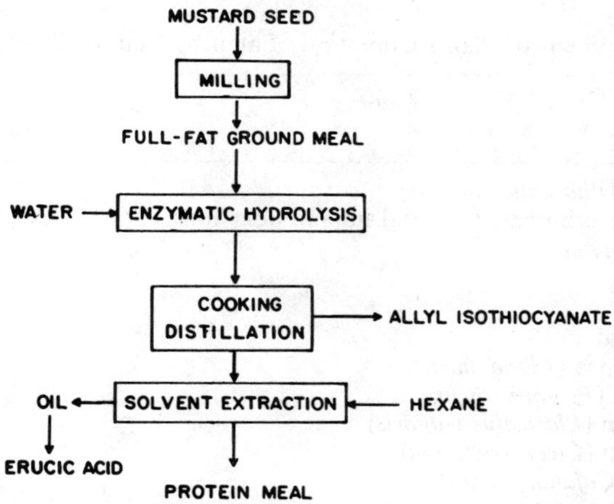

Fig. 4. Flow-sheet for the enzymatic detoxification of mustard seed. Adapted from Mustakas *et al.*[30]

render the meal toxic,[3] but also provides a good source of oil which possesses special properties as a lubricant. The final product of this process is a bland meal which contains almost 50% protein and less than 0·01% allyl isothiocyanate. Feeding tests with rats have shown that the meal produced by this method did not inhibit growth, and a 50:50 mixture with soybean meal gave as good growth as soybean meal alone.[31] Furthermore, preliminary cost estimates indicate that the costs of industrializing this process are nearly the same as for a comparable soybean plant.[30] Butt *et al.*[32] have reported a modification of the Mustakas procedure which involved the use of an enzyme preparation from germinated mustard seed to accelerate the autolysis step.

A microbial fermentation technique has also been employed to destroy the glucosinolates in rapeseed meal. Staron[33] has reported the biological destruction of glucosinolates through the action of the fungus *Geotrichum candidum* in rapeseed press cake. Not only was there an improvement in nutritive value, but there was also an increase in the solubility of the rapeseed proteins.

3. CYANOGENIC GLYCOSIDES

It has been known for over 150 years that a wide variety of plants are potentially toxic because they contain glycosides which release HCN upon

Table 1
Cyanide Content of Plants Commonly Eaten by Man or Domestic Animals[38]

Plant	*HCN Yield* (*mg 100 g*$^{-1}$)
Lima bean (*Phaseolus lunatus*)	
Samples incriminated in fatal human poisoning	210–312
Normal levels	14·4–16·7
Sorghum	250
Cassava	113
Linseed meal	53
Black-eyed pea (*Vigna sinensis*)	2·1
Garden pea (*Pisum sativum*)	2·3
Kidney bean (*Phaseolus vulgaris*)	2·0
Bengal gram (*Cicer arietinum*)	0·8
Red gram (*Cajanus cajans*)	0·5

hydrolysis; several reviews[34–37] deal with this subject. These glycosides themselves are not toxic, but the HCN which is released upon hydrolysis by an endogenous enzyme may be toxic because its principal target is cytochrome oxidase, a key enzyme necessary for the survival of aerobic organisms. Those plants which contain high levels of cyanogenic glycosides and are commonly eaten by man or domestic animals include cassava (*Manihot esculenta* or *utilissima*), lima beans (*Phaseolus lunatus*), sorghum (*Sorghum vulgare*) and linseed (*Linum usitatissimum*). See Table 1 for potential HCN content of some commonly consumed plants.

3.1. Cassava

Cassava is a staple food item commonly eaten throughout the tropics where it is eaten as a boiled vegetable or prepared into a variety of dishes by traditional methods.[39–41] The principal cyanogenic glycoside in cassava is linamarin which is a β-glucoside of acetone cyanohydrin (Fig. 5). The liberation of HCN is effected by maceration of the root which brings the substrate (linamarin) into intimate contact with an endogenous β-glucosidase called linamarase. Strictly speaking, linamarase refers to the enzyme which hydrolyzes the β-glycosidic bond of linamarin to yield glucose and an α-hydroxynitrile (2-cyano-2-propanol). The latter may decompose spontaneously or may be acted upon by a second enzyme, hydroxynitrilase, producing acetone and HCN.

Since the peel is particularly rich in the glycoside,[42] some reduction in potential toxicity is achieved by removal of the peel and thorough washing

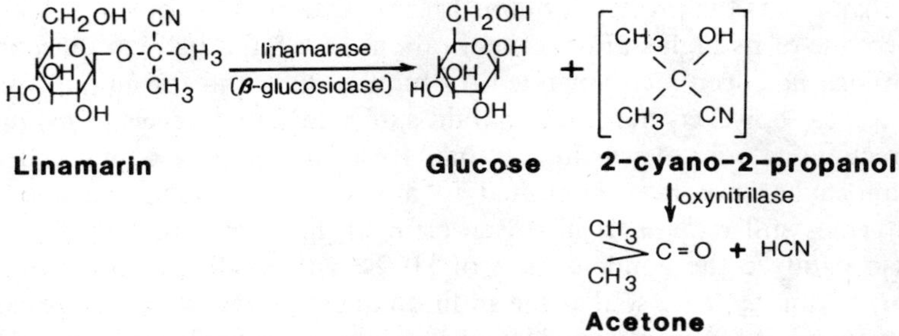

Fig. 5. Enzymatic hydrolysis of linamarin, the principal cyanogenic glycoside of cassava and lima beans.

of the pulp from the root with running water. A further reduction in toxicity is achieved by the application of heat (boiling, roasting or sun-drying) which serves to inactivate linamarase and to volatilize any HCN that may have been released. The leaves of the cassava plant have also been used as food in West Africa, New Guinea and central Java. Although rich in the cyanogenic glycoside, about 98% of the cyanide can be removed by pulping and boiling.[43]

In the preparation of the native Nigerian dish called gari, the pulp of the root is allowed to undergo fermentation by *Corynebacterium manihot*. A marked reduction in cyanide results from this fermentative process which, at one time, was attributed to the hydrolytic effect of the organic acids produced during the fermentation.[44,45] This is then followed by frying which serves to eliminate almost completely all of the liberated cyanide. More recent work[46–48] has demonstrated the unusual stability of linamarin in dilute acid even at 100°C, thus making it unlikely that detoxification is due solely to chemical hydrolysis by the weak acids produced by microbial fermentation. It now appears that the detoxification resulting from the gari fermentation is in fact due mainly to the action of linamarase.[49] The use of an exogenous source of linamarase derived from fungi (*Aspergillus sydowi*, *Fusarium equiseti* or *Penicillium streckii*) has been proposed for enhancing the detoxification of linamarin in the preparation of gari.[50,51] Before the use of these fungal preparations of linamarase can be recommended, however, it is important to establish the absence of any toxic substances that may be produced by these fungi.

Fermented cassava meal has also been shown to be free of any toxic effects when incorporated into feeds for broilers,[52] chicks[53] or swine.[54] Presumably the same mechanism of detoxification is involved as in the preparation of gari.

Tapioca is the product obtained from cassava after thorough washing. Because of its high starch content (about 65%) and its high digestibility, tapioca has been recommended for human diets and for animal feeding. Tapioca, however, may still contain sufficiently high levels of cyanide to preclude its use at levels higher than 10% for poultry feeding.[55] Because tapioca has not been subjected to any heat treatment, the hydrolytic enzymes still remain. It has been claimed, however, that these enzymes also catalyze the condensation of HCN with aldehydic compounds to form cyanohydrins, so that the addition of glucose to unprocessed cassava causes the disappearance of HCN.[56] It has therefore been suggested that in the preparation of foods from unprocessed cassava liberal amounts of glucose be added.

Despite the apparent effectiveness of traditional methods of preparing cassava dishes in removing cyanide, the consumption of cassava in certain parts of Africa has been associated with at least two pathological conditions, namely ataxic neuropathy[57-59] and goiter.[60] Ataxic neuropathy is believed to involve demyelination of the optic nerve tissue resulting in defective vision (amblyopia). Elevated levels of thiocyanate in the blood plasma have been observed in individuals who suffer from this disease. Thiocyanate is the end-product arising from the body's attempt to detoxify cyanide (see Fig. 6). The mechanism whereby cyanide or its metabolic product, thiocyanate, is involved in the pathogenesis of ataxic neuropathy remains to be elucidated. On the other hand, since thiocyanate is a well established goitrogen, it is not surprising that a high incidence of goiter and cretinism is so prevalent in those parts of Africa where cassava constitutes such an important part of the diet.

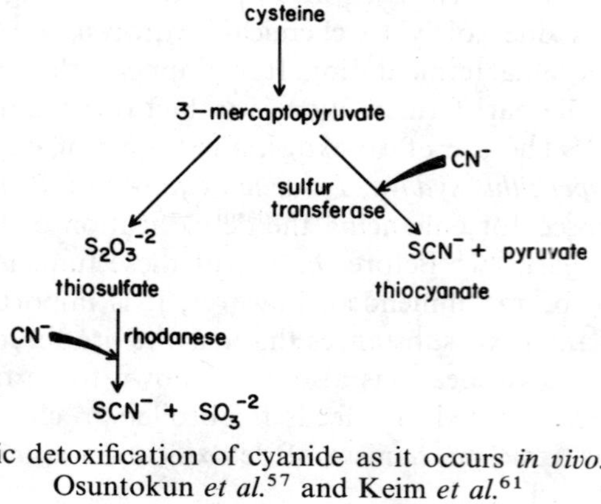

Fig. 6. Metabolic detoxification of cyanide as it occurs *in vivo*. Adapted from Osuntokun *et al.*[57] and Keim *et al.*[61]

3.2. Lima Beans

Serious outbreaks of poisoning in man and animals have frequently been attributed to the consumption of certain varieties of lima beans (*Phaseolus lunatus*), particularly those which are highly colored.[37,38] As shown in Table 1, those varieties of lima beans which have been implicated in cases of fatal human poisoning contain over 15 times the level of HCN that is found in most varieties (white) of lima beans which are normally used in the diet. The particular cyanogenic glycoside involved in this toxicity is identical to the one present in cassava (linamarin), although it is sometimes referred to in the older literature as phaseolunatin. Rahman *et al.*[62] have described a simple procedure for reducing the cyanide content of lima beans to a negligibly low, non-toxic level. Water is added to the ground bean flour to a level of 25–30% which favors the intimate contact of the glycoside with the enzyme resulting in the ensuing release of HCN. Subsequent drying and cooking of the flour with water for 20 min reduces the cyanide content to a level where it is physically impossible to eat enough to reach the estimated toxic dose of 50 mg of HCN for man.

3.3. Fruit and Nut Kernels

The kernels of both almonds and apricots are used for making marzipan, a common ingredient in baking and confectionery products. The principal cyanogenic glycoside in these particular seeds is amygdalin in which the aglycone is D-mandelonitrile and the sugar component is gentiobiose (see Fig. 7). The endogenous enzyme responsible for the hydrolysis of amygdalin is most commonly referred to as emulsin. This enzyme, although inactive towards linamarin, is similar in action to linamarase in that hydrolysis occurs by a two-stage process, the first step involving hydrolysis by a β-glycosidase to yield gentiobiose and mandelonitrile,

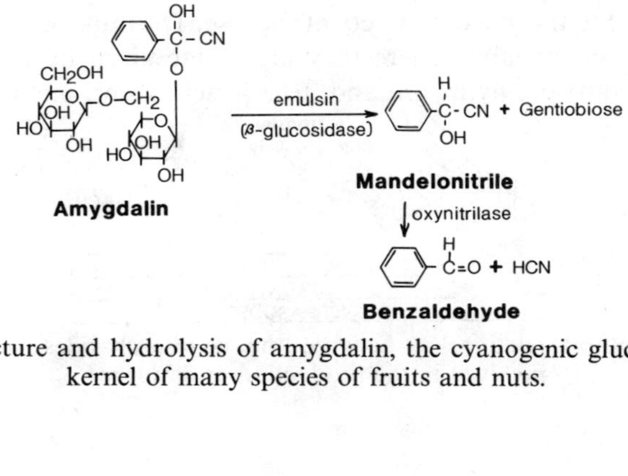

Fig. 7. Structure and hydrolysis of amygdalin, the cyanogenic glucoside in the kernel of many species of fruits and nuts.

and a second enzymatic step in which mandelonitrile is converted to benzaldehyde and HCN. In the conventional preparation of marzipan the cyanogenic glucoside is removed to a large extent by simply soaking the ground kernels in water for 24–48 h (the 'debittering' process).[63] This process is wasteful in both time and water and also poses a pollution problem created by the release of HCN into the wash water. Also the benzaldehyde which is released is insoluble in water and dissolves in the kernel oil where it is readily oxidized to benzoic acid. The latter may then reach levels higher than permitted by food regulations. In order to avoid some of these problems Schab and Yannai[64] developed a debittering procedure which involved soaking coarsely ground apricot kernels in acidified water, pH 6·5, for 2 h at 55°C. Complete hydrolysis of amygdalin was effected under these conditions, and the products of hydrolysis could be removed by steam distillation. Marzipan made from apricot kernels which had been processed in this fashion was found to be as organoleptically acceptable as the product made from almonds prepared in the conventional fashion.

4. FLATULENCE-PRODUCING OLIGOSACCHARIDES

One of the important factors limiting the use of legumes in the human diet is the production of flatulence associated with their consumption.[65–69] Flatulence is generally attributed to the fact that humans are not endowed with the enzymes (α-galactosidase and β-fructosidase) necessary to hydrolyze certain oligosaccharides which contain α-galactosidic and β-fructosidic linkages. As shown in Fig. 8, these oligosaccharides (raffinose, stachyose and verbascose) are related by having one or more α-1,6-D-galactopyranosyl groups in their structure, where one of the α-galactose units is linked to the glucose moiety of sucrose. The intact oligosaccharides enter the lower intestine where they are metabolized by the microflora into carbon dioxide, hydrogen and, to a lesser extent, methane. It is the

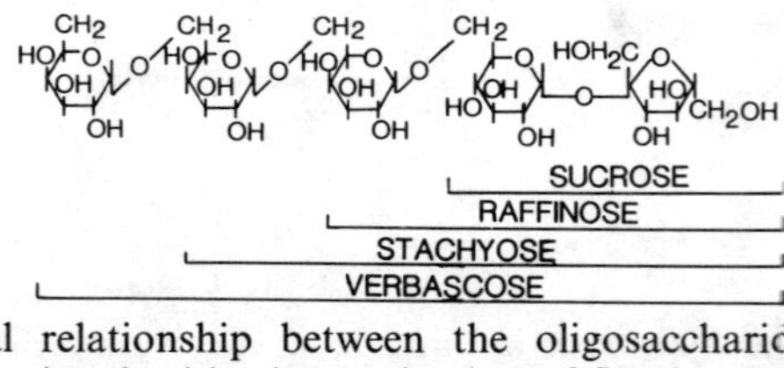

Fig. 8. Structural relationship between the oligosaccharides believed to be involved in the production of flatulence.

production of these gases which leads to the characteristic features of flatulence, namely nausea, cramps, diarrhea, abdominal rumbling, and the social embarrassment which accompanies the ejection of rectal gas. Attempts to eliminate these offending oligosaccharides from legumes have taken advantage of the endogenous enzymes present in the plant tissue or the introduction of exogenous sources of enzymes capable of hydrolyzing α-1,6-galactosidic linkages.

A significant reduction in oligosaccharide content of dry beans (*Phaseolus vulgaris*) can be achieved by maximizing their hydrolysis by enzymes normally present in the plant tissue.[70,71] For example, Becker *et al.*[71] were able to lower the oligosaccharide content of California small white beans by allowing a slurry of the raw, ground bean to undergo autolysis at pH 5·2 at 55°C for various periods of time. The effect of these treatments in reducing the flatulence effect in rats, as measured by the production of hydrogen gas, is shown in Fig. 9. This simple procedure of soaking the beans has also proved effective in reducing the raffinose and stachyose content of several other varieties of *P. vulgaris* as well as soybeans and a number of other legumes.[68,72,73] It should be appreciated, however, that when soaking alone is employed, it is difficult to assess how much of this reduction in oligosaccharides is caused by leaching or by autolysis by endogenous enzymes.

By allowing beans to germinate it becomes possible to take advantage of the increase in the activity of enzymes capable of degrading oligosaccharides. Thus, Kim *et al.*[72] found that a combination of soaking, germination for 4 h and re-soaking produced a 70% reduction in the

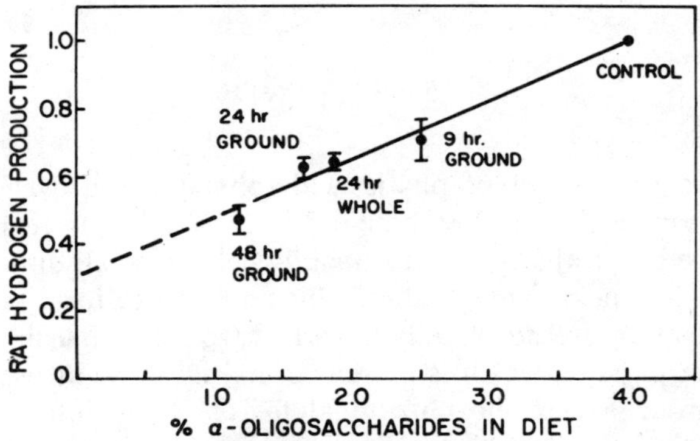

Fig. 9. Flatulence in rats produced by California small white beans which had been allowed to autolyze at pH 5·2 and 55°C for various periods of time. Taken from Becker *et al.*[71]

oligosaccharide content of soybeans. The germination of commonly consumed Indian legumes including chick pea or Bengal gram (*Cicer arietinum*), black gram (*Phaseolus mungo*), red gram (*Cajanus cajan*) and the broad bean (*Vicia faba*) for 48 h resulted in the almost complete disappearance of raffinose, stachyose and verbascose.[74,75]

The preparation of traditional soybean foods such as tempeh involves fermentation, in which the beans are inoculated with mold spores from *Rhizopus oligosporus*. Tempeh has been found to have little flatus activity,[65] presumably due to the hydrolytic enzymes elaborated by the mold. The inoculation of suspensions of soybean flour with cultures of various strains of the yeast *Saccharomyces* has also been employed to remove flatus-forming carbohydrates.[76,77]

The treatment of soymilk by an enzyme preparation from *Aspergillus saitoi*, which possessed both α-galactosidase and invertase activities, resulted in the complete hydrolysis of its constituent oligosaccharides.[78] However, the effectiveness of this treatment in reducing flatulence was not evaluated. Crude enzyme preparations from yeast or mold have also been employed for the removal of oligosaccharides from soybeans[79] or soymilk.[80] The pre-digestion of soybeans with papain followed by a lactic acid fermentation with *Lactobacillus acidophilus* or *L. bulgaricus* has been reported to result in the complete elimination of raffinose and stachyose.[81] Immobilized preparations of α-galactosidase derived from microbial sources have likewise proved to be effective in hydrolyzing the oligosaccharides of soybeans when employed either as continuous flow reactors[82,83] or in the form of a hollow-fiber dialyzer.[84]

5. PHYTATE

Phytic acid or the salt of phytic acid (phytate) is a cyclic compound (inositol) containing six phosphate radicals (see Fig. 10). Its physiological significance lies in the fact that it readily chelates such di- and tri-valent metal ions as calcium, magnesium, zinc and iron to form poorly soluble compounds that are not readily absorbed from the intestines. Thus, the presence of phytate, which is present primarily in plant tissue, has long been regarded as an anti-nutritional factor which interferes with the bioavailability of minerals essential for optimal health. Further details on the nutritional significance of phytates and its interactions in food systems may be found in several reviews.[85–89]

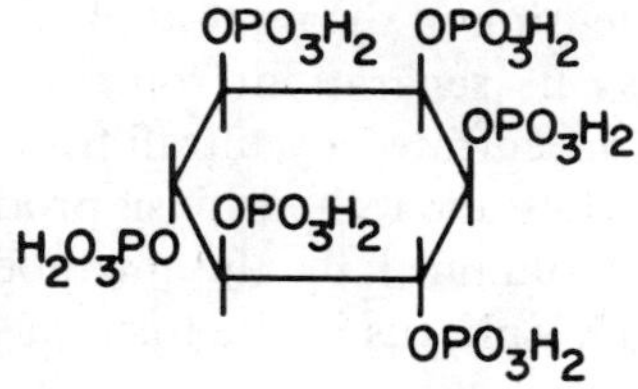

Fig. 10. Structure of phytic acid (myoinositol 1,2,3,4,5,6-hexadihydrogen phosphate).

The ability of phytate to bind metal ions is lost when the phosphate groups are removed from inositol by hydrolysis through the action of the enzyme phytase. Although phytase activity has been shown to be present in the small intestine of various experimental animals,[90] its presence in the intestines of humans remains a controversial issue.[91,92] Even if phytase were present, its activity does not appear to be sufficiently great to prevent phytate from interfering with the utilization by man of calcium, zinc and phosphorus from diets which contain proteins of plant origin.[92] As in the case of other toxicants which have thus far been considered, the phytate content of many plant foodstuffs can be reduced by taking advantage of the endogenous phytase which accompanies the phytate in separate compartments of the plant tissue, by subjecting the plant material to a fermentation process which serves to provide an exogenous source of the enzyme, or, in some cases, by adding preparations of phytase derived from other sources.

5.1. Effect of Autolysis

The phytate content of California small white beans (*P. vulgaris*) can be greatly reduced by allowing a suspension of raw beans to autolyze at pH 5·2 for 20–48 h at 35–55°C; these are essentially the same conditions which have proven so effective for reducing the flatulence-causing oligosaccharides in these beans.[70,71] This same treatment has also proved effective for reducing the phytate content of mung bean, soybean, lima bean and wheat.[93] The bean phytase responsible for the autolytic hydrolysis of phytate has been purified and characterized.[94]

Soybeans also contain high levels of phytate which can interfere with mineral utilization, and much of this phytate remains bound to the protein even after concentrating or isolating the protein derived therefrom.[95] A low phytate-containing soybean protein isolate can be prepared from soybean flour, however, by allowing the endogenous phytase to act on the phytate in a 6% suspension of the flour at pH 5 at 65°C.[96] Hydrolysis of the phytate facilitates its separation from the bulk of the soybean protein, which is then concentrated by ultrafiltration using a membrane which is permeable to phytate and its hydrolysis products but impermeable to protein. The product obtained by this method contains over 90% protein and only 0·3% phosphorus in the form of phytate.

5.2. Effect of Adding Phytase

The phytate content of rapeseed could be totally eliminated by adding wheat phytase to a 5% dispersion of rapeseed meal or a protein concentrate at pH 5·15, followed by a 2 h incubation at 57°C and subsequent dialysis against distilled water for 7 days.[97] The residual phytate content of autolyzed California small beans could also be further reduced by the addition of wheat phytase.[70]

5.3. Effect of Germination

Germination has been shown to produce an increase in the phytase activity of mature seeds with a concomitant reduction in the phytate content of navy beans (*Phaseolus vulgaris*),[98,99] black-eyed beans (*Vigna sinensis*),[99] green gram (*Phaseolus aureus*),[100] horse gram (*Dolichos biflorus* or *Macrotyloma uniflorum*),[100,101] red gram (*Cajanus cajan*),[100] moth bean (*Vigna aconitifolium*),[101] faba bean (*Vicia faba*),[102] rapeseed (*Brassica campestris*)[103] and sorghum.[104] Details concerning the experimental conditions of germination necessary to effect significant reductions in the phytate content of these seeds may be found in the aforementioned citations.

5.4. Effect of Fermentation

Traditional legume-based foods which have undergone fermentation likewise have reduced levels of phytate, presumably due to the action of phytase elaborated by microorganisms. Tempeh, a soybean food produced by fermentation with the mold *Rhizopus oligosporus*, has one-half[105] to two-thirds[106] of the phytate content of the soybeans prior to fermentation. Idli, a blend of black gram (*Phaseolus mungo*) and rice which has been subjected to the fermentative action of yeast and *Leuconostoc mesenteroides*, undergoes a 35–40% reduction in phytate.[107] The treatment of soybean meal with the culture filtrate of *Aspergillus ficcum* caused a 90% release of the phytate-bound phosphorus and reduced the chick's requirement for calcium by at least one-third.[108] A culture filtrate of this organism has also been used to reduce the phytate content of

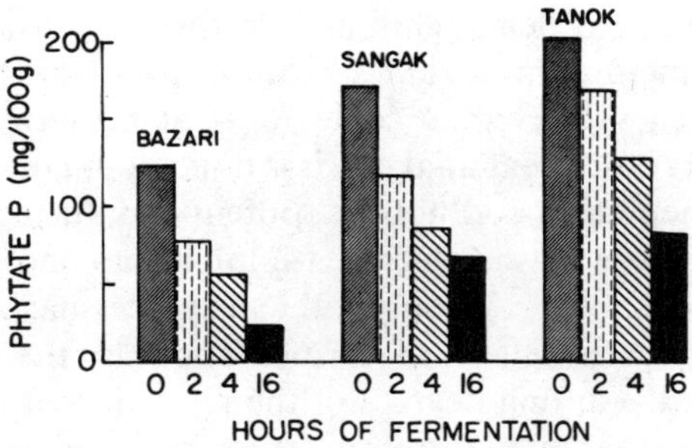

Fig. 11. Destruction of phytate in whole-wheat flours by the action of yeast. Bazari, 75–85% extraction rate (ER); Sangak, 85–90% ER; Tanok, 95–100% ER. Taken from Reinhold.[112]

solvent-extracted cottonseed meal with a concomitant increase in the availability of phosphorus and zinc for the growing chick.[109] This treatment also resulted in a substantial increase in the metabolizable energy of the meal for chicks, an effect which was attributed to the hydrolysis of phytate and its release from the protein, thereby making the latter more digestible.

It is believed that a high incidence of mineral imbalance among the rural population of Iran is due to their life-long consumption of unleavened bread made from a high-extraction-rate flour with a high phytate content.[110,111] The use of yeast for leavening bread dough, on the other hand, produces a significant reduction in phytate content of bread[112,113] and enhances the absorption of zinc by the rat intestine *in vitro*[114] or by human subjects *in vivo*.[113] The destruction of phytate in three different whole-wheat flours commonly used in Iran (by yeast fermentation) is illustrated in Fig. 11. A significant reduction in phytate also occurs in yeast-leavened bread made from a soy-fortified wheat flour.[115]

6. OTHER ANTI-NUTRITIONAL FACTORS

6.1. Protease Inhibitors and Lectins

As has already been pointed out in connection with phytate and flatus-causing oligosaccharides, fermentation has proved to be a very effective way of reducing these particular undesirable constituents in various food or feed products. The improvement in the nutritive value generally associated with the fermentation of legume-based foods[116,117] may also be due to the partial destruction of proteinaceous components such as protease inhibitors and hemagglutinins (lectins) which are the principal anti-nutritional factors in legumes.[1] Since most of these fermented legumes have been subjected to some form of heat treatment prior to fermentation and subsequent analysis, it is difficult to determine the extent to which the fermentation and not the protein-denaturing effect of heat is responsible for the low levels of protease inhibitors and lectins generally found in such foods.[116,117] In only a few instances has an attempt been made to determine the effect of fermentation in the absence of the application of heat. Varying degrees of the reduction of trypsin inhibitor activity by natural fermentation prior to heat treatment have been reported for chick peas or Bengal gram (*Cicer arietinum*),[118,119] cow peas (*Vigna sinensis* or *catjung*)[119] and soy idli (a blend of soybeans and rice).[116] In the absence of further studies it can only be assumed that the decreases in trypsin inhibitor activity observed in these cases were due to proteolytic attack by enzymes secreted by the organisms involved in the fermentation. Ramakrishnan in his review[116] refers to the fact that *Leuconostoc mesenteroides*, one of the organisms isolated from soy idli, was capable of hydrolyzing the lectins present in soybeans, kidney beans, navy beans and black beans. He suggests that the use of this organism for the fermentation of legumes may provide an effective means for eliminating the toxic effects of the lectins which they contain.

6.2. Toxic Glucosides

In addition to the glycosides which have thus far been considered, namely, the goitrogenic and cyanogenic glucosides and the flatus-producing oligosaccharides, at least two other plants are known to contain toxic glucosides. These are the safflower and jojoba plant.

Safflower meal is commonly used in animal feeds, but is not suitable for human food because of its bitterness and cathartic effect. The latter property has been attributed to two phenolic glucosides which have been

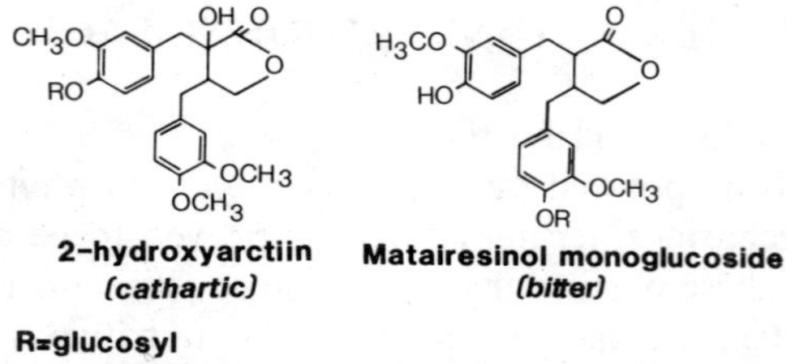

Fig. 12. Structures of the phenolic glucosides responsible for the bitterness and cathartic activity of safflower meal.

identified as 2-hydroxyarctiin and matairesinol monoglucoside whose structures are shown in Fig. 12. Treatment of safflower meal with β-glucosidase at pH 5, 37°C, for 2 days has been shown to be an effective means of hydrolyzing these glucosides resulting in almost complete elimination of its bitterness and cathartic activity.[120]

The jojoba plant, *Simmondsia californica*, is an evergreen plant that grows wild in the arid regions of south-western United States. Its economic importance lies in the fact that the female plants produce seeds that contain a liquid wax comparable in properties to sperm whale oil. The meal remaining after extraction of the oil contains about 27% protein but

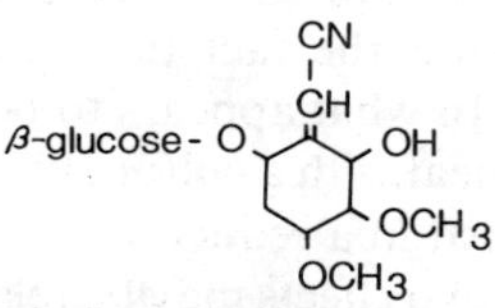

Fig. 13. Structure of simmondsin, the toxic principle of the jojoba plant.

is very poorly utilized by animals. The toxic principle largely responsible for the toxicity of the jojoba meal has been identified as a glucoside of a multiply-substituted cyclohexane bearing a cyanomethylene substituent.[121] This compound has been given the name simmondsin whose structure is shown in Fig. 13. The toxicity of simmondsin is apparently related to the cyano-aglycone which can be split from the glucose conjugate in the gut or stomach of monogastric animals or by microorganisms in the rumen of sheep and cows.[122] Treatment of moistened jojoba meal with selected strains of *Lactobacillus acidophilus* for 21 days at 26°C produces a 95% reduction in the level of the toxicant.[123] Enzymes produced by this organism apparently modify the cyano group of simmondsin thereby rendering it non-toxic to mice, poultry, sheep and cattle. The precise enzyme (or enzymes) responsible for detoxification is not known, but the possible action of a nitrilase has been suggested.[123]

6.3. Gossypol

It is well known that cottonseed meal, unless properly processed or derived from a glandless variety, is poorly utilized as an animal feed because of the presence of gossypol (Fig. 14).[124] Treatment of cottonseed meal with a suspension of spores from a fungus (a strain of *Diplodia*) caused a marked reduction in free gossypol which was accompanied by a

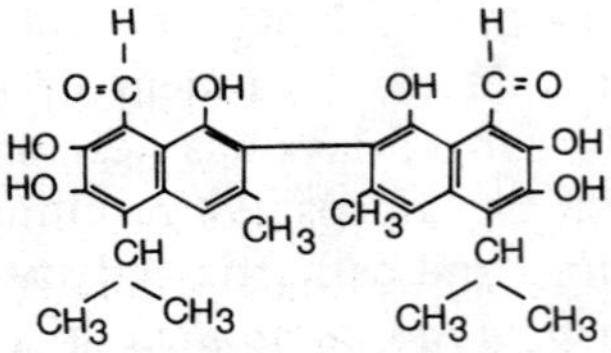

Fig. 14. Structure of gossypol present in the pigment gland of the cottonseed.

corresponding increase in protein-bound gossypol.[125] Rat feeding studies confirmed the fact that the fungal-treated cottonseed meal was essentially devoid of toxicity compared to the untreated meal. The conclusion that this reduction in toxicity is caused by the conversion of free to bound gossypol is compatible with the fact that, unlike free gossypol, bound gossypol is non-toxic.[124] In what appears to be a related observation, the treatment of cottonseed meal with a culture filtrate of *Aspergillus ficcum* as a source of phytase resulted in a reduction in the level of free gossypol as well as phytate.[109] These two effects may be related if one can assume that, as a result of the action of phytase, protein is released from a phytate–protein complex and thus becomes more readily available for binding gossypol.

6.4. Lathyrogenic Factor

Lathyrism, a neurological disease associated with the consumption of *Lathyrus sativus*, is not an uncommon occurrence in India, and has been attributed to β-*N*-oxalyl-L-α,β-diaminopropionic acid (Fig. 15).[126] In his review on Indian fermented foods, Ramakrishnan[116] reported that a non-pathogenic *Bacillus* isolated from rat intestine was capable of growing in a medium containing this neurotoxin and that this organism possessed enzymes capable of degrading this toxin to CO_2 and H_2O. Whether this organism can be used to ferment *L. sativus* to produce a non-toxic item of human food in a country sorely in need of alternative sources of protein seems worthy of further investigation.

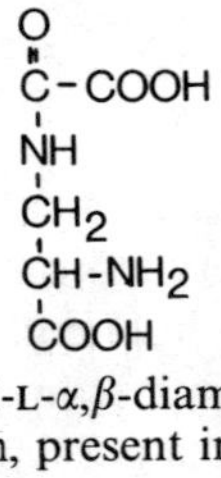

FIg. 15. Structure of β-*N*-oxalyl-L-α,β-diaminopropionic acid, the causative factor of lathyrism, present in *Lathyrus sativus*.

7. CONCLUSIONS

In this chapter an attempt has been made to cite examples of how toxic substances which are natural constituents of many plant tissues may be removed or at least reduced through the action of specific enzymes. These enzymes may be present endogenously to release the toxic substances from an innocuous precursor; processing techniques or traditional modes of food preparation are then employed to eliminate the toxicants so formed. Such is the case with the goitrogenic glycosides of rapeseed and the cyanogenic glycosides of cassava, lima beans and fruit kernels. A most promising approach to the problem of detoxification, and one which will probably be used more commonly in the future, is the use of microbial enzymes which selectively attack natural toxicants. This method has been employed successfully to reduce the flatulence-producing factors in beans, the phytate content of soybeans and wheat, and the gossypol in cottonseed. The use of immobilized cells or the enzymes which they produce represents an area of technological advancement in the processing of foods[127,128] that will no doubt find wider application in the future for the enzymatic detoxification of foods.

REFERENCES

1. Liener, I. E. (Ed.), *Toxic Constituents of Plant Foodstuffs*, 2nd Edn, Academic Press, New York, 1980.
2. Ory, R. L. (Ed.), *Antinutrients and Natural Toxicants in Foods*, Food and Nutrition Press, Westport, Connecticut, 1981.
3. Maheshwari, P. N., Stanley, D. W. and Gray, J. I., *J. Food Protection*, **44**, 459 (1981).
4. Bell, J. M., *Can. J. Agric. Sci.*, **35**, 242 (1955).
5. Miller, R. W., Van Etten, C. H., McCrew, C., Wolff, I. A. and Jones, Q. J., *J. Agric. Food Chem.*, **10**, 426 (1962).
6. Tookey, H. L., Van Etten, C. H. and Daxenbichler, M. E., in *Toxic Constituents of Plant Foodstuffs*, Liener, I. E. (Ed.), 2nd Edn, Academic Press, New York, 1980, pp. 103–42.
7. Eapen, K. E., Tape, N. W. and Sims, R. P. A., *J. Am. Oil Chem. Soc.*, **45**, 194 (1968).
8. Eapen, K. E., Tape, N. W. and Sims, R. P. A., *J. Am. Oil Chem. Soc.*, **46**, 52 (1969).
9. Agren, G. and Eklund, A., *J. Sci. Food Agric.*, **23**, 1451 (1972).
10. Kozlowska, H., Sosulski, F. W. and Youngs, C. G., *Can. Inst. Food Sci. Technol. J.*, **5**, 149 (1972).
11. Van Etten, C. H., Daxenbichler, M. E., Peters, J. E., Wolff, I. A. and Booth, A. N., *J. Agric. Food Chem.*, **13**, 24 (1965).

12. Kirk, L. D., Mustakas, G. C., Griffin, E. L. Jr and Booth, A. N., *J. Am. Oil Chem. Soc.*, **48**, 845 (1971).
13. Mustakas, G. C., Kirk, L. D., Griffin, E. L. Jr and Clanton, D. C., *J. Am. Oil Chem. Soc.*, **45**, 53 (1968).
14. Oginsky, E. L., Stein, A. E. and Greer, M. A., *Proc. Soc. Exp. Biol. Med.*, **119**, 360 (1965).
15. Marangos, A. and Hill, R., *Proc. Nutr. Soc.*, **33**, 90A (1974).
16. Greer, M. A. and Deeney, J. M., *J. Clin. Invest.*, **38**, 1465 (1959).
17. Poulsen, E., *Nord. Vet. Med.*, **10**, 487 (1958).
18. Akiba, Y. and Matsumoto, T., *Nippon Chikusan Gakkai Ho*, **50**, 73 (1979).
19. Belzile, R. J. and Bell, J. M., *Can. J. Animal Sci.*, **51**, 165 (1966).
20. Goering, K. J., US Pat. 2,987,399 (1963).
21. Mukherjee, K. D., Afzalpurkar, A. B. and El-Nockrashy, A. S., *Fette, Seifen, Anstrichm.*, **78**, 306 (1976).
22. Mukherjee, K. D., Mangold, H. K. and El-Nockrashy, A. S., *Nutr. Metab.*, **23**, 1 (1978).
23. Goering, K. J., Thomas, O. O., Beardsley, D. R. and Curran, W. A. Jr, *J. Nutr.*, **72**, 210 (1960).
24. Vaccarino, C., *Rev. Ital. Sostanze Grasse*, **52**, 91 (1975).
25. Vaccarino, C., *Rev. Ital. Sostanze Grasse*, **52**, 195 (1975).
26. Vaccarino, C., Toscano, M. A. and Tripodo, M. M., *Rev. Ital. Sostanze Grasse*, **52**, 291 (1976).
27. Tookey, A. L., Van Etten, C. H., Peters, J. E. and Wolff, I. A., *Cereal Chem.*, **42**, 507 (1965).
28. Mustakas, G. C., Kirk, L. D. and Griffin, E. L. Jr, *J. Am. Oil Chem. Soc.*, **39**, 372 (1962).
29. Mustakas, G. C., Griffin, E. L. Jr, Gastrock, E. A., D'Aquin, E. L., Keating, E. J. and Paton, E. L., *Biotech. Bioeng.*, **5**, 27 (1963).
30. Mustakas, G. C., Kirk, L. D., Sohns, V. E. and Griffin, E. L. Jr, *J. Am. Oil Chem. Soc.*, **42**, 33 (1965).
31. Mustakas, G. C., *Chemurgic Digest*, Jan./Feb., 9 (1964).
32. Butt, J. M., Yunus, O. and Shah, F. H., *Pakistan J. Sci. Ind. Res.*, **9**, 130 (1966).
33. Staron, T., *Aliment. Vie*, **62**, 165 (1974).
34. Conn, E. E., *J. Agric. Food Chem.*, **17**, 519 (1969).
35. Conn, E. E., in *Toxicants Occurring Naturally in Foods*, Nat. Acad. Sci., Washington, DC, 1973, p. 309.
36. Montgomery, R. D., in *Toxic Constituents of Plant Foodstuffs*, Liener, I. E. (Ed.), 2nd Edn, Academic Press, New York, 1980, p. 143.
37. Cheeke, P. R. and Shull, L. R., *Natural Toxicants in Feeds and Poisonous Plants*, Avi Publishing Co., Westport, Connecticut, 1985, p. 173.
38. Montgomery, R. D., *Am. J. Clin. Nutr.*, **17**, 103 (1965).
39. Oyefess, J. A., *Indian J. Nutr. Diet.*, **13**, 77 (1976).
40. Ketiker, A. O., Akinyele, I. O., Keshinro, O. O. and Akinnawo, O. O., *Food Chem.*, **3**, 221 (1978).
41. Maduagwu, E. N. and Umoh, I. B., *Toxicol. Lett.*, **10**, 245 (1982).
42. Wood, T., *J. Sci. Food Agric.*, **16**, 300 (1965).
43. Seigler, D. S., *Phytochemistry*, **14**, 9 (1975).
44. Collard, P. and Levi, S., *Nature*, **183**, 620 (1959).

45. Akinrele, I. A., *J. Sci. Food Agric.*, **15**, 589 (1964).
46. Wood, T., *J. Sci. Food Agric.*, **17**, 85 (1966).
47. Okafor, N., *J. Appl. Bacteriol.*, **42**, 279 (1977).
48. Maduagwu, E. N., *Toxicol. Lett.*, **15**, 335 (1983).
49. Ikediobi, C. O. and Onyike, E., *Agric. Biol. Chem.*, **46**, 1667 (1982).
50. Ikediobi, C. O. and Onyike, E., *Process Biochem.*, **17**, 2 (1982).
51. Ikediobi, C. O., Ogundu, E. C. and Ukoha, A. I., *Process Biochem.*, **20**, 99 (1985).
52. Pido, P. P., Adeyanju, S. A. and Adegbola, A. A., *Poultry Sci.*, **58**, 427 (1979).
53. Wyllie, D., Mtui, M., Oloya, J. D. and Kategile, J. A., *Nutr. Rep. Int.*, **30**, 1127 (1984).
54. Obioha, F. C. and Anikwe, P. C. N., *Nutr. Rep. Int.*, **26**, 961 (1982).
55. Vogt, H., *World's Poult. Sci. J.*, **22**, 113 (1966).
56. Oke, O. L., *Nature*, **212**, 1055 (1966).
57. Osuntokun, B. O., Durowoju, J. E., McFarlane, H. and Wilson, J., *Brit. Med. J.*, **3**, 647 (1968).
58. Osuntokun, B. O., *Brit. J. Nutr.*, **24**, 797 (1970).
59. Osuntokun, B. O., Aladetoyinbo, A. and Adauja, A. D. G., *Lancet*, **II**, 372 (1970).
60. Ermans, A. M., Mbulamoko, N. M., Delange, F. and Ahluwalia, R., Int. Develop. Res. Center, publ. IDRC-136e, Ottawa, Canada, 1980.
61. Keim, P., Kohanski, R. A. and Heinrikson, R. L., in *Antinutrients and Toxicants in Foods*, Ory, R. L. (Ed.), Food and Nutrition Press, Westport, Connecticut, 1981, p. 331.
62. Rahman, S. A., De, S. S. and Subrahmanyan, V., *Current Sci.*, **11**, 351 (1947).
63. Fincke, A., in *Handbuch der Lebensmittelchemie*, Schormuller, J. and Acker, L. (Eds), Springer-Verlag, Berlin, 1967, p. 719.
64. Schab, R. and Yannai, S., *J. Food Sci. Technol.*, **10**, 57 (1973).
65. Calloway, D. H., Hickey, C. A. and Murphy, E. L., *J. Food Sci.*, **36**, 251 (1971).
66. Rackis, J. J., *Adv. Chem. Ser.*, **15**, 207 (1975).
67. Olson, A. C., Becker, R., Miers, J. C., Gumbmann, M. R. and Wagner, J. R., in *Protein Nutritional Quality of Foods and Feeds*, Friedman, M. (Ed.), Marcel Dekker, New York, 1975, p. 551.
68. Rackis, J. J., *J. Am. Oil Chem. Soc.*, **58**, 503 (1981).
69. Olson, A. C., Gray, G. M., Gumbmann, M. R., Sell, C. R. and Wagner, J. R., in *Antinutrients and Natural Toxicants in Foods*, Ory, R. L. (Ed.), Food and Nutrition Press, Westport, Connecticut, 1981, p. 275.
70. Kon, S., Olson, A. C., Frederick, D. P., Eggling, S. B. and Wagner, J. R., *J. Food Sci.*, **38**, 215 (1973).
71. Becker, R., Olson, A. C., Frederick, D. P., Kon, S., Gumbmann, M. R. and Wagner, J. R., *J. Food Sci.*, **39**, 766 (1974).
72. Kim, W. J., Smit, J. B. and Nakayama, T. O. M., *Lebensm.-Weiss. Technol.*, **6**, 201 (1973).
73. DaSilva, R. S. F., Ida, E. I., Perre da Silva, M. A. and Fabre-Sanches, S., *Cienc. Technol. Aliment.*, **1**, 107 (1981).
74. Reddy, N. R., Salumkhe, D. K. and Sharma, R. P., *J. Food Sci.*, **45**, 1161 (1980).

75. Jood, S., Mehta, U., Singh, R. and Bhat, C. M., *J. Agric. Food Chem.*, **33**, 268 (1985).
76. Hansen, O. K., Ger. Pat. 2,524,753 (1975).
77. Dasek, J., Shepherd, D. and Wood, R. D., Belg. Pat. 872,336 (1979).
78. Sugimoto, H. and Van Buren, J. P., *J. Food Sci.*, **35**, 655 (1970).
79. Sherba, S. E., S. African Pat. 6,902,503 (1970).
80. Cruz, R., Batistela, J. C. and Wosiacki, G., *J. Food Sci.*, **46**, 1196 (1982).
81. Hernandez, J. L., Adris, J., deRank, E. F., Farias, R. and Samman, N., *J. Am. Oil Chem. Soc.*, **58**, 510 (1981).
82. Thananukul, D., Tanaka, M., Chichester, C. O. and Lee, T.-C., *J. Food Sci.*, **41**, 173 (1976).
83. Reynolds, J. H., *Biotechnol. Bioeng.*, **16**, 135 (1974).
84. Smiley, K. L., Hensley, D. E. and Gasdorf, H. J., *Appl. Environ. Microbiol.*, **31**, 615 (1976).
85. Oberleas, D., in *Toxicants Occurring Naturally in Foods*, Nat. Acad. Sci., Washington, DC, 1973, p. 363.
86. Erdman, J. W. Jr, *J. Am. Oil Chem. Soc.*, **56**, 736 (1979).
87. Cheryan, M., *Crit. Rev. Food Sci. Nutr.*, 297 (1980).
88. Maga, J. A., *J. Agric. Food Chem.*, **30**, 1 (1982).
89. Reddy, N. R., Sathe, S. K. and Salunkhe, D. K., *Adv. Food Res.*, **28**, 1 (1982).
90. Spitzer, R. R. and Phillips, P. H., *J. Nutr.*, **30**, 183 (1945).
91. McCance, R. A. and Widdowson, E. M., *Biochem. J.*, **29B**, 2694 (1935).
92. Bitar, K. and Reinhold, J. G., *Biochim. Biophys. Acta*, **268**, 442 (1972).
93. Schwimmer, S. and Burr, H. K., *J. Food Sci.*, **42**, 108 (1977).
94. Chang, R. and Schwimmer, S., *J. Food Biochem.*, **1**, 45 (1977).
95. Okubo, K., Myers, D. V. and Iacobucci, G. A., *Cereal Chem.*, **53**, 513 (1976).
96. Okubo, K., Waldrop, A. B., Iacobucci, G. A. and Myers, D. V., *Cereal Chem.*, **52**, 263 (1975).
97. Serraino, M. R. and Thompson, L. U., *J. Agric. Food Chem.*, **32**, 38 (1984).
98. Lolas, G. M. and Markakis, P., *J. Food Sci.*, **42**, 1094 (1977).
99. Tabekhia, M. M. and Luh, B. S., *J. Food Sci.*, **45**, 406 (1980).
100. Biri, J., Parvatham, R. and Santhini, K., *Indian J. Nutr. Diet.*, **18**, 87 (1981).
101. Borade, V. P., Kakam, S. S. and Salunkhe, D. K., *Qual. Plant. Plant Foods Hum. Nutr.*, **34**, 151 (1984).
102. Eskin, N. A. M. and Wiebe, S., *J. Food Sci.*, **48**, 270 (1983).
103. Thompson, L. U. and Serraino, M. R., *J. Food Sci.*, **50**, 1200 (1985).
104. Glennie, C. W., Cilliers, J. J. L. and Geyer, H. L., *Nutr. Rep. Int.*, **32**, 343 (1985).
105. Sutardi and Buckle, K. A., *J. Food Sci.*, **50**, 260 (1985).
106. Sudarmaji, S. and Markakis, P., *J. Sci. Food Agric.*, **28**, 381 (1977).
107. Reddy, N. R. and Salunkhe, D. K., *J. Food Sci.*, **45**, 1708 (1980).
108. Nelson, T. S., Shieh, T. R., Wodzinski, R. J. and Ware, J. H., *Poultry Sci.*, **47**, 1842 (1968).
109. Rojas, S. W. and Scott, M. L., *Poultry Sci.*, **48**, 819 (1969).
110. Reinhold, J. G., *Am. J. Clin. Nutr.*, **24**, 1204 (1971).

111. Reinhold, J. G., Hedayati, H., Lahingarzadeh, A. and Nasr, K., *Ecol. Food Nutr.*, **2**, 157 (1973).
112. Reinhold, J. G., *J. Am. Diet. Assoc.*, **66**, 38 (1975).
113. Navert, B., Sandstrom, B. and Cederblad, A., *Brit. J. Nutr.*, **53**, 47 (1985).
114. Reinhold, J. G., Parsa, A., Karinian, N., Hammick, J. W. and Ismail-Beigi, F., *J. Nutr.*, 976–82 (1974).
115. Ranhotra, G. S., Loewe, R. J. and Puyat, L. V., *J. Food Sci.*, **39**, 1023 (1974).
116. Ramakrishnan, C. V., *Baroda J. Nutr.*, **6**, 1 (1979).
117. Reddy, N. R., Pierson, M. D., Sathe, S. K. and Salunkhe, D. K., *Crit. Rev. Food Sci. Nutr.*, 335 (1982).
118. Rajalakshim, R. and Vanaja, K., *Brit. J. Nutr.*, **21**, 467 (1967).
119. Zemora, A. F. and Fields, M. L., *J. Food Sci.*, **44**, 234 (1979).
120. Lyon, C. K., Gumbmann, M. R., Betschart, A. A., Robbins, D. J. and Saunders, R. M., *J. Am. Oil Chem. Soc.*, **56**, 560 (1979).
121. Booth, A. N., Elliger, C. A. and Waiss, A. Jr, *Life Sci.*, **15**, 1115 (1974).
122. Verbiscar, A., Banigan, T. F., Weber, C. W., Reid, B. L., Trei, J. E., Nelson, E. A., Raffauf, R. F. and Kosersky, D., *J. Agric. Food Chem.*, **28**, 571 (1980).
123. Verbiscar, A., Banigan, T. F., Weber, C. W., Reid, B. L., Swingle, R. S., Trei, J. E. and Nelson, E. A., *J. Agric. Food Chem.*, **29**, 296 (1981).
124. Berardi, L. C. and Goldblatt, L. A., in *Toxic Constituents of Plant Foodstuffs*, 2nd Edn, Liener, I. E. (Ed.), Academic Press, New York, 1980, p. 184.
125. Baugher, N. L. and Campbell, T. C., *Science*, **164**, 1526 (1969).
126. Padmanabhan, G., in *Toxic Constituents of Plant Foodstuffs*, 2nd Edn, Liener, I. E. (Ed.), Academic Press, New York, 1980, p. 239.
127. Pitcher, W. H. Jr, *Immobilized Enzymes for Food Processing*, CRC Press, Boca Raton, Florida, 1980.
128. Laskin, A. I. (Ed.), *Enzymes and Immobilized Cells in Biotechnology*, Butterworth Publishers, Stoneham, Massachusetts, 1985.

113. [illegible] Freeman, H. [illegible] and [illegible] K., [illegible] Vitr. [illegible]
114. [illegible]
115. [illegible] Gadamboud [illegible] 3, 71 (1985)
116. [illegible] McCarthy, [illegible] Kathleen, N., Hammock, K. W. and [illegible]
117. [illegible] Vorsp. [illegible] 52 (1975)
118. [illegible] and Lloyd, [illegible] Soc. 39, 1022 (1975)
119. [illegible] Appl., 6, 4(15) [illegible]
120. [illegible] M. D. Sathe, [illegible] and Schmerling, [illegible] Proc. Soc. [illegible]
121. [illegible] 46 (1965)
122. [illegible] Sci., 47 [illegible]
123. [illegible] Robinson, D. [illegible] Sci. 58, 26 (1979)
124. [illegible] Nature [illegible] 18, 145 (1963)
125. [illegible] C. W. [illegible] E. Nelson, [illegible] J. Agric. Food Chem., 28, 571 (1980)
126. [illegible] R., Swingle, K. [illegible] J. Agric. Food Chem., 29, 39 (1981)
127. [illegible] and Goldstein [illegible] Constituents of Plant Foods [illegible] Academic Press, New York, 1981, p. 181
128. [illegible] and [illegible] Science 164, 1326 (1969)
129. [illegible] Constituents of Plant Foodstuffs, 2nd edn. [illegible] New York, 1980, p. 239
130. [illegible] Food [illegible] CRC Press, Boca Raton, Florida [illegible]
131. [illegible] (Eds.) [illegible] AVI Publishing [illegible] Westport, Mass. [illegible] 1983

7

Organic Acids by Fermentation, especially Citric Acid

P. E. MILSOM

John & E. Sturge Ltd, Selby, North Yorkshire, UK

1. INTRODUCTION

Aliphatic hydroxy acids containing one or more carboxyl groups are an important group of additives having extensive uses in the food industry. Such compounds include citric acid, malic acid, lactic acid, tartaric acid, gluconic acid. These acids all occur naturally in the plant and animal kingdoms, some more widely than others. Citric and *l*-malic acid are intermediates in the oxidation of acetate via the so-called citric acid cycle, a metabolic system occurring in most living cells. Lactic acid is a product of the oxidation of glucose in muscle. It is also produced by the starter cultures such as *Streptococcus lactis* and *Lactobacillus bulgaricus* which are used in the dairy industry to acidify the milk used in the manufacture of cheese and yoghurt respectively. Gluconic acid is a product of glucose oxidation in many living systems, particularly microorganisms. Only tartaric acid is of limited distribution, occurring only in the grape and tamarind.

Whilst all these acids are aliphatic hydroxy acids they differ in their chemical behaviour and in their organoleptic effects. The sour taste of an acid is in part related to the titrable acidity but this is not the whole story. The acid taste may be sharp or bland, and it is possible to list organic acids in an order which grades them in this sense.[1] Thus tartaric acid is considered to be the sharpest and lactic acid the most bland; citric and malic come in between, in that order.

Whilst the acids under consideration in this chapter occur in fruits and therefore in wines and ciders, they do not directly contribute to the flavour of the fruit. Citric acid has no lemon flavour nor does malic acid taste of apples. However, the sharpness of the taste of the fruit is affected by the predominant acid present, that is citric and malic acids respectively. Arnold[2] gives a table showing the acids present in various fruits. From this it is clear that citric acid and malic acid are the principal food acids. Malic acid is the foremost acid in apples, pears, stone fruit, grapes and bananas, whereas citric acid is dominant in citrus fruits, raspberry, strawberry, currants, tomato and pineapple. Blackberries contain large amounts of iso-citric acid which is not a commercial product and is not treated in this article as a fermentation acid (but see under citric acid from yeast). Although grapes contain much tartaric acid, the principal acid is malic acid as mentioned above. During the maturing of wine the malic acid is progressively converted to lactic acid in a secondary fermentation giving rise to a smoother flavour.

Food acids are used to adjust the acid flavour in soft drinks, fruit juices, canned fruit, cider and wine. It might be thought that the acid used would match that already present in the fruit in question. However, there are other considerations such as the power of citric acid to inhibit the development of metal-catalysed off-flavours and colour deterioration. In some cases blends of citric acid and lactic or malic acids are used. Although lactic acid and malic acid might be considered the appropriate materials for adjustment of acidity in wines and ciders respectively, in many cases citric acid is used in wines, and mixtures of citric and malic acids in ciders.

In jams, citric acid (or sometimes malic acid, where this is the natural acid of the fruit in question) is used to help adjust the flavour and to bring the pH to the correct value for pectin setting. In table jellies citric, or possibly malic acid, is used to bring the gelatin to its isoelectric point.

In sugar confectionery, acid is added to bring about inversion of the sugar and in some cases to give 'bite' to the product. In some speciality products, malic, lactic or tartaric acid may be used, but more usually citric acid is the acidulent of choice. Buffered lactic acid is often used in the preparation of soft candies.

In dry mixes such as effervescent powders, citric or malic acids, calcium hydrogen lactate or sometimes tartaric acid are used. In dry cake mixes gluconolactone or calcium hydrogen lactate are often employed as latent acids activated when water is added.

Citric acid or citrates are used as additives to combat oxidative deterioration in flavour or colour in a wide range of foodstuffs. The property which makes citric acid particularly suitable for this duty is its ability to complex heavy metals such as copper and nickel which catalyse oxidation reactions. Amongst the foodstuffs protected by addition of citric acid or one of its derivatives are fats, animal or vegetable oils, and fish. Citric acid or alkali citrates are not soluble in fats and oils, and the use of monostearyl citrate may be preferred. Fruit and vegetables are dipped in citric acid solution before canning or freezing. This treatment reduces enzymic browning and loss of vitamin C.

Sodium citrate assists in the emulsification of processed cheese and is widely used for this purpose. Calcium lactate is a so-called protein plasticiser and is employed in the preparation of dried milk powders for baby foods and also in foamed protein products.

2. PRODUCTION VOLUMES AND METHODS OF MANUFACTURE OF ORGANIC ACIDS

The approximate annual production tonnages of the acids being discussed in this chapter are as follows: citric 350 000 tonnes; malic 25 000 t; tartaric 40 000 t; gluconic 50 000 t; lactic 20 000 t. It is thus clear that citric acid is easily the most important product in this group. Moreover, if food applications only are considered, the position of citric acid is, if anything, strengthened. Approximately 75% of citric acid on the market is used in food, confectionery and beverages whereas, for instance, only a small proportion of gluconic acid is sold as gluconolactone for food use, the remainder being employed as the sodium salt in bottle-washing and other industrial applications. Again, only perhaps 50% of the tartaric acid on the market is used in food but most of the malic and lactic acid is devoted to this purpose.

In a chapter devoted to the production of food acids by fermentation, a discussion of the manufacture of tartaric acid would be inappropriate because all available material is made from wine residues. During the production of wine from grapes, tartaric acid precipitates or crystallises as potassium hydrogen tartrate (algols, wine lees) or calcium hydrogen tartrate (sablons). These materials are worked up to yield tartaric acid, cream of tartar (potassium hydrogen tartrate) or Rochelle salt (sodium potassium tartrate) which are the main forms in which tartaric acid appears in commerce. In fact, the supply of tartaric acid is declining because of changes in the wine industry, and this fact has caused the price to rise to levels where it is used only when substitutes such as citric acid cannot be employed. It is true that synthetic tartaric acid is available by chemical hydroxylation of maleic acid. However, this process produces DL-tartaric acid whereas the wine residues contain the L(+) stereoisomer.

Apart from any possible difficulties in obtaining regulatory approval for the use of the racemic DL-tartaric acid in food, its solubility is very much less than that of the L(+) isomer, making it difficult to substitute the DL acid for the L(+) acid in existing manufacturing routines.

The case of malic acid shows certain similarities but the analogy does not hold completely. L(−)-Malic acid is an article of commerce but it is more expensive than the DL product. The optically active variety can be made by fermentation of glucose or by the action of the enzyme fumarase on fumaric acid (see below). The racemic acid is made by hydration of maleic and/or fumaric acid by water under pressure. However, DL-malic acid is not less soluble than the L(−) variety and its use in food has been approved by the FDA, the European Community regulatory authorities and MAFF in the UK. As a result L(−)-malic acid produced biotechnologically is used only in a few special applications, the bulk of the market being satisfied by the synthetic racemic acid.

Gluconic acid and its lactone can be produced by the oxidation of glucose solutions using air or oxygen and an organic catalyst. However, at the present time, most or all of the available gluconic acid is produced by fermentation of glucose by *Aspergillus niger* or species of *Gluconobacter*.

Lactic acid can likewise be produced synthetically by hydrolysis of lactonitrile which itself is produced either as a by-product of acrylonitrile manufacture or by the reaction of acetaldehyde with hydrogen cyanide. Either of these chemical processes leads to DL-lactic acid. The fermentation route using *Lactobacillus delbrückii* is said to produce about 3% of D(−) acid in addition to 97% of the L(+) isomer.[3] Contaminant organisms, if present, e.g. *Lactobacillus plantarum*, may elaborate enzymes causing racemisation resulting in the production of a much greater proportion of the D(−) acid. *Lactobacillus lactis* produces only the D(−) isomer. Commercial food grade fermentation lactic acid is variable in optical rotation but usually corresponds to 65–75% of the L(+) isomer.

In the light of the above considerations the remainder of this chapter will describe the fermentative production of citric, lactic and L(−)-malic acid, together with gluconic acid and its δ-lactone. Citric acid, because it takes such a large share of the market, will be given the most detailed treatment.

A matter of terminology perhaps needs to be cleared up. Fermentation was a term originally applied to the anaerobic process such as that used to produce ethanol. Today the word is used to denote both anaerobic and aerobic processes such as the lactic acid fermentation and the citric acid fermentation respectively.

3. CITRIC ACID

Citric acid (2-hydroxy-1,2,3-propanetricarboxylic acid) [77-92-9] has the following structure:

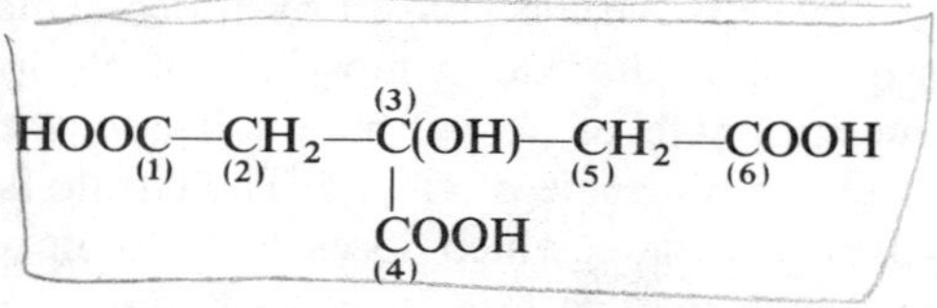

It crystallises from water in two forms, namely anhydrous and monohydrate. When crystallisation occurs at a temperature exceeding 36·6°C the anhydrous form is produced. At lower temperatures the monohydrate is formed. The monohydrate was at one time the only form commercially available but increasing amounts of the anhydrous acid are now used. It is essential in dry-mix formulations and has a lower tendency to cake on storage.

Up until about 1920 all commercial citric acid was obtained from the juice of lemons or limes. It was isolated by precipitation as the insoluble calcium salt followed by regeneration of the acid using sulphuric acid. This is the classical citric acid recovery process (see below).

In the second decade of this century supply difficulties and increasing prices led to the development of fermentation processes for the production of citric acid, and by 1930 a number of manufacturers were using this method, including La Citrique Belge, Chas. Pfizer Inc. and John & E. Sturge (Citric) Limited. These processes derived directly or indirectly from the work of Currie,[4] and employed the mould *Aspergillus niger*, growing on the surface of a medium consisting of a solution of sucrose and inorganic salts. Developments in this process led to the use of beet molasses as the main medium constituent.

The last 40 years have seen the development of the submerged citric acid fermentation utilising glucose or beet or cane molasses as substrate. At first the organism employed was still *A. niger* but more recently certain yeasts have been used instead. A process using yeasts and n-paraffins as substrate proved to be an aberration, not only because of steeply rising oil prices but also because of problems with regulatory bodies, particularly in Italy and Japan. The process utilising n-paraffins as substrate will not be discussed in this chapter. However, the literature has been reviewed[5] and a plant built to use n-paraffins to produce sodium citrate has been described.[6]

3.1. Mechanism of Citric Acid Accumulation in Microorganisms

The metabolic pathways by which simple mono- and di-saccharides are converted to citric acid are included in Fig. 1. It will be observed that both the Embden–Meyerhof glycolysis and the citric acid cycle are involved.

The citric acid cycle is the mechanism by which acetate is oxidised to two molecules of carbon dioxide in most, if not all, living cells. In the normal way one molecule of acetate is condensed with a molecule of oxaloacetate to give a molecule of citrate. The citrate is metabolised via iso-citrate to oxalosuccinate which loses a molecule of CO_2 to yield α-ketoglutarate, which in turn loses a further molecule of CO_2 to give succinate. The succinate is further metabolised via fumarate and malate to regenerate oxaloacetate. Now if the citric acid cycle is interrupted at citric acid, no oxaloacetate is regenerated to condense with acetyl coenzyme A and so form citrate, and it is necessary to bring into play another so-called anapleurotic reaction to produce oxaloacetate. The anapleurotic reaction, at least in *A. niger*, is the carboxylation of pyruvate by the enzyme pyruvate carboxylase.

The evidence that this reaction does indeed occur in *A. niger* during citrate accumulation was provided by Cleland and Johnson[7] who used the high-yielding *A. niger* strain Wisconsin 72-4 acting on glucose-3:4-^{14}C. After exposing a solution of this labelled glucose for a short time to a culture of *A. niger* previously brought to a high citrate yielding condition, they were able to recover labelled citric acid. The ^{14}C activity was found almost quantitatively in carbon 4. About 15% of the original specific activity was located in carbon 6, but no other carbon exhibited any significant labelling.

When fructose-1,6-bisphosphate is split during glycolysis, both the 3 and 4 carbons in the original glucose become the carbon of the aldehyde group in 3-phosphoglyceraldehyde which itself becomes the carbon of the carboxyl group of pyruvate. The labelling in carbons 4 and 6 in citrate in this experiment provided strong circumstantial evidence for the carboxylation of pyruvate as the anapleurotic reaction in *A. niger*. The lower specific activity in carbon 6 was attributed to the equilibration of the CO_2 released from one molecule of pyruvate with unlabelled CO_2 present in the cells prior to its incorporation into oxaloacetate by the action of pyruvate carboxylase on a second molecule of pyruvate. The fixation of CO_2 by *A. niger* under citric acid production conditions and the presence of pyruvate carboxylase have been established.[8,9] Also demonstrated is the presence of all the enzymes of the Embden–Meyerhof glycolysis and of citrate synthetase.[10,11]

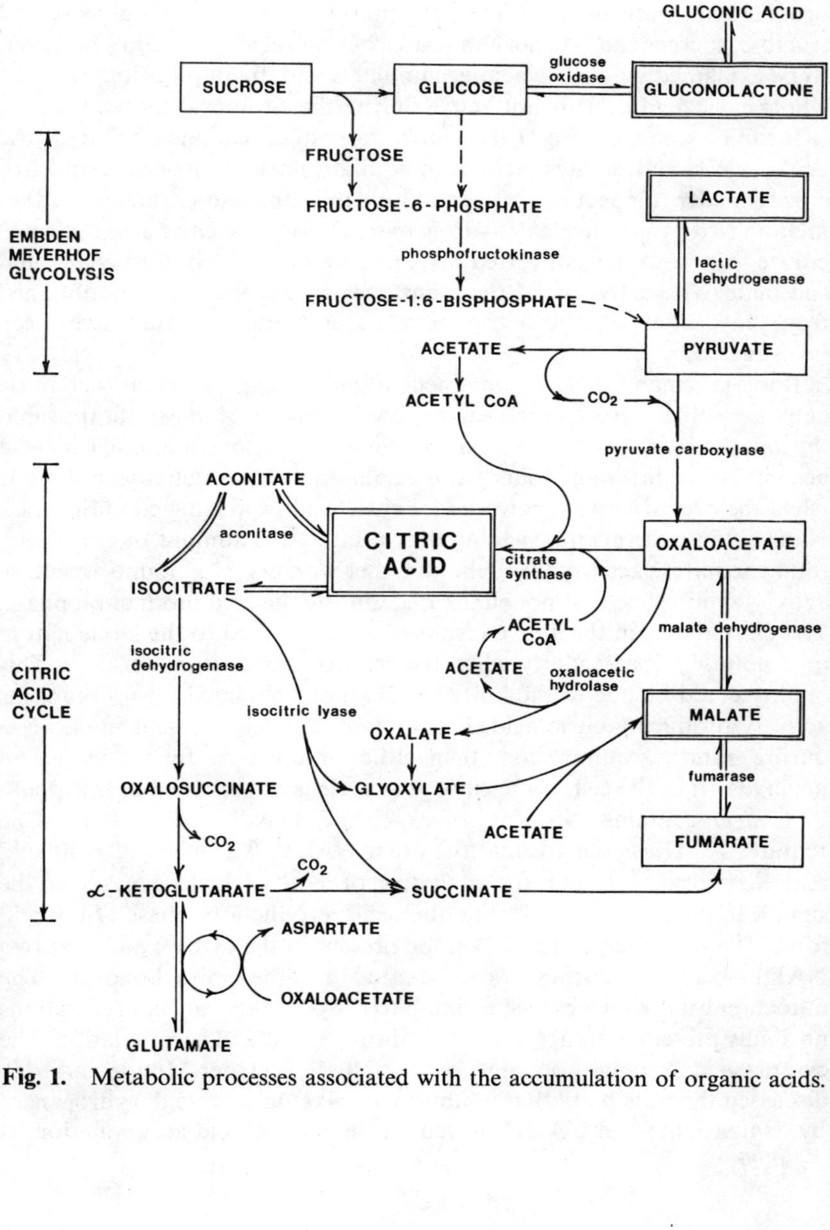

Fig. 1. Metabolic processes associated with the accumulation of organic acids.

From the above it is clear that all the enzymes needed to convert glucose to citric acid have been demonstrated in *A. niger*. Since this mould produces an invertase and since hexokinase will act on both glucose and fructose, sucrose can also act as a source of citric acid. It remains, however, to be explained why citric acid accumulates and, by implication, why the whole citric acid cycle is not active during the citric acid fermentation.

It can be seen from Fig. 1 that, under normal conditions when the citric acid cycle is active, citric acid is in equilibrium with aconitate and iso-citrate under the action of aconitase, and that iso-citrate is further metabolised by iso-citrate dehydrogenase. Outside the citric acid cycle iso-citrate can also be converted by iso-citrate lyase to glyoxylate and succinate. The activities of these enzymes in cell-free extracts obtained from mycelium in the citric acid accumulation phase have been investigated.

Ramakrishnan *et al.*[12] obtained results suggesting that aconitase activity, whilst present in the early growth phase of *A. niger*, disappeared during the acid production phase. Neilson,[13] using a non-citric acid accumulating medium, found appreciable aconitase activity, but none when the organism was grown under citric acid producing conditions. La Nauze,[14] in experiments yielding only a moderate amount of citric acid, found aconitase activity throughout. Other workers[15–17] found aconitase activity diminished, but not eliminated, during the acid production phase. The discrepancy in the above results may be ascribed to the labile nature of aconitase,[14] particularly when the cell extracts contain citrate.[15] This lability could be due to the low pH of extracts obtained from organisms actively producing citric acid. If aconitase is indeed present in *A. niger* during citrate accumulation, then either it must be for some reason inactive within the cell or the citric acid cycle is not arrested at this point.

A. niger contains two types of iso-citrate dehydrogenase (ICD). One requires NAD as cofactor and the other NADP.[18] Täufel and Behnke[15] and Szczodrak[17] found the activities of both dehydrogenases to be significantly diminished during the acid production phase. Mattey[19] found the NAD-requiring ICD to be present in the cytosol and that two NADP-specific enzymes were located in the mitochondria. The mitochondrial enzymes were inhibited by citrate at concentrations normally present within *A. niger* cells during citric acid accumulation. The cytoplasmic enzyme was not so inhibited. Meixner-Monori *et al.*[20] discussed the possibility that inhibition of α-ketoglutarate dehydrogenase by oxaloacetate and NADH played a role in citric acid accumulation by *A. niger*.

Kubicek and Röhr[21] state that iso-citric lyase is not present in *A. niger* cultures under citric acid producing conditions.

If we consider the regulation of the glycolysis system, the carboxylation of pyruvate and the action of citrate synthase on oxaloacetic acid and acetyl coenzyme A to produce citric acid, we find that most of the glycolytic enzymes, pyruvate carboxylase and citrate synthase are poorly regulated.[22] However, in normal circumstances phosphofructokinase (PFK) is inhibited by citrate giving a measure of feed-back control over the whole system.[22,23] This inhibition by citrate is of obvious importance in the citric acid fermentation because, unless it is relieved in some way, little product could accumulate. Two possibilities have been advanced. Firstly, Bowes[24] showed that, whereas the PFK in cell-free extracts from a low citrate producing strain was inhibited by added citrate, that from a high yielding strain was not. Secondly, Habison *et al.*[23] found that elevated levels of ammonium ion in the cells would also relieve the inhibition of PFK by citrate. These observations are not necessarily in conflict because the extract from the high yielding strain could possibly have contained more NH_4 than the extract from the low yielding strain (see also Habison *et al.*[23]).

The mechanism for the great accumulation of citric acid in *A. niger* grown under the correct conditions is thus not wholly clear. Although a fast flux of carbon through the Embden–Meyerhof system, coupled with virtually unregulated pyruvate carboxylase and citrate synthase, seems well established, the precise reason for the lack of operation of the citric acid cycle which would normally catabolise citrate has yet to be demonstrated. The activity of aconitase seems to be diminished during citric acid accumulation; citric acid inhibits the NADP dependent ICD and oxaloacetate inhibits α-ketoglutarate dehydrogenase. It is possible that all three effects have roles to play in citric acid production.

Now all strains of *A. niger* under all conditions of growth do not accumulate citric acid in the quantities required for an economic commercial process. Two factors are of prime importance, namely selection of good strains of mould and the maintenance of the correct conditions of fermentation, especially in terms of nutrition.

3.2. Nutritional Considerations Affecting the Production of Citric Acid by *A. niger*

If unlimited growth of *A. niger* is permitted by use of a rich culture medium, the amount of citric acid accumulated in the medium is minimal. If, however, growth is restricted in appropriate ways, citric acid is produced in amounts equivalent to 70–90% w/w on sucrose supplied. In the hypothetical ideal situation in which no mould growth occurs and no carbon dioxide or other by-product is produced, the yield would be 123%. Thus, under favourable conditions, up to about 75% of sucrose supplied appears as citric acid.

In addition to carbohydrate, *A. niger* requires a supply of nitrogen, phosphate, lesser amounts of potassium, magnesium, and sulphate, as well as low concentrations of trace metals such as zinc, copper, iron and manganese. *A. niger* grows well on a medium in which nitrogen is supplied as ammonium or nitrate ions, but organic nitrogen compounds can also be used effectively by the mould.

In order to accumulate citric acid, growth may be restricted by limiting the supply either of nitrogen[25,26] or of phosphate.[21] Kristiansen and Sinclair[25] considered the exhaustion of nitrogen to be a necessary prerequisite for citric acid production. They proposed that, following nitrogen exhaustion from the medium, growth slowed drastically and available nitrogen within the cells was carried by streaming towards the hyphal tips. The remainder of the cells in the hyphal filament became storage cells except for those at the origin of the filament which were devoid of function.[26] The storage cells produced citric acid as a carbon storage material. It is interesting to compare this hypothesis with the demonstrated mechanism for the accumulation of fat in oleaginous yeasts and moulds. Fats accumulate in large amounts upon nitrogen exhaustion when the organisms are grown in a medium having a high carbon to nitrogen ratio. In these circumstances the citric acid cycle is blocked at isocitrate dehydrogenase because of AMP depletion. Citrate accumulates in the mitochondria whence it is transferred to the cytosol in exchange for malate. However, in these fat accumulating organisms, the citric acid is then broken down by ATP citrate lyase giving acetyl coenzyme A which is the principal fat precursor.[27] If indeed there is an analogy between citrate and fat accumulation in fungi, it must be presumed that *A. niger* contains no ATP citrate lyase under citric acid producing conditions.

Shu and Johnson[28] and Kubicek and Röhr,[21] on the other hand, have shown that citrate production is possible in the presence of an excess of nitrogen in the medium provided that phosphate is limiting. The mechanism giving rise to this effect has not been elucidated but it is clear that such an alternative mechanism is necessary to explain the production of citric acid using nitrogen-rich beet molasses, a substrate favoured by many citric acid manufacturers.

Limitation of nitrogen or of phosphate is, however, not the only constraint which must be applied to a culture of *A. niger* if citric acid is to be produced. It is also necessary to restrict the supply of certain of the trace metals, namely iron and manganese. Ferrous ions are known activators of aconitase, and therefore a deficit of iron is plausibly a requisite for the accumulation of citrate if the citric acid cycle is interrupted at this step. Shu and Johnson[28] found that a small amount of iron was required for maximum citrate production. Excess of iron was deleterious. The optimum iron concentration was dependent on the level of zinc ions in the medium. However, La Nauze[14] added a 20-fold excess of iron and reduced the yield level by only 25%.

The level of manganese appears to have a much greater importance, and the maximum permitted levels, although varying somewhat from strain to strain of *A. niger*, are much less. Indeed Shu and Johnson,[29] using their high yielding 72-4 strain cultured in a sucrose/salts medium, found that sufficient manganese could be carried over in the spores used for the inoculum to reduce substantially the yield of citric acid. These authors solved this problem by growing the sporulation culture in a low-manganese medium. Kubicek *et al.*[30] found an increase in the amino acid pool and in the intracellular ammonium ion concentration in *A. niger* cultures accumulating citric acid under conditions of manganese deficit. Habison *et al.*[31] showed that physiological concentrations of ammonium ions such as are induced by a deficiency of manganese can relieve the feed-back inhibition of PFK by citrate (see also latter part of Section 3.1). It is interesting also to note, in view of the analogy between fat and citrate accumulation reported above, that manganese deficiency reduces fat synthesis in *A. niger*.[32] A deficiency of manganese appeared to increase the permeability of the *A. niger* cells,[30] and there is also evidence[33] that lack of manganese results in an impairment of protein turnover which could affect the concentration of some enzymes. The effect of manganese deficiency is thus manifold. It ensures a rapid flow of carbon through glycolysis by unblocking the action of PFK, reduces the loss of carbon into fat and may affect the synthesis of enzymes (see also Neilson[13]).

3.3. Fermentation Conditions Affecting Citric Acid Accumulation by *A. niger*

The most important conditions governing a successful citric acid fermentation are probably temperature, pH and aeration. pH considerations are rather complicated. Because citric acid is being produced all the time during the acid producing phase, the pH will drift downwards during a batch fermentation whatever the starting value

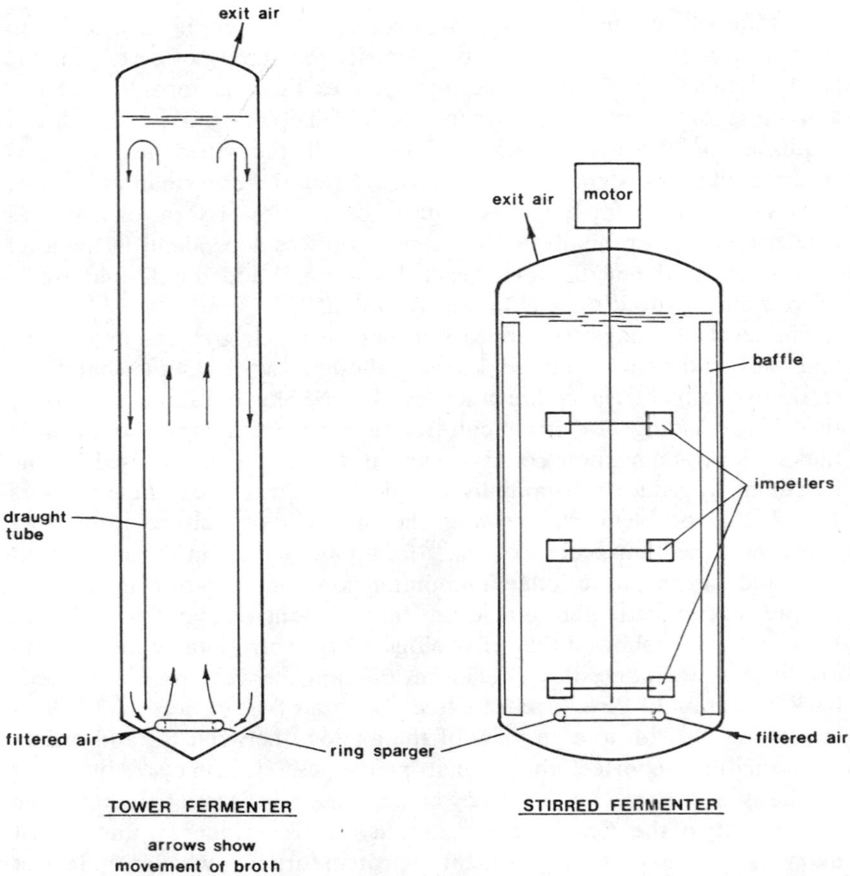

Fig. 2. Alternative designs of fermenter for use in citric acid production.

unless this is very low. The initial pH used will depend on the substrate. If a molasses solution is being employed, higher initial pH values are required because *A. niger* does not germinate or grow well at low pH in this medium. This is due to the presence of acetate in molasses.[34] Where the substrate is sucrose or glucose plus inorganic salts an initial pH of 3·5 can be used.[35] In the so-called surface process (to be described below) very little can be done to control pH once the fermentation has started. In the submerged process this is not so. Kristiansen and Sinclair[26] investigated the effect of pH in a submerged fermentation in the continuous culture mode and reported the optimum pH for citric acid production to be 3·4.

Provision of oxygen by aeration is a significant problem only in the submerged fermentation. Dissolution of oxygen in the medium may be promoted by agitation in a stirred baffled fermenter. Alternatively, the air may be dispersed finely at the base of a tall, narrow fermenter (tower fermenter), thus making use of a high hydrostatic pressure at the base of the vessel (Fig. 2). Kristiansen and Sinclair[26] found growth rate to be almost unaffected by dissolved oxygen tension (DOT) in the range investigated. However, the rate of citric acid production increased linearly with increased DOT. Kubicek *et al.*[36] examined the effect of DOT on the citric acid fermentation in batch fermentations in a stirred fermenter of 35 litres working volume. They found the minimum DOT for citric accumulation was about 25 mbar. The rate of citric acid production was shown to increase linearly with the DOT of the medium in the range 40–150 mbar. Interruptions of air supply resulted in a cessation of citric acid production.

3.4. Strains of *A. niger* and their Improvement

Not all wild type strains of *A. niger* are suitable for citric acid production even under optimum conditions of fermentation. Strains may be isolated from appropriate geographic locations and screened for citric acid production using shake-flask cultures. Promising cultures may then be improved by mutation. Suspensions of spores are treated with mutagens, e.g. UV light, X-rays, *N*-methyl-*N'*-nitro-*N*-nitrosoguanidine or other chemicals. The treated suspensions are plated on a suitable solid medium and single colonies isolated. These colonies after further propagation are screened in shake-flask cultures. The best isolations may be exposed to further treatment using the same or another mutagen. Thus Musilkova *et al.*[37] improved the yield of a strain of *A. niger* from 35–40% to 70–80% by a combination of such treatments. Junior *et al.*[38] selected mutants for resistance to benlate. Azevedo and Bonatelli[39] used the parasexual cycle to generate diploid strains and their haploid derivatives. However, no derived strains were better than the parent. Junior and Azevedo[40] found that diploids so produced exhibited a lower degree of variability than the parent.

3.5. Manufacturing Processes using *A. niger*

Most of the world production of citric acid is centred on Europe and the United States, although some is manufactured in India, China and South America. In the USA there are two manufacturers, namely Chas. Pfizer Inc. and Miles Laboratories, the latter now being a subsidiary of Bayer of

West Germany. In Europe the acid is produced by Joh. A. Benckiser of West Germany, La Citrique Belge (a subsidiary of Hofmann La Roche), John & E. Sturge Ltd in England (a subsidiary of RTZ) and Pfizer in the Republic of Ireland. There is also production in France, Czechoslovakia, Austria and the USSR.

Both the older surface process and the more recently developed submerged process are still used. Smeets states that the surface process will not necessarily become obsolete. The surface process requires more manpower but less energy than the submerged method. Also higher concentrations of substrate can be employed in the surface method, although the submerged fermentation is faster.[41] Most manufacturers still employ a strain of *A. niger* in their fermentations but one major company is reported to be using a strain of yeast growing on a carbohydrate-based medium.

3.5.1. The Surface Process

Nowadays the raw material used in the surface fermentation is almost exclusively beet molasses. Cane molasses has not been used successfully in this process. Because molasses contains too high a level of trace metals, i.e. zinc, copper, iron and manganese (see above), it is necessary to treat the medium with a reagent to remove them. Usually sodium ferrocyanide is used. The suitability of beet molasses for citric acid production varies considerably from year to year depending on the season. There are also large variations between sugar factories. Those using the so-called Quentin process for the recovery of sucrose produce particularly intractable molasses. Fermentation tests are the only reliable guide to performance in the process. For use the molasses is diluted to a sugar concentration of 14–20% w/v and heated to eliminate harmful contaminant microorganisms. The ferrocyanide may be added at this point together with supplementary nutrients such as phosphate and/or zinc. The medium is then run into trays of aluminium.[42] The trays are assembled on racks in fermentation chambers.[42]

The medium is inoculated with spores obtained from the selected culture of *A. niger* grown on a solid medium.[43,44] The chambers are ventilated with filtered air.[42,43,45] The air serves primarily to control the temperature of the fermentation and secondarily to supply oxygen. The spores germinate and grow to cover the medium with mycelium. The appearance of the mycelium during the fermentation has been described.[43]

The fermentation is complete within 9–12 days, after which the mycelium and fermented liquor are separated. The mycelium is washed to remove citric acid and the washings are added to the main liquor.[43] Very little has been published concerning the surface fermentation because manufacturers like to keep their knowledge secret. Some of the references given are rather old and processes may well have been modified since.

3.5.2. The Submerged Process

In the submerged process the fungal mycelium is not permitted to float on the surface of the medium but is dispersed throughout the liquid phase. In this situation a primary requirement is to maintain a concentration of dissolved oxygen in the solution to facilitate the respiration of the mould[36] (see also end of Section 3.3).

Advantages of the submerged method include the possible use of a wider range of substrate and better control of the fermentation. Substrates usable in the submerged process include glucose, sucrose, beet molasses and cane molasses. Other sources such as whey permeate[46] and apple pomace[47] have been mentioned. Hossain *et al.*[48] state that sucrose is the best substrate for citric acid production. Glucose, fructose and lactose are the next best in that order.

As in the surface process, the suitability of molasses for submerged citric acid production varies considerably. This applies to the cane-derived product as much as to that from beet, and practical trials are necessary to establish fermentability. Molasses is a very complex mixture of many components,[49] and attempts have been made to relate the presence of toxic components to fermentation performance.[50,51] Advantages have been claimed for certain medium pretreatments, e.g. the use of bentonite[52] or calcium phosphate plus gel filtration.[53] The addition of methanol[54] or salts of sulphur oxyacids[55] has been claimed to assist the production of high citric yields from molasses.

The form of growth of the mycelium of *A. niger* is said greatly to influence fermentation performance. On the macroscopic scale many authors have claimed that the formation of small dense smooth pellets of mycelium is essential for good citric acid yields.[56] For instance Svensk Sockerfabriks describe a method for the production of such pellets,[57] and Clark[58] illustrates the kind of pellets necessary for good yields of citric acid. Others consider a free filamentous type of growth to be preferable.[59–61]

On the microscopic scale it is claimed that the hyphae should appear abnormally short, stubby, forked and bulbous, and that there should be numerous swollen oval to spherical shaped cells throughout the mycelial structure.[62] This state of affairs can be ensured by maintaining the iron concentration below 1 ppm.[62] To some extent the effect of excess iron can be offset by additions of cupric ion.[63]

If molasses is used as the source of carbohydrate, sufficient nitrogen may be already present to enable growth and citric acid production. Additions of other salts such as phosphate may, however, be necessary. As in the surface process, ferrocyanide must be added to lower the trace metal concentration in the medium. Medium preparation will be broadly similar to that described for the surface process but greater precautions must be taken to ensure its sterility. If the medium is sterilised outside the fermenter, the latter must be sterilised separately, e.g. by the use of steam at not less than 121°C.

If glucose is used as the carbohydrate source it may be necessary to remove metal ions by the use of cation exchangers.[64] If sucrose is used as the substrate, metal ions may also be removed by the use of ferrocyanide.[65] In both cases a full range of nutrient salts must be provided.

Submerged citric acid fermentations may be initiated by the direct addition of spores of *A. niger* to the sterilised fermentation medium contained within the fermenter. Alternatively the spores may be introduced into a smaller fermenter of perhaps 10% of the volume of the main vessel. After a period of growth the vegetative inoculum so produced is transferred to the main fermenter.

Whether a tower fermenter or a stirred fermenter is employed will depend on the process in use. A pellet type process is most easily carried out in a tower fermenter[56] whereas filamentous mycelium is more readily handled in a stirred vessel. The lower viscosity of the pellet broth allows easier circulation and oxygen transfer in the tower type vessel (Fig. 2).

Citric acid broth is very corrosive and fermenters and pipe-lines must be constructed of a high grade stainless steel.

The course of the fermentation may be followed by taking measurements of relevant parameters. Sugar and citric acid levels in the medium may be estimated by chemical or enzymic methods. pH, dissolved oxygen and medium temperature are measured using appropriate sterilisable sensors installed in the fermenter. Oxygen uptake and carbon dioxide output are determined by instrumental analysis of the gas flowing out of the fermenter. Siebert and Hustede[66] show typical curves obtained using the above measurements in a molasses based fermentation in a stirred vessel. These authors also describe the control of the fermentation using a computer which receives data relating to the above-mentioned parameters as well as several other important variables, i.e. redox potential, temperature of cooling water, stirrer speed, gas pressure in the fermenter etc. In turn the computer calculates derivative functions and modulates stirrer speed, air flow rate and other controllable variables so as to keep the values of the derivative functions within previously determined optimum values. By this means an improved performance of the plant is obtained.

When citric acid ceases to accumulate at an economic rate the fermentation is terminated and the liquor separated from the mycelium using a rotary vacuum filter, a disc-bowl centrifuge or other suitable device (Fig. 3).

3.5.3. Submerged Process using Yeast in Carbohydrate-based Media

In recent years it has been found that certain species of yeast cultivated on carbohydrate media are able to accumulate citric acid. Thus Takeda Chemical Industries published a patent[67] claiming the use of species of several genera, namely *Candida, Debaromyces, Hansenula, Torulopsis, Pichia, Saccharomyces, Kloekera* and *Trichisporon*, for citric acid production. It was, however, soon discovered that the citric acid produced contained very significant quantities of iso-citric acid, and methods were sought to reduce the amount of this impurity. For instance, Takeda Chemical Industries[67] claimed that mutants sensitive to fluoroacetate produced very much smaller amounts of iso-citrate. A feature of the yeast-based process for citric acid manufacture was the necessity to maintain the pH in a neutral range. This meant that the citrate was produced mainly as a salt, usually the sodium or calcium salt, thus complicating the recovery process. Pfizer[68] discovered a strain of *Candida* (now re-classified as *Yarrowia*) *lipolytica* requiring neutral conditions only during the initial growth phase, after which the pH could be permitted to fall. Subsequently Pfizer obtained a further patent in which species of certain genera of osmophilic yeasts were used to produce good yields of citric acid from high concentrations of cane molasses.[69] It is understood that beet molasses can also be used. The advantages of the yeast-based process are said to be high productivity (fermentations are finished in only half the time of an *A. niger* fermentation) and relative insensitivity to molasses variations.

3.5.4. Recovery of Citric Acid

In the classical recovery process, clarified fermented liquor is treated with lime to precipitate the insoluble tricalcium salt. The precipitate is carefully washed to remove soluble impurities and treated with sulphuric acid to

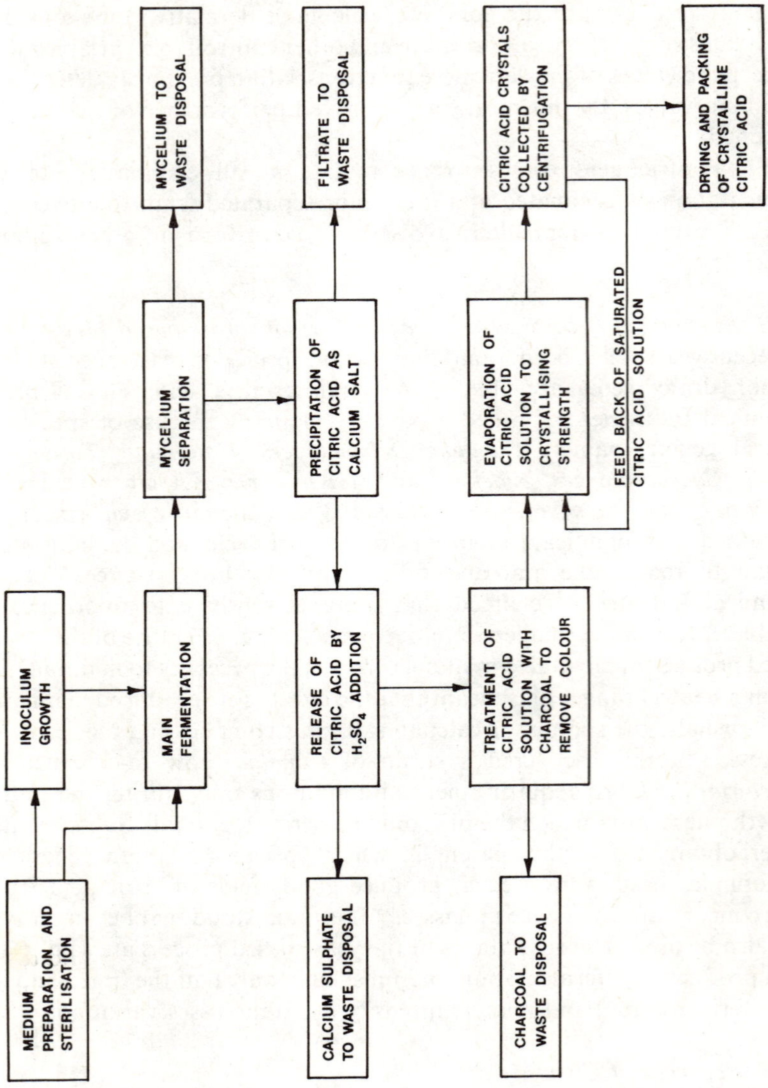

Fig. 3. Diagram showing stages in citric acid manufacture by submerged fermentation.

precipitate calcium sulphate and regenerate the citric acid. The solution so obtained is concentrated, treated with ion exchangers and charcoal and finally crystallised. Evaporative crystallisation yields the anhydrous material whereas cooling crystallisation gives the monohydrate (Fig. 3). Further crystallisation from water produces a high-grade material suitable for food use.

Solvent extraction methods for the recovery of citric acid have been described. These are most suitable where the fermentation medium is composed of sucrose or glucose and inorganic salts. Media based on molasses contain solvent-extractable impurities. Surface active components of molasses render the separation of the solvent and aqueous phases rather difficult.

A number of different solvents have been proposed. Les Usines de Melle have claimed the use of 2-butanol[69] and tributyl phosphate.[70] Baniel *et al.*[71] describe an extraction with a secondary or tertiary amine having at least twenty carbon atoms in the molecule admixed with a diluting organic solvent. This process has been approved by the FDA and is understood to be in use by Miles Laboratories. Wennersten[72] has studied the extraction of citric acid using amines.

3.5.5. Possible Future Developments in Citric Acid Production

Most, if not all, citric acid production utilises batch processing. Batch processes suffer from the defect that at the commencement of each batch a sufficient number of cells of the organism must be grown to carry out the conversion of the sugar to citric acid. This part of the fermentation occupies valuable fermenter time. Theoretically, a continuous process in which fresh medium is continually added to the fermenter, and a mixture of cells and fermented liquor is withdrawn at the same rate, would overcome this defect. In practice, however, such continuous processes are best suited to the production of cells or of a directly growth-related product. The citric acid process does not fit this requirement very well. Furthermore, in order that the process shall be economic, high concentrations of sugar must be employed and, in addition, this sugar must be substantially utilised necessitating the use of several fermenters in series, each exhausting the medium to a lower level. This complication tends to nullify the potential advantages of the continuous fermentation mode for the citric fermentation.

Another possibility for the future is the immobilised enzyme reactor. In an idealised system a series of immobilised enzyme reactors would carry out in turn the ten or so steps between sucrose and citric acid.

Unfortunately the exact mechanism of citric acid accumulation is not known. Additionally some of the enzymes are very unstable and/or need cofactors for their operation. Such cofactors could be lost to the substrate or need to be continually regenerated. In the next century this type of production may be possible. A more immediate prospect is the use of immobilised cells which contain all the necessary enzymes, cofactors and cofactor regeneration systems.

Production of citric acid by means of immobilised or entrapped cells. So far as is known, only laboratory work has been carried out.

Borglum and Marshall[73] of Miles Laboratories have reviewed the situation as it appeared to them in 1983. They considered cell immobilisation by three possible methods, namely adsorption on to solid supports, entrapment in gel matrices, and covalent attachment to inert supports. They studied *A. niger* cells entrapped in beads either of agar or of κ-carrageenan gel. Citric acid was produced from glucose using these gels but at a reduced rate compared with the performance of *A. niger* pellets. A problem identified in this work was the continued growth of the organism leading to the appearance of free mycelium in the medium surrounding the beads. Borglum and Marshall have assigned a patent application to Miles Laboratories covering the gelling of κ-carrageenan containing viable microorganisms using an epichlorhydrin alkylene polyamine polymer.[74] Vaija *et al.*[75] immobilised pellets of *A. niger* in calcium alginate gel beads and studied the production of citric acid in a continuous system. They obtained a maximum rate of citric acid production of 70 mg g^{-1} h^{-1} which was five times higher than with un-immobilised pellets. Surprisingly, some iso-citric acid was also produced.

Eikmeier and Rehm[76] immobilised spores of *A. niger* in beads of calcium alginate and allowed these to grow by immersing the beads in a growth medium. In order to prevent outgrowth of mycelium from the beads, the growth medium was nitrogen limited.[77] Outgrowth in the fermentation stage, which was carried out in an air-lift fermenter, was prevented by increasing the ratio of particle volume to medium volume.[76] A 50% yield of citric acid on sugar consumed was obtained.

Several workers have investigated citric acid production by entrapped yeasts. Maddox and Kingston[78] studied *Saccharomycopsis* (*Yarrowia*) *lipolytica* entrapped in a polyacrylamide gel. Average production rates were 40–50 mg citric acid $litre^{-1}$ h^{-1}. Lim and Choi[79] used *Candida* (*Yarrowia*) *lipolytica* immobilised within calcium alginate beads. Briffaud and Engasser[80] used a diploid strain of *Saccharomycopsis* (*Yarrowia*) *lipolytica* grown on wood chips in a packed column.

While these results are promising, they are far from adequate and it is clear that further studies are required before a commercial process could be envisaged.

4. L(−)-MALIC ACID

L(−)-Malic acid (ethan-1-ol-1,2-dicarboxylic acid, hydroxybutanedioic acid) [97-67-6] has the following structure:

$$\mathrm{HOOC{-}CH_2{-}\overset{*}{C}H(OH){-}COOH}$$

The asterisk indicates the chiral centre. The acid crystallises from water as the anhydrous acid.

A number of ways to manufacture L(−)-malic acid have been described. The first of these is a straightforward fermentation of glucose. For instance Kyowa Hakko Kogyo Co. described the use of species of *Aspergillus*, namely *A. parasiticus*, *A. flavus* and *A. oryzae*.[81] Yields of up to 58% based on glucose supplied were obtained.

The preparation of fumaric acid by fermentation of glucose, e.g. by *Rhizopus arrhizus*, has long been known. However, the route by which it is formed has not been properly established. There is some evidence that the fumarate is formed via oxaloacetate and malate (see Fig. 1), that is by the citric acid cycle operating in reverse. The malic acid fermentation may be regarded as a variant of the fumarate fermentation in which the metabolism for whatever reason stops at malate. Takao and Hotta[82] carried out a fumaric acid fermentation with *R. arrhizus* and part way through added a culture of *Proteus vulgaris*, an organism known to have a high fumarase activity (see Fig. 1). Yields of L(−)-malic acid of over 80% were obtained.

From this it is only a step to a process in which fumaric acid (readily available from maleic anhydride) becomes the main carbon source for malic acid production. Thus Degen *et al.*[83] used a strain of *Paracolobactrum aerogenoides* to ferment a mixture of glucose (for cell growth) and fumaric acid. A 75% yield of L(−)-malic acid based on fumarate consumed was obtained. *P. aerogenoides* could be also used as a source of fumarase which could be employed after isolation in the conversion of fumarate to malate.[83] Morisi *et al.*[84] obtained the fumarase either from *P. aerogenoides* or from species of *Pseudomonas* and immobilised the enzyme within cellulose triacetate filaments. The immobilised enzyme was then used to convert fumarate to L(−)-malate.

Kyowa Hakko[85] used fumarase containing cells of *Lactobacillus* species to convert fumarate to L(−)-malate. The novel feature of this process was that the starting material was a suspension of the sparingly soluble calcium salt of fumaric acid. However, since calcium malate is even less soluble than the fumaric acid salt, the effect was to force the equilibrium over to the malate salt and virtually 100% yields of malate were obtained.

Takata *et al.*[86,87] immobilised cells of *Brevibacterium flavum* in κ-carrageenan and used the resultant biocatalyst to convert fumaric acid to malic acid. They found that succinic acid was formed as a by-product and this unwanted material could be eliminated if the immobilised cells were first treated with bile extract.

Schindler[88] has proposed the use of moulds from various genera but especially *Aspergillus wentii.* Nutrient salts and sucrose were added to a solution formed by neutralising 125 g of fumaric acid suspended in 1 litre of water using a mixture of sodium and potassium hydroxides. After sterilisation the medium was inoculated with a culture of *A. wentii.* After 3 days the concentration of L(−)-malic acid was 132 g litre^{-1}. The mycelium could be re-used in a nutrient-free solution of fumarate salts, giving excellent conversion in 20 h.

The L(−)-malic acid produced in any of the above-mentioned processes may be recovered as the insoluble calcium salt. This salt may be re-converted to malic acid using sulphuric acid, and after removal of the calcium sulphate the resultant solution may be evaporated and crystallised. Unconverted fumaric acid, which is much less water-soluble than malic acid, may be removed from the system by filtration before the precipitation of the calcium salt.[88]

5. LACTIC ACID

Lactic acid (2-hydroxypropionic acid, 2-hydroxypropanoic acid) [50-21-5] has the following structure:

$$CH_3-\overset{*}{C}H(OH)-COOH$$

The asterisk indicates the chiral centre.

Lactic acid is produced from mono- or di-saccharides via the Embden–Meyerhof glycolysis. Under anaerobic conditions the pyruvic acid so produced is reduced to lactic acid by the enzyme lactic dehydrogenase (Fig. 1). The properties of lactic dehydrogenase have been studied in the bacterium *Lactobacillus plantarum* which, as has been mentioned above (latter part of Section 2), produces a racemic mixture of the L(+) and D(−) stereoisomers. Dennis and Kaplan found[89] that *L. plantarum* elaborated two stereospecific lactate dehydrogenases, one producing the L(+) isomer and the other the D(−) form. Because pyruvate is in equilibrium with both enantiomers of lactic acid under the influence of these two enzymes, racemisation of lactic acid can take place. Another possible mechanism for the racemisation of lactic acid has been reported by Hiyama *et al.*[90] who isolated a lactate racemase from *Lactobacillus sake*. Organisms which produce only one stereoisomer of lactic acid must contain only one active lactate dehydrogenase and no active racemase.

5.1. Lactic Acid Fermentation

In the production of lactic acid by anaerobic fermentation, strains of the so-called homolactic organisms such as *Lactobacillus delbrückii* or *L. bulgaricus* are usually used. These organisms form only lactic acid from the carbohydrate supplied. The homolactic fermenters should be distinguished from heterolactic fermenters which produce acetic acid as well as lactic acid and are used in the preparation of sauerkraut, pickled cucumbers and similar products. Normally a pure strain of organism is used for lactic acid production but some authors have claimed that mixed strains give a faster fermentation.[91]

The organism most usually used for lactic acid production is *Lactobacillus delbrückii.* This species is able to use sucrose or glucose as a carbon source and operates best at 50°C.[3] Where lactose or whey is the substrate, *L. bulgaricus* is employed because *L. delbrückii* cannot metabolise lactose. The optimum temperature for *L. bulgaricus* is about 45°C.

The choice of medium in the lactic acid fermentation has a great effect on the process as a whole. It is well known that the recovery of lactic acid from the fermentation broth is more difficult than the fermentation. The more complex the medium the more difficult it is to produce a pure product.[92] Thus the cheapest raw material, e.g. beet molasses, causes the most problems in recovery, whilst the most expensive, sucrose, gives the least. In the US, glucose or enzyme hydrolysed starch are the most commonly used substrates.[93] In Europe, beet molasses or sucrose have been employed. Other substrates which have been suggested are whey,[94] sorghum[95] and cellulose.[96,97]

Lactic acid bacteria have a requirement for B vitamins.[98] These are often added as malt sprouts.[93] A source of phosphate and inorganic nitrogen may also be added. The medium is maintained at a pH in the range 5·5–6·0 by additions of calcium carbonate. This pH control is necessary to ensure a satisfactory rate of fermentation.[92,93] At lower pH values the fermentation is progressively slower and at pH 4·5 or lower no lactic acid is formed.

The medium is charged to fermenters which are large tanks equipped with a slow stirrer to keep the bacteria and calcium carbonate in suspension. The fermenters may be of stainless steel but in some plants are made of wood. Vessels of up to 100 000 litres capacity have been reported. The medium may be unsterilised or at best pasteurised. Because of the relatively high temperatures employed there is little danger of infections. Thermophilic *Clostridia*, however, which produce butanol and butyric acid, have caused problems in some plants.

A culture of the organism is multiplied by transfer through a series of fermenters of progressively increasing size until a volume of 5% of the main fermenter charge is reached. The addition of this culture to the medium in the main fermenter starts the fermentation which lasts for 5 days or more depending on the organism, medium and/or conditions.[93]

Although the fermentation is anaerobic, the cultures are not unduly sensitive to oxygen and the carbon dioxide released by the neutralisation of the accumulating lactic acid by the added calcium carbonate assists in the exclusion of air.

Yields of lactic acid in the range 80–90% of the carbohydrate supplied have been reported.[99]

5.2. Possible Advances in the Lactic Acid Fermentation

5.2.1. Use of Fed Batch and Continuous Culture

Voelskow and Sukatsch[100] have described a fed batch process for producing D(−)-lactic acid from whey by a strain of *L. bulgaricus*. Starting with a medium containing 60 g whey litre^{-1}, further additions of whey were made to make up for the lactose used until the total lactose supplied corresponded to 130 g litre^{-1}. Final concentrations of D(−)-lactic acid were up to 115 g litre^{-1}.

The earliest mention of a continuous lactic acid fermentation proper was that of Childs and Walesby.[101] In this process the pH was controlled by additions of a calcium or sodium carbonate. Production of lactic acid was at the rate of about 3·7 g $litre^{-1}$ h^{-1}. Other workers have used ammonia as a pH control agent. Vlck *et al.*[102] using glucose as a substrate and *L. delbrückii* with cell recycle obtained a volumetric productivity of 76 g $litre^{-1}$ h^{-1} with a concentration of lactic acid in the effluent of 35 g $litre^{-1}$. Major and Bull[103] studied a process without biomass recycle over a wide range of dilution rates and found a maximum volumetric productivity of 8·93 g $litre^{-1}$ h^{-1} at dilution rates between 0·35 and 0·40 h^{-1}. Hanson and Tsao[104] produced a model for batch fermentations of glucose by *L. delbrückii* and used this to predict the behaviour of the system in a continuous mode.

As far as is known, no continuous process has yet been operated commercially.

5.2.2. Use of Immobilised Organisms

A number of investigations have been made into the possibility of using immobilised organisms for the production of lactic acid. Bacteria entrapped in various types of gel have the potentiality for repeated re-use either in batch or continuous mode.

Tanabe Seiyaku Co.[105] claimed the use of *Streptococcus thermophilus* grown within gels of carrageenan which produced lactic acid when immersed in a glucose-based medium. Tipayang and Kozaki[106] immobilised cells of *Lactobacillus vaccinostercus* in a calcium alginate gel to give a system stable for a month which produced lactic acid from xylose. Mitsubishi Chemical Industries claimed a system in which *Lactobacillus thermophilus* was immobilised in carrageenan.[107]

Tuli *et al.*[108] experimented with two different matrices in which to immobilise cells of *Lactobacillus casei.* Agar was found to be more satisfactory than polyacrylamide. A temperature range of 40–50°C and a pH range of 4·5–6·0 were established as optimal. 90% conversion of lactose to lactic acid was observed.

Stenroos *et al.*[109] entrapped *L. delbrückii* in calcium alginate gel and used the beads so obtained in both recycle batch and continuous modes. pH was controlled using calcium carbonate. A maximum lactic acid yield of 97% was obtained, more than 90% of which was the L(+) isomer. In continuous culture the half-life of the system was said to be 100 days.

Whilst these results can be described as promising, as far as is known no immobilised system has yet been used in commercial production.

5.3. Recovery of Lactic Acid from Fermentation Broth

Lactic acid can be obtained as a pure crystalline solid melting at about

54°C. However, special methods are needed to obtain the crystalline material, for instance molecular distillation techniques.

Because lactic acid contains both a carboxyl and a hydroxyl group, it can condense with itself to form esters such as lactoyl-lactic acid and dilactide.[110] When a solution of lactic acid is evaporated, part of the lactic acid is converted to these compounds and to polymers containing more than two lactic acid units, so that a somewhat viscous non-crystallising syrup is obtained.

At the end of a fermentation carried out with calcium carbonate as a pH control agent, the bacteria and any other solid materials are removed by filtration giving a solution of calcium lactate. Sulphuric acid is added and the calcium sulphate so formed removed by filtration. If the fermentation medium has been based on sucrose and the latter has been substantially consumed, recovery is relatively easy.[92] After a preliminary evaporation, colouring matter may be removed by charcoal[92,111] or bleached with hydrogen peroxide.[92] If any iron is present giving an off-colour, this can be precipitated with ferrocyanide.[93,112] Finally the solution may be concentrated again by evaporation to give a food grade material containing about 80% lactic acid. A higher purity material may be obtained by extracting the partly evaporated lactic acid solution with a solvent, e.g. isopropyl ether,[93] followed by re-extraction into water. Alternatively the lactic acid may be esterified with methanol or ethanol and the ester purified by distillation.[92,93] The ester is then hydrolysed to give free lactic acid.

If a crude carbohydrate, e.g. molasses, is used in the fermentation medium it becomes essential to use a solvent extraction and/or esterification procedure to obtain a satisfactory food-grade product. Many solvents other than isopropyl ether have been proposed for extraction of lactic acid from aqueous solutions. Grinstead (Dow Chemical Company) has filed a series of patents for a wide variety of solvents claimed to be useful for this purpose.[e.g.113,114] Solvents such as fusel oil,[115] C_9 aliphatic alcohols[116] and long-chain aliphatic secondary or tertiary amines[71] have also been proposed for the extraction of lactic acid.

Purification of lactic acid may also be carried out using ion-exchangers.[92,117] Other methods for the extraction of lactic acid from filtered, decalcified fermenter broths include reverse osmosis[118,119] and electrodialysis.[120] It is not known whether these methods have found their way into commercial production.

6. GLUCONIC ACID AND GLUCONO-δ-LACTONE

6.1. Conversion of Gluconic Acid to Glucono-δ-lactone

As previously stated, gluconic acid finds its chief use in foodstuffs in the form of glucono-δ-lactone. This substance is in fact the primary product of glucose oxidation by oxygen whether the catalyst be an inorganic one or an enzyme.

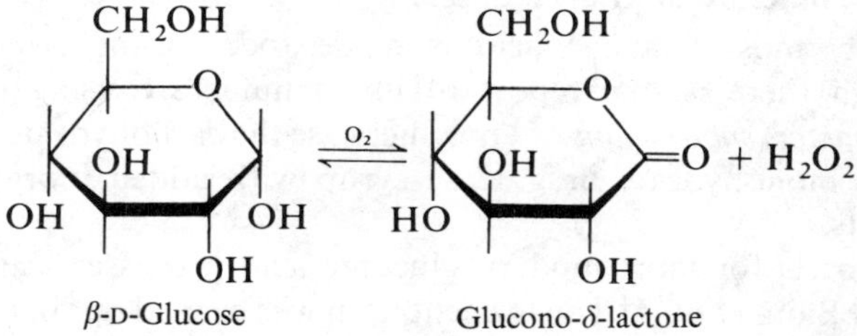

The enzyme catalyst can either be glucose oxidase, e.g. in *A. niger* (Fig. 1), or glucose dehydrogenase, e.g. in *Gluconobacter* (*Acetobacter*) *suboxidans*. However, the lactone once produced in an aqueous system is hydrolysed either chemically or with the aid of a gluconolactonase,[121] to yield a mixture of gluconic acid and lactone. Because the enzymic oxidation of glucose is an equilibrium reaction, it is necessary, if the reaction is to proceed to completion, that the gluconolactone be removed from the sphere of action. This is normally done by carrying out the fermentation at a pH near neutrality, pH control being normally carried out by additions of calcium carbonate or sodium hydroxide. In this situation the end product of the fermentation is either calcium or sodium gluconate. These salts are, in fact, commercial products. However, to obtain the lactone, gluconic acid must be regenerated. The calcium salt may be treated with sulphuric acid to liberate the acid, the calcium sulphate being filtered off. Alternatively the calcium ions may be removed by ion exchange. Ion exchange is the preferred method of liberating gluconic acid from its sodium salt.

Glucono-δ-lactone may be crystallised from a supersaturated solution of gluconic acid by seeding it with crystals of the lactone. The temperature of the solution should be in the range 30–70°C because at lower temperatures gluconic acid crystallises and at higher temperatures glucono-γ-lactone is obtained. Yamauchi and Shimizu[122] have claimed that the best temperature of crystallisation of glucono-δ-lactone depends on the concentration of the solution. Thus a solution containing 1·5 kg of gluconic acid/glucono-δ-lactone at a concentration of 78·3% was seeded and water removed *in vacuo* at 2 g h^{-1} at 47–48°C. After about 18 h the glucono-δ-lactone crystals were removed and the mother-liquor was treated in the same way to give an overall yield of over 75%.

Glucono-δ-lactone is a white crystalline powder melting at 153°C.

6.2. Manufacture of Gluconic Acid by Fermentation

Probably most gluconic acid is made today using *Aspergillus niger*, although there is one report of its manufacture using *Gluconobacter* (*Acetobacter*) *suboxydans*.[123] In either case the carbohydrate source will be glucose monohydrate or glucose syrup with added inorganic or other nutrients.

The basis for most modern gluconic acid processes was described in 1952 by Blom *et al.*[124] The fermentation was carried out in a stainless steel stirred fermenter of about 1000 litres capacity using a medium containing glucose, corn steep liquor, magnesium sulphate, potassium phosphate, urea, ammonium phosphate and sulphuric acid to pH 4·5. The medium was sterilised batchwise or by means of a continuous steriliser. Before inoculation the pH was adjusted to 6–6·5 and it was maintained at this value during the fermentation by automatic addition of sodium hydroxide.

Spores of *A. niger* were germinated in a smaller inoculum fermenter, before transfer to the main fermenter. Fermentation time with a 10% v/v inoculum was about 20 h, at which time the glucose was substantially all consumed.

The initial glucose concentration in these trials was of the order of 200 g $litre^{-1}$. In order to make the fermentation process more economical and to reduce evaporation costs in recovery, it is desirable to increase the initial glucose level. If this is done two problems arise. At the higher glucose levels growth of the organism is less rapid. Furthermore, the sodium gluconate produced reaches the limits of its solubility and tends to crystallise. These difficulties were resolved in a patent application by Ziffer *et al.*[125] In the first place the glucose was added in stages, the first and any subsequent glucose supplements being added unsterilised. Secondly, sodium hydroxide additions for pH control were discontinued part way through the fermentation, giving a final yield of a mixture of gluconic acid and sodium gluconate which shows no tendency to crystallise out. From the point of view of glucono-δ-lactone production, an additional advantage is apparent, namely that only part of the product is in the form of sodium gluconate. Since glucono-δ-lactone is made from gluconic acid, the lower sodium content of the broth produced by this method will reduce recovery costs.

6.3. Possible Future Developments

Hatcher has shown[126] that, at the end of a classical gluconic acid fermentation as described above, the mycelium of *A. niger* can be filtered off and re-used to convert a new charge of nutrient-free glucose. This process can be repeated once more. Further use of the mycelium is said to be possible if certain (unspecified) nutrients are added to the glucose.

These results clearly imply that the use of immobilised mycelium of *A. niger* is a possibility. However, this system has not been studied to any extent, although Hartmeier and Döppner[127] have reported the immobilisation of *A. niger* mycelium containing the glucose oxidase system. Seiskari *et al.*[128] have immobilised living cells of *Gluconobacter oxydans* on a fibrous nylon carrier and studied the production of gluconic acid in an aerated reactor. An advantage of using *Gluconobacter* is that conversion of glucose to gluconic acid can take place at low pH values, avoiding the need for the continuous neutralisation which is required with the *A. niger* system. However, the glucose oxidising enzyme in *Gluconobacter* ssp. is glucose dehydrogenase which needs a cofactor. This cofactor, namely NADP, must be regenerated if oxidation is to proceed. This is done by the normal cell metabolic system and explains the need for entrapped living cells.

The system of Hartmeier and Döppner[127] is in essence an immobilisation of glucose oxidase and catalase on the *A. niger* cells which do not subsequently grow. It is an extension of earlier work by Hartmeier and Tegge[129] in which extracted glucose oxidase and catalase are immobilised on hardened gelatin particles using glutaraldehyde. Others have immobilised glucose oxidase/catalase on various supports such as porous glass,[130] activated carbon[131] and nickel–silica alumina.[132] The entrapment of glucose oxidase/catalase in polyacrylamide[133] and 2-hydroxyethyl methacrylate[134] gels has been described.

$$\text{Glucose} + O_2 \xrightleftharpoons[\text{glucose oxidase}]{} \text{Gluconolactone} + H_2O_2$$

$$2H_2O_2 \xrightarrow[\text{catalase}]{} 2H_2O + O_2$$

It will be observed from the above equations that the oxidation of the glucose by oxygen in the presence of glucose oxidase produces not only gluconolactone but also hydrogen peroxide. Thus the utilisation of the oxygen is not efficient unless catalase is present. Most aerobic cells contain catalases and *A. niger* is no exception, the catalase accompanying the glucose oxidase when the latter is extracted from the cells. The purpose of the catalase in the cell is to eliminate hydrogen peroxide formed by oxidase enzymes. The hydrogen peroxide would otherwise be lethal to the cell.

In immobilised glucose oxidase/catalase systems the catalase serves to protect the glucose oxidase which is very sensitive to hydrogen peroxide. It is essential that the catalase be bound to the carrier in close proximity to the glucose oxidase to prevent damage to the latter and loss of activity in the system as a whole.[133] In the method of Hartmeier and Döppner[127] additional catalase may be immobilised on the outside of the cells.

If the immobilised glucose oxidase/catalase is contained in a fixed bed, air or oxygen cannot be used as the oxidising agent because the oxygen is insufficiently soluble and gaseous air or oxygen will disturb the bed. The use of hydrogen peroxide introduced into the glucose stream as it enters the packed bed has been proposed. The use of immobilised glucose oxidase/catalase in stirred aerated tanks has been studied.[129] Richter and Heinecker[135] have studied the use of air, oxygen and hydrogen peroxide as 'oxygen carriers' and report that there are problems to be solved including inhibitory effects, and loss of productivity with time.

At the present time it is thought that no immobilised enzyme reactors are being used commercially for gluconic acid production. However, if the problems can be solved, enzyme reactors may well in time oust the conventional fermentation method.

7. CONCLUDING REMARKS

The present-day production methods used in the manufacture of citric acid, L(−)-malic acid, lactic acid and glucono-δ-lactone have been described. Future developments in the manufacture of these acids can fall into three categories.

(1) In the immediate future the use of so-called genetic engineering will be confined to the production of single protein molecules and thus could possibly be of benefit to the production of glucono-δ-lactone from glucose or L(−)-malic acid from fumaric acid. Where a number of steps separate the carbon source and the product, the application of genetic engineering is likely to be delayed until the exact sequence of metabolic steps and the

enzyme which catalyses them have been properly defined. When this has been done, site-directed mutations may be a possibility.

(2) The possible use of immobilised enzymes and cells in so-called bioreactors has been separately discussed above under each topic. The general conclusion reached is that promising results have been obtained but that many difficulties remain to be overcome. It is expected that in the next two decades some organic acids will be produced commercially by immobilised enzymes or cells in bioreactors.

(3) Until suitable bioreactors have been developed, the efficient use of power in promoting mixing and oxygen solution in fermenters will be a principal preoccupation of biochemical engineers. Much remains to be done in this field and it is to be expected that great improvements in energy utilisation in fermenters will become apparent within the next decade.

REFERENCES

1. Arnold, M. H. M., *Acidulants for Food and Beverages*, Food Trade Press, London, 1975, p. 58.
2. Arnold, M. H. M., *ibid.*, p. 9.
3. Nanninga, G. L., *Antonie v. Leeuwenhoek*, **49**, 86 (1983).
4. Currie, J. N., *J. Biol. Chem.*, **31**, 15–37 (1917).
5. Stottmeister, U., Behrens, U., Weissbrodt, E., Barth, G., Franke-Rinker, D. and Schulze, E., *Z. Allg. Mikrobiol.*, **22**(6), 399–424 (1982).
6. Ferrara, L., De Cesari, L. and Salvini, E., *Chimica e Industria*, **59**(3), 202–6 (1977).
7. Cleland, W. W. and Johnson, M. J., *J. Biol. Chem.*, **208**, 679–89 (1954).
8. Waronick, C. L. and Johnson, M. J., *J. Biol. Chem.*, **235**(1), 9–15 (1960).
9. Bloom, S. J. and Johnson, M. J., *J. Biol. Chem.*, **237**(9), 2718–20 (1962).
10. Damodaran, M., Jagannathan, V. and Singh, K., *Enzymologia*, **17**, 199–206 (1955).
11. Ramakrishnan, C. V. and Martin, S. M., *Can. J. Biochem. Phys.*, **32**, 434–9 (1954).
12. Ramakrishnan, C. V., Steel, R. and Lentz, C. P., *Arch. Biochem. Biophys.*, **55**(1), 270–3 (1955).
13. Neilson, N. E., *J. Bacteriol.*, **71**, 356–61 (1956).
14. La Nauze, J. M., *J. Gen. Microbiol.*, **44**, 73–81 (1965).
15. Täufel, K. and Behnke, U., *Nährung*, **10**, 443–51 (1966).
16. Usami, S., Saegusa, N. and Tateishi, T., *Hakko Kyokaishi*, **29**(8), 399–404 (1971).
17. Szczodrak, J., *Eur. J. Appl. Microbiol. Biotechnol.*, **13**, 107–12 (1981).
18. Ramakrishnan, C. V. and Martin, S. M., *Arch. Biochem. Biophys.*, **55**, 403–7 (1955).
19. Mattey, M., *FEMS Microbiol. Lett.*, **2**, 71–4 (1977).
20. Meixner-Monori, B., Kubicek, C. P., Habison, A., Kubicek-Pranz, E. M. and Röhr, M., *J. Bacteriol.*, **161**, 265–71 (1985).

21. Kubicek, C. P. and Röhr, M., *Eur. J. Appl. Microbiol.*, **4**, 167–75 (1977).
22. Röhr, M. and Kubicek, C. P., *Process Biochem.*, **16**(4), 34–7, 44 (1981).
23. Habison, A., Kubicek, C. P. and Röhr, M., *FEMS Microbiol. Lett.*, **5**, 39–42 (1979).
24. Bowes, I., Ph.D. Thesis, University of Strathclyde, Glasgow, 1980.
25. Kristiansen, B. and Sinclair, C. G., *Biotech. Bioeng.*, **20**, 1711–22 (1978).
26. Kristiansen, B. and Sinclair, C. G., *Biotech. Bioeng.*, **21**, 297–315 (1979).
27. Evans, C. T. and Ratledge, C., *Biotechnol. Genet. Eng. Rev.*, **3**, 349–75 (1985).
28. Shu, P. and Johnson, M. J., *J. Bacteriol.*, **56**, 577–85 (1948).
29. Shu, P. and Johnson, M. J., *J. Bacteriol.*, **54**, 161–7 (1947).
30. Kubicek, C. P., Hampel, W. and Röhr, M., *Arch. Microbiol.*, **123**, 73–9 (1979).
31. Habison, A., Kubicek, C. P. and Röhr, M., *Biochem. J.*, **209**, 669–76 (1983).
32. Orthofer, R., Kubicek, C. P. and Röhr, M., *FEMS Microbiol. Lett.*, **5**, 403–6 (1979).
33. Kubicek, C. P., Röhr, M. and Ma, H., *Third European Congress on Biotechnology*, Vol. II, Verlag Chemie, Munich, 1984.
34. Fencl, Z. and Leopold, J., *Folia Microbiol.*, **4**, 7–15 (1959).
35. Noguchi, Y. and Johnson, M. J., *J. Bacteriol.*, **82**, 538–41 (1961).
36. Kubicek, C. P., Zehentgruber, O., El-Kalak, H. and Röhr, M., *Eur. J. Appl. Microbiol.*, **9**, 101–15 (1980).
37. Musilkova, M., Ujcova, E., Seichert, L. and Krumphanzl, V., *Kvasny Prum.*, **29**, 230–2 (1983).
38. Junior, R. B., Azevedo, J. L. and Valent, G. U., *Rev. Brasil. Genet.*, **5**, 483–92 (1982).
39. Azevedo, J. L. and Bonatelli, R., Overproduction of microbial products, *FEMS Symp.* 1982, **13,** 439–50 (1982).
40. Junior, R. B. and Azevedo, J. L., *Biotechnol. Lett.*, **4**, 761–6 (1982).
41. Smeets, F., *Antonie v. Leeuwenhoek*, **49**, 86–7 (1983).
42. Ellis, W. J. and Gresford, G. B., BIOS Target No. C22/541(a), British Intelligence Objectives Sub-Committee, HMSO, London, 1946.
43. Ledingham, G. A., Eamer, R. A. and Hrudka, G. E., BIOS Trip No. 3185, Technical Information and Documents Unit, London, 1947.
44. Karklin, R. Y., Rumba, A. A. and Azanda, V. K., UK Pat. Appl. 2,076,018A (1981).
45. Schmitz, von R., *Chemie-Technik*, **5**, 271–2 (1976).
46. Hossain, M., Brooks, J. D. and Maddox, I. S., *New Zealand J. Dairy Sci. Technol.*, **18**, 161–8 (1983).
47. Hang, Y. D. and Woodams, E. E., *Biotechnol. Lett.*, **6**, 763–4 (1984).
48. Hossain, H., Brooks, J. D. and Maddox, I. S., *Appl. Microbiol. Technol.*, **19**, 393–7 (1984).
49. Tressl, von R., Jakob, R., Kossa, T. and Bronn, W. K., *Branntweinwirtschaft*, **116**, 117–19 (1976).
50. Zakowska, Z. and Jaloka, A., *Acta Biotechnol.*, **4**, 171–8 (1984).
51. Banerjee, N. and Viswanathan, L., *Proc. Ann. Conv. Sugar Technol. Assoc. India*, **41**, G75–80 (1976).
52. Nasim, S., Dasgupta, J., Khan, A. W. and Vora, V. C., *Indian J. Microbiol.*, **21**, 343–4 (1981).
53. Panda, T., Kundu, S. and Majumdar, S. K., *Process Biochem.*, 183–7 (1984).

54. Moyer, A. J., *Appl. Microbiol.*, **1**, 1–13 (1953).
55. Ilczuk, Z., *Eur. J. Appl. Microbiol. Biotechnol.*, **17**, 69–72 (1983).
56. Sodeck, G., Modl, J., Kominek, J. and Salzbrunn, W., *Process Biochem.*, Oct. /Nov. 9–11 (1981).
57. Svensk Sockerfabriks AB, UK Pat. 951,629 (1964).
58. Clark, D. S., *Can. J. Microbiol.*, **8**, 133–6 (1962).
59. Takahashi, J., Hidaka, H. and Yamada, K., *Agric. Biol. Chem.*, **29**, 331–6 (1965).
60. Musilkova, M., Ujcova, E., Seichert, L. and Fencl, Z., *Folia Microbiol.*, **27**, 328–32 (1982).
61. Seichert, L., Ujcova, E., Musilkova, M. and Fencl, Z., *Folia Microbiol.*, **27**, 333–4 (1982).
62. Snell, R. L. and Schweiger, L. B., UK Pat. 653,808 (1951).
63. Miles Laboratories Inc., UK Pat. 742,972 (1956).
64. Miles Laboratories Inc., UK Pat. 669,733 (1952).
65. Aktiengesellschaft Jungbunzlauer Spiritus- und Chemische Fabrik, UK Pat. 1,342,311 (1974).
66. Siebert, D. and Hustede, H., *Chem.-Ing.-Tech.*, **54**, 549–69 (1982).
67. Takeda Chemical Industries Ltd, Jap. Pat. 20707/68 (1968).
68. Pfizer Inc., UK Pat. 1,304,753 (1973).
69. Les Usines de Melle, UK Pat. 874,030 (1961).
70. Les Usines de Melle, UK Pat. 936,339 (1963).
71. Baniel, A. M., Blumberg, R. and Hajdu, K., UK Pat. 1,426,018 (1976).
72. Wennersten, R., *J. Chem. Tech. Biotechnol.*, **33B**, 85–94 (1983).
73. Borglum, G. B. and Marshall, J. J., *Appl. Biochem. Biotechnol.*, **9**, 117–30 (1984).
74. Miles Laboratories Inc., Eur. Pat. Appl. 0,052,829 (1982).
75. Vaija, J., Linko, Y.-Y. and Linko, P., *Appl. Biochem. Biotechnol.*, **7**, 51–4 (1982).
76. Eikmeier, H. and Rehm, H. J., *Appl. Microbiol. Biotechnol.*, **20**, 365–70 (1984).
77. Eikmeier, H., Westmeier, F. and Rehm, H. J., *Appl. Microbiol. Biotechnol.*, **19**, 53–7 (1984).
78. Maddox, I. S. and Kingston, P. J., *Biotechnol. Lett.*, **5**, 795–8 (1983).
79. Lim, D. J. and Choi, C. Y., *Hwahak Konghak*, **22**, 213–23 (*Chem. Abstr.*, **101**, 189,644).
80. Briffaud, J. and Engasser, J. M., *Cell Immobilisées Colloq.*, 197–210 (1979).
81. Kyowa Hakko, Kogyo Co. Ltd, UK Pat. 844,029 (1961).
82. Takao, S. and Hotta, K., *J. Ferment. Technol.*, **54**, 197–204 (1976).
83. Degen, L., Oddo, N. and Olivieri, R. (Snamprogetti spA), UK Pat. 1,419,668 (1975).
84. Morisi, F., Mosti, R. and Bettonte, M. (Snamprogetti spA), UK Pat. 1,426,137 (1976).
85. Kyowa Hakko Kogyo Co. Ltd, UK Pat. 874,580 (1961).
86. Takata, I., Yamamoto, K., Tosa, T. and Chibata, I., *Enzyme Microb. Technol.*, **2**, 30–6 (1980).
87. Tanabe Seiyako Co. Ltd., UK Pat. 1,506,112 (1978).
88. Schindler, F. (Chemische Werke Hüls AG), Ger. Pat. 3310849A1 (1984).
89. Dennis, D. and Kaplan, N. O., *J. Biol. Chem.*, **235**, 810–18 (1960).

90. Hiyama, T., Fukui, S. and Kitihara, K., *J. Biochem.* (Tokyo), **64**, 99–107 (1968).
91. Tiwara, K. P., Pandey, A. and Mishra, N., *Proc. Nat. Acad. Sci. India*, Sect. A, **47**, 130–2 (1977).
92. Buchta, K., *Chem. Z.*, **98**, 532–8 (1974).
93. Lockwood, L. B., in *Microbial Technology*, Vol. 1, 2nd Edn, Peppler, H. J. and Perlman, D. (Eds), Academic Press, New York, 1979, pp 373–5.
94. Teuber, M., *Wiss. Umwelt.*, 135–40 (1981).
95. Samuel, W. A. and Lee, Y. Y., *Biotech. Bioeng.*, **22**, 757–77 (1980).
96. Wang, D. I. C., Cooney, C. L., Demain, A. L., Gomez, R. F. and Sinskey, A. J., Report 1977, C00-4198-4, 241 pp. From *Energy Res. Abs.*, **3**, Abstr. No. 39663 (1978).
97. Abe, S., Tagaki, M. and Yamada, K. (Ajinomoto Co. Inc.), Jap. Pat. 78 99397 (1978).
98. Tiwara, K. P., Singh, S. P. and Pandey, A., *Proc. Indian Nat. Acad.*, Part B, **43**, 219–22 (1977/78).
99. Miall, L. M., in *Economic Microbiology*, Rose, A. H. (Ed.), Vol. 2, Academic Press, London, 1978, pp 95–9.
100. Voelskow, H. and Sukatsch, D. (Hoechst AG), Eur. Pat. 72010 (1983).
101. Childs, C. G. and Walesby, B., *Process Biochem.*, **1**(8), 441–4 (1966).
102. Vick, R. T. B., Mandel, D. K., Dea, D. K., Blanch, H. W. and Wilke, C. R., *Biotechnol. Lett.*, **5**, 665–70 (1983).
103. Major, N. C. and Bull, A. T., *Biotechnol. Lett.*, **7**, 401–5 (1985).
104. Hanson, T. P. and Tsao, G. T., *Biotechnol. Bioeng.*, **14**, 233–52 (1972).
105. Tanabe Seiyaku Co. Ltd, Jap. Pat. 57,144,985 (1982).
106. Tipayang, P. and Kozaki, M., *J. Ferment. Technol.*, **60**, 595–8 (1982).
107. Mitsubishi Chemical Industries Co. Ltd, Jap. Pat. 82,110,192 (1982).
108. Tuli, A., Sethi, R. P., Khanna, P. K., Marwaha, S. S. and Kennedy, J. F., *Enzyme Microb. Technol.*, **7**, 164–8 (1985).
109. Stenroos, S.-L., Linko, Y.-Y. and Linko, P., *Biotechnol. Lett.*, **4**, 159–64 (1982).
110. Holten, C. H., *Lactic Acid*, Verlag Chemie, Weinheim, 1971, pp 192–231.
111. Nikulina, I. D., Smirnov, V. A., Bondarenko, A. N. and Antipenko, M. V., *Tr. Leningrad Nauch.-Issled. Inst. Pishch. Prom.*, No. 1, 146–8 (1971) (*Chem. Abstr.*, **78**, 41842e).
112. Nikulina, I. D., Smirnov, V. A., Bondarenko, A. N. and Antipenko, M. V., *Tr. Leningrad Nauch.-Issled. Inst. Pishch. Prom.*, No. 1, 120–36 (1971) (*Chem. Abstr.*, **78**, 41538).
113. Grinstead, R. R., US Pat. 3,980,704 (1976).
114. Grinstead, R. R., US Publ. Pat. Appl. B 467,328 (1976).
115. Bansal, T. K., Vishawanathan, M. S. and Ghosh, T. K., *Indian Chem. Eng.*, **18**, 24–8 (1976).
116. Sporidi, G., Ger. Pat. 1,493,196 (1970).
117. Napierala, W. and Siminski, M., *Przem. Ferment. Rolny*, **16**, 4–10 (1972).
118. Matsuura, T., *J. Appl. Polym. Sci.*, **17**, 3663–8 (1973).
119. Smith, B. R., MacBean, R. D. and Cox, G. C., *Aust. J. Dairy Technol.*, **32**, 23–6 (1977).
120. Jacquemet, J. C., Ger. Offen. 1,957,395 (1970).
121. Dumontier, M. and Hanss, M., *Biochimie*, **56**, 1291–2 (1974).

122. Yamauchi, T. and Shimizu, K., Jap. Pat. 75,123,674 (1975).
123. Huchette, M. and Devos, F. (Roquette Frères), Fr. Pat. 2,054,829 (1971).
124. Blom, R. H., Pfeifer, V. F., Moyer, A. J., Traufler, D. H., Conway, H. F., Crocker, C. K., Farison, R. E. and Hannibal, D. V., *Ind. Eng. Chem.*, 435–40 (1952).
125. Ziffer, J., Gaffney, A. S., Rothenberg, S. and Cairney, T. J. (assigned to Pabst Brewing Co., now owned by Joh. A. Benckiser, West Germany), UK Pat. 1,249,347 (1971).
126. Hatcher, H. J. (Economics Laboratory Inc.), US Pat. 3,669,840 (1972).
127. Hartmeier, W. and Döppner, T., *Biotechnol. Lett.*, **5**, 743–8 (1983).
128. Seiskari, P., Linko, Y.-Y. and Linko, P., *Appl. Microbiol. Biotechnol.*, **21**, 356–60 (1985).
129. Hartmeier, W. and Tegge, G., *Starch/Stärke*, **31**, 348–53 (1979).
130. Weibel, M. K. and Bright, H. J., *Biochem. J.*, **124**, 801–7 (1971).
131. Cho, Y. K. and Bailey, J. E., *Biotech. Bioeng.*, **20**, 1651–5 (1978).
132. Herring, W. M., Laurence, R. L. and Kittrell, J. R., *Biotech. Bioeng.*, **14**, 975–84 (1972).
133. Boehringer Mannheim GmbH, UK Pat. 1,373,562 (1974).
134. Hinberg, I., Kapoulas, A., Korus, R. and O'Driscoll, K., *Biotech. Bioeng.*, **16**, 159–68 (1974).
135. Richter, G. and Heinecker, H., *Starch/Stärke*, **31**, 418–22 (1979).

Index